CAMBRIDGE TRACTS IN MATHEMATICS

General Editors

121 Lévy Processes

Jean Bertoin

Laboratoire de Probabilités
Université Pierre et Marie Curie

Lévy Processes

CAMBRIDGE
UNIVERSITY PRESS

CAMBRIDGE
UNIVERSITY PRESS

The Edinburgh Building, Cambridge CB2 8RU, UK

Cambridge University Press is part of the University of Cambridge.

It furthers the University's mission by disseminating knowledge in the pursuit of education, learning and research at the highest international levels of excellence.

www.cambridge.org
Information on this title: www.cambridge.org/9780521646321

First published 1996
First paperback edition 2007
Sixth printing 2009

A catalogue record for this publication is available from the British Library

Library of Congress Cataloguing in Publication data
Bertoin, Jean.
Lévy processes / Jean Bertoin.
p. cm. – (Cambridge tracts in mathematics; 121)
Includes bibliographical references.
ISBN 0 521 56243 0 (hardback)
1. Lévy processes. I. Title II. Series.
QA274.73B48 1996
519.2'82 – dc20 95–48226 CIP

ISBN 978-0-521-64632-1 Paperback
ISBN 978-0-521-56243-0 Hardback

Contents

Preface

Lévy processes can be thought of as random walks in continuous time, that is they are stochastic processes with independent and stationary increments. The state space may be a fairly general topological group, but in this text, we will stick to the Euclidean framework. The best known and most important examples are the Poisson process, Brownian motion, the Cauchy process, and more generally stable processes. Lévy processes concern many aspects of probability theory and its applications. In particular, they are prototypes of Markov processes (actually, they form the class of space-time homogeneous Markov processes) and of semimartingales; they are also used as models in the study of queues, insurance risks, dams, and more recently in mathematical finance. From the viewpoint of functional analysis, they appear in connection with potential theory of convolution semigroups.

Historically, the first researches go back to the late 20's (that is when the foundations of modern probability theory were laid down) with the study of infinitely divisible distributions. Their general structure has been gradually discovered by de Finetti, Kolmogorov, Lévy, Khintchine and Itô; it is described by the celebrated Lévy-Khintchine formula which points out the correspondence between infinitely divisible distributions and processes with independent and stationary increments. After the pioneer contribution of Hunt in the mid-50's, the spreading of the theory of Markov processes and its connection with abstract potential theory has had a considerable impact on Lévy processes; see the works of Doob, Dynkin, Blumenthal and Getoor, Skorohod, Kesten, Bretagnolle,

Port and Stone, Berg and Forst, Kanda, Hawkes, $\cdots$. At the same time, the fluctuation theory for random walks developed chiefly by Spitzer, Feller and Borovkov via analytic methods has been extended to continuous time by approximations based on discrete time skeletons; and many important properties of the sample paths of Lévy processes have been noted by Rogozin, Taylor, Fristedt, Pruitt, and others. Path transformations such as reflexion, splitting or time reversal form another set of useful techniques that were applied initially (in continuous time) to Brownian motion. Their importance for Lévy processes was recognized first by Millar, Greenwood and Pitman, who presented a direct approach to fluctuation theory. Further developments in this setting were made quite recently by Bertoin, Doney and others. Local times have received a lot of attention in the last ten years or so; the most impressive result in that field is perhaps the characterization by Barlow and Hawkes of the class of Lévy processes which possess jointly continuous local times; see also the recent works of Marcus and Rosen in the symmetric case. To complete this brief overview, we stress that the so-called general theory of processes also has many important applications to Lévy processes, in particular concerning stochastic calculus and limit theorems.

Several books contain sections or chapters on Lévy processes (e.g. Lévy (1954), Itô (1961), Gihman and Skorohod (1975), Jacod and Shiryaev (1987), Sato (1990, 1995), Skorohod (1991), Rogers and Williams (1994), ...); see also the surveys by Taylor (1973), Fristedt (1974) and Bingham (1975). The purpose of this monograph is to present an up-dated and concise account of the theory, which may serve as a reference text. I endeavoured to make it as self-contained as possible; the prerequisite is limited to standard notions in probability and Fourier analysis.

Here is a short description of the content. A chapter of preliminaries introduces the notation and reviews some elementary material on infinitely divisible laws, Poisson processes, martingales, Brownian motion and regularly varying functions. The core of the theory of Lévy processes in connection with the Markov property and the related potential theory is developed in chapters I and II. The theory of general Markov processes is doubtless one of the most fascinating fields of probability, but it is also one of the most demanding. Nonetheless, the special case of Lévy processes is much easier to handle, thanks to techniques of Fourier analysis and the spatial homogeneity. We stress that no prior knowledge of Markov processes is assumed. Chapter III is devoted to the study of subordinators, which form the class of increasing Lévy processes; a special emphasis is given to the properties of their sample paths. Subordinators also have a key part in chapter IV, where we introduce Itô's theory of the excursions of a Markov process away from a point, and in chapter V, where we investigate the local times of Lévy processes. The fluctuation theory is presented in chapter VI, following the Greenwood-Pitman approach based on excursion theory. Chapter VII is devoted to Lévy processes with no positive jumps, for which fluctuation theory becomes remarkably simple. Some path transformations are described, which extend well-known identities for Brownian motion due to Williams and Pitman. Finally, several consequences of the scaling property of stable processes are presented in chapter VIII. Each chapter ends with exercises, which provide additional information on the topic for the interested reader, and with comments, where credits and further references are given. To avoid duplication with the existing literature on semimartingales, we did not include material on stochastic calculus or limit theorems; we refer to Jacod (1979), Protter (1990) and Jacod and Shiryaev (1987) for detailed expositions.

We use the following labels. Roman numbers refer to chapters, arabic numbers to statements, and numbers between parentheses to equations or formulas. For instance, Proposition V.2 designates the proposition with label 2 of chapter V, and (III.10) the equation with label (10) in chapter III. The roman number refering to a chapter is omitted within the same chapter.

This text is partially based on a 'cours de troisième cycle' taught in the Laboratoire de Probabilités de L'Université Pierre-et-Marie-Curie. My work was greatly eased by the position I had at this time in the Centre National de la Recherche Scientifique. I should like to thank warmly my colleagues in the Laboratoire de Probabilités, and to express my deep gratitude to Nick Bingham, Ron Doney, Daniel Revuz and Hrvoje Sikic, who read preliminary versions of the manuscript and corrected uncountable errors, misprints and misuses of the English language.

O

Preliminaries

1. Notation

In this section, we set down notation which will be used throughout the text.

We denote by $\mathbb{R}^d$ the d-dimensional Euclidean space, equipped with the standard scalar product $\langle \cdot, \cdot \rangle$ and the Euclidean norm $|\cdot|$. It is endowed with the Borel sigma-field $\mathscr{B}(\mathbb{R}^d)$ and the Lebesgue measure dx. The abbreviation a.e. refers to 'almost everywhere' with respect to the Lebesgue measure. The lower and upper bounds of a subset A of the nonnegative half-line $[0, \infty)$ are denoted by $\inf A$ and $\sup A$, respectively, with the convention that $\inf \emptyset = \infty$ and $\sup \emptyset = 0$. We say that a function $f : [0, \infty) \to [0, \infty]$ is increasing if $f(s) \leq f(t)$ for all $0 \leq s \leq t$. If the preceding condition holds with $\leq$ replaced by $<$, we say that f is strictly increasing. We use Landau's notation $f = o(g)$, $f = O(g)$ and $f \sim g$ for $\lim(f/g) = 0$, $\limsup(f/g) < \infty$ and $\lim(f/g) = 1$, respectively.

Next, we introduce the so-called canonical notation for right-continuous substochastic (i.e. possibly defective) processes having left limits. Specifically, take an isolated point ∂ which will serve as cemetery. Consider

$$\Omega = D([0, \infty), \mathbb{R}^d \cup \{\partial\}),$$

the set of paths $\omega : [0, \infty) \to \mathbb{R}^d \cup \{\partial\}$ with lifetime

$$\zeta(\omega) = \inf\{t \geq 0 : \omega(t) = \partial\}$$

which are right-continuous on $[0, \infty)$, have a left limit denoted by $\omega(s-)$ for any $s \in (0, \infty)$, and stay at the cemetery point ∂ after the lifetime $\zeta(\omega)$. This space is endowed with Skorohod's topology, for which we refer to chapter VI in Jacod and Shiryaev (1987). In particular, Ω is a Polish space, that is it is metric-complete and separable. We shall not use Skorohod's topology directly, but it is crucial to work on a Polish space to apply fundamental theorems of probability theory, such as the existence of conditional laws. The Borel sigma-field of Ω is denoted by $\mathscr{F}$.

We then introduce the coordinate process $X = (X_t, t \geq 0)$, where

$$X_t = X_t(\omega) = \omega(t).$$

We also write $\zeta = \zeta(\omega)$ for the lifetime of X and

$$X_{s-} = X_{s-}(\omega) = \omega(s-) \quad , \quad \Delta X_s = X_s - X_{s-}$$

respectively for the left limit and the jump at time $s \in (0, \zeta)$. The family of mappings $\theta_t : \Omega \to \Omega$ and $\mathrm{k}_t : \Omega \to \Omega$ $(t \geq 0)$, specified by

$$\theta_t \omega(s) = \omega(t+s) \qquad (s \geq 0)$$

and

$$\mathrm{k}_t \omega(s) = \begin{cases} \omega(s) & \text{if } s < t\,, \\ \partial & \text{otherwise} \end{cases}$$

are called the translation and the killing operators, respectively.

Suppose that $\mathbb{P}$ is a probability measure on $(\Omega, \mathscr{F})$, and Y is a random variable, say taking values in $\mathbb{R}^d$. We denote the expectation of Y under $\mathbb{P}$ by $\mathbb{E}(Y)$ whenever it makes sense. We then write $\mathbb{E}(Y, \Lambda_1, \cdots, \Lambda_k)$ for $\mathbb{E}(\mathbf{1}_\Lambda Y)$ with $\Lambda = \Lambda_1 \cap \cdots \cap \Lambda_k$, where $\Lambda_1, \cdots, \Lambda_k \in \mathscr{F}$, and $\mathbb{E}(Y \mid \mathscr{G})$ for the conditional expectation given some subfield $\mathscr{G}$. Finally, we denote either by $\mathbb{P}(Y \in \cdot)$ or by $\mathbb{P}(Y \in dy)$ the distribution of Y under $\mathbb{P}$. We say that a family $(\mathbb{P}(\cdot | Y = y), y \in \mathbb{R}^d)$ of laws on $(\Omega, \mathscr{F})$ is a version of the conditional law $\mathbb{P}$ given Y if the mapping $y \to \mathbb{P}(\cdot | Y = y)$ is measurable, $\mathbb{P}(Y = y | Y = y) = 1$ for all $y \in \mathbb{R}^d$, and

$$\mathbb{P}(\Lambda) = \int_{\mathbb{R}^d} \mathbb{P}(\Lambda \mid Y = y) \mathbb{P}(Y \in dy), \qquad \Lambda \in \mathscr{F}.$$

We refer e.g. to chapter III in Dellacherie and Meyer (1975) for the existence of conditional laws.

2. Infinitely divisible distributions

Consider a probability measure μ on $\mathbb{R}^d$, and its characteristic function

$$\mathscr{F}\mu(\lambda) = \int_{\mathbb{R}^d} \exp\{\mathrm{i}\langle \lambda, x \rangle\} \mu(dx) \quad (\lambda \in \mathbb{R}^d).$$

The law μ is called *infinitely divisible* if for any positive integer n, there exists a probability measure μ_n with characteristic function $\mathscr{F}\mu_n$ such that $\mathscr{F}\mu = (\mathscr{F}\mu_n)^n$. In other words, μ can be expressed as the n-th convolution power of μ_n. The simplest examples of infinitely divisible laws are Dirac point masses, Gaussian and stable distributions, and in dimension $d = 1$, Poisson and Gamma distributions.

Assume now that μ is infinitely divisible. Then its characteristic function never vanishes and can be expressed as follows. There is a unique continuous function $\Psi : \mathbb{R}^d \to \mathbb{C}$, called the *characteristic exponent* of μ, such that $\Psi(0) = 0$ and

$$\mathscr{F}\mu(\lambda) = \exp\{-\Psi(\lambda)\} \quad (\lambda \in \mathbb{R}^d).$$

We see that if μ_1 and μ_2 are two infinitely divisible laws with respective characteristic exponents Ψ_1 and Ψ_2, then the convolution $\mu_1 \star \mu_2$ is again infinitely divisible with characteristic exponent $\Psi_1 + \Psi_2$.

The starting point of many studies of infinitely divisible laws is the famous Lévy-Khintchine formula (see for instance section 7.6 in Chung (1968), or chapter XVII in Feller (1971)) which determines the class of characteristic functions corresponding to infinitely divisible laws.

Lévy-Khintchine formula *A function $\Psi : \mathbb{R}^d \to \mathbb{C}$ is the characteristic exponent of an infinitely divisible probability measure on $\mathbb{R}^d$ if and only if there are $a \in \mathbb{R}^d$, a positive semi-definite quadratic form Q on $\mathbb{R}^d$, and a measure Π on $\mathbb{R}^d - \{0\}$ with $\int (1 \wedge |x|^2)\Pi(dx) < \infty$ such that*

$$\Psi(\lambda) = \mathrm{i}\langle a, \lambda\rangle + \frac{1}{2}Q(\lambda) + \int_{\mathbb{R}^d} \left(1 - \mathrm{e}^{\mathrm{i}\langle \lambda, x\rangle} + \mathrm{i}\langle \lambda, x\rangle \mathbf{1}_{\{|x|<1\}}\right)\Pi(dx) \qquad (1)$$

for every $\lambda \in \mathbb{R}^d$.

The parameters a, Q, and Π appearing in (1) are determined by Ψ, and their probabilistic meanings will be clarified in section I.1. The measure Π is called the *Lévy measure* of μ and the quadratic form Q the *Gaussian coefficient* . We mention that some authors use a slightly different expression for the Lévy-Khintchine formula. Specifically, the cut-off function $\mathbf{1}_{\{|x|<1\}}$ is replaced by a bounded smooth function which is equivalent to 1 at the origin, the most common being $(1 + |x|^2)^{-1}$. Such a change in the choice of the cut-off function does not alter the Lévy measure and the Gaussian coefficient, but the parameter a has to be replaced by

$$a' = a + \int_{\mathbb{R}^d} x \left(\frac{1}{1 + |x|^2} - \mathbf{1}_{\{|x|<1\}}\right) \Pi(dx).$$

3. Martingales

Consider a probability space $(\Omega, \mathscr{F}, \mathbb{P})$ endowed with a filtration $(\mathscr{F}_t)_{t\geq 0}$, i.e. an increasing family of sub-fields, which fulfils the usual conditions. That is each $\mathscr{F}_t$ is $\mathbb{P}$-complete and $\mathscr{F}_t = \bigcap_{s>t} \mathscr{F}_s$ for every t. A real-valued stochastic process $M = (M_t, 0 \leq t < \infty)$ is a *martingale* if

$$\mathbb{E}(M_t \mid \mathscr{F}_s) = M_s, \qquad 0 \leq s \leq t.$$

(It is implicit here that $\mathbb{E}(|M_t|) < \infty$ for all t.) We say that M is right-continuous if its sample paths are right-continuous a.s., and *uniformly integrable* if there exists an increasing function $f : [0, \infty) \to [0, \infty)$ with $x = o(f(x))$ as x goes to ∞, such that $\sup\{\mathbb{E}(f(|M_t|)) : t \geq 0\} < \infty$.

We assume from now on that M is a right-continuous martingale. The following key results are due to Doob, and we refer to chapter VI of Dellacherie and Meyer (1980) for a complete account.

Maximal inequality *For every $t > 0$, we have*

$$\mathbb{E}\left(\sup\{|M_s|^2 : 0 \le s \le t\}\right) \le 4\mathbb{E}(|M_t|^2).$$

A nonnegative random variable T is called a *stopping time* if for every $t \ge 0$, $\{T \le t\} \in \mathcal{F}_t$.

Optional sampling theorem *Suppose that T is a stopping time, a.s. finite.*

(i) *The stopped process $(M_{T\wedge t}, t \ge 0)$ is again a martingale.*

(ii) *Suppose moreover that M is uniformly integrable. Then $\mathbb{E}(M_T) = \mathbb{E}(M_0)$.*

Convergence theorem *Suppose that M is uniformly integrable. Then $\lim_{t\to\infty} M_t = M_\infty$ exists a.s. and in $L^1(\mathbb{P})$, and $M_t = \mathbb{E}(M_\infty \mid \mathcal{F}_t)$ for all t.*

4. Poisson processes

The proofs of the results stated in this section and the next can be found in section XII.1 in Revuz and Yor (1994).

The Poisson distribution with parameter (or intensity) $c > 0$ is the probability measure on integers which assigns mass $e^{-c}c^k/(k!)$ at point $k \in \mathbb{N}$. Its characteristic function is

$$\sum_{k=0}^{\infty} e^{i\lambda k} e^{-c}\frac{c^k}{k!} = \exp\{-c(1-e^{i\lambda})\} \quad , \quad \lambda \in \mathbb{R}.$$

The Poisson distribution is infinitely divisible, and the results of section I.1 below guarantee the existence of a unique (in law) increasing right-continuous process N with stationary independent increments, called a Poisson process of parameter (or intensity) c, such that for each $t > 0$, N_t has a Poisson distribution with parameter ct. One can also construct N directly as follows. Consider a probability measure $\mathbb{P}$ and a sequence $\tau_1, \cdots, \tau_n, \cdots$ of independent exponential variables with parameter c, that is $\mathbb{P}(\tau_i > s) = e^{-cs}$ for $s \ge 0$. Introduce the partial sums $S_n = \tau_1 + \cdots + \tau_n$, $n \in \mathbb{N}$, so that S_n has the Gamma(c, n) distribution,

$$\mathbb{P}(S_n \in ds) = \frac{c^n}{(n-1)!} s^{n-1} e^{-cs}\, ds \qquad (s \ge 0) .$$

Then consider the right-continuous inverse $N_t = \sup\{n \in \mathbb{N} : S_n \leq t\}$ ($t \geq 0$), so that for every $t \geq 0$ and $k \in \mathbb{N}$,

$$\mathbb{P}(N_t = k) = \mathbb{P}(S_k \leq t, S_{k+1} > t) = \int_0^t \frac{c^k}{(k-1)!} s^{k-1} e^{-cs} e^{-c(t-s)}\, ds$$
$$= e^{-ct}(ct)^k/(k!)\,.$$

On the other hand, it follows easily from the so-called lack-of-memory property of the exponential law that for every $0 \leq s \leq t$, the increment $N_{t+s} - N_t$ has the Poisson distribution with parameter cs and is independent of the sigma-field generated by $(N_u, u \leq t)$.

Next, let $(\mathcal{G}_t)$ be a filtration which satisfies the usual conditions. We say that N is a $(\mathcal{G}_t)$-Poisson process if N is a Poisson process which is adapted to $(\mathcal{G}_t)$ and for every $s, t \geq 0$, the increment $N_{t+s} - N_t$ is independent of $\mathcal{G}_t$. In particular, N is a $(\mathcal{G}_t)$-Poisson process if $(\mathcal{G}_t)$ is the natural filtration of N.

There are three important families of martingales related to a $(\mathcal{G}_t)$-Poisson process. First, one says that a process $H = (H_t, t \geq 0)$ is *predictable* if it is measurable in the sigma-field generated by the left-continuous adapted processes. If H is a real-valued predictable process with $\mathbb{E}(\int_0^t |H_s| ds) < \infty$ for all $t \geq 0$ and if $N = (N_t, t \geq 0)$ is a $(\mathcal{G}_t)$-Poisson process with parameter $c > 0$, then the *compensated integral*

$$M_t = \int_0^t H_s dN_s \; - \; c \int_0^t H_s ds \quad (t \geq 0)$$

is a $(\mathcal{G}_t)$-martingale. If moreover $\mathbb{E}(\int_0^t H_s^2 ds) < \infty$, then

$$M_t^2 - c \int_0^t H_s^2 ds \quad (t \geq 0)$$

is also a martingale. Finally, if H is predictable and bounded, then the same holds for the exponential process

$$\exp\left\{ \int_0^t H_s dN_s + c \int_0^t (1 - e^{H_s}) ds \right\} \quad (t \geq 0)\,.$$

Here, the various integrals with dN_s as integrator are taken in the sense of Stieltjes.

We conclude this section by recalling a well-known criterion for the independence of Poisson processes.

Proposition 1 *Let $N^{(i)}, i = 1, \cdots, d$, be $(\mathcal{G}_t)$-Poisson processes. They are independent if and only if they never jump simultaneously, that is for* *every i, j with i ≠ j*

$$N_t^{(i)} - N_{t-}^{(i)} = 0 \text{ or } N_t^{(j)} - N_{t-}^{(j)} = 0 \quad \text{for all } t > 0, \text{ a.s.},$$

where $N_{t-}^{(k)}$ *stands for the left limit of* $N^{(k)}$ *at time t.*

It is crucial in Proposition 1 to assume that the $N^{(i)}$ are Poisson processes in the same filtration. Otherwise, it is easy to construct Poisson processes which never jump simultaneously and which are not independent.

5. Poisson measures and Poisson point processes

Let E be a Polish space and ν a sigma-finite measure on E. We call a random measure φ on E *a Poisson measure with intensity* ν if it satisfies the following. For every Borel subset B of E with $\nu(B) < \infty$, $\varphi(B)$ has a Poisson distribution with parameter $\nu(B)$, and if $B_1, \cdots, B_n$ are disjoint Borel sets, the variables $\varphi(B_1), \cdots, \varphi(B_n)$ are independent. Plainly, φ is then a sum of Dirac point masses.

One can construct Poisson measures as follows. First, assume that the total mass of ν is finite, and put $c = \nu(E)$. Let $\xi_1, \cdots, \xi_n, \cdots$ be a sequence of independent identically distributed random variables with common law $c^{-1}\nu$ and a Poisson variable N with parameter c independent of the ξ_n's. The random measure

$$\varphi = \sum_{j=1}^{N} \delta_{\xi_j},$$

where δ_ϵ stands for the Dirac point mass at $\epsilon \in E$, is a Poisson measure with intensity ν. If ν is merely sigma-finite, there exists a partition $(E_n, n \in \mathbb{N})$ of E into Borel sets such that $\nu(E_n) < \infty$ for every integer n. Then we can construct a sequence φ_n of independent Poisson measures with respective characteristic measures $\mathbf{1}_{E_n}\nu$, and $\varphi = \sum_n \varphi_n$ is a Poisson measure with intensity ν.

We then consider the product space $E \times [0, \infty)$, the measure $\mu = \nu \otimes dx$, and a Poisson measure φ on $E \times [0, \infty)$ with intensity μ. It is easy to check that a.s., $\varphi(E \times \{t\}) = 0$ or 1 for all $t \geq 0$. This enables us to represent φ in terms of a stochastic process taking values in $E \cup \{\Upsilon\}$, where Υ is an isolated additional point. Specifically, if $\varphi(E \times \{t\}) = 0$, then put $e(t) = \Upsilon$. If $\varphi(E \times \{t\}) = 1$, then the restriction of φ to the section $E \times \{t\}$ is a Dirac point mass, say at (ϵ, t), and we put $e(t) = \epsilon$. We can now express the Poisson measure as

$$\varphi = \sum_{t \geq 0} \delta_{(e(t),t)}.$$

The process $e = (e(t), t \geq 0)$ is called a *Poisson point process* with characteristic measure ν. We denote its natural filtration after the usual completion by $(\mathcal{G}_t)$.

For every Borel subset B of E, we call

$$N_t^B = \text{Card}\{s \leq t : e(s) \in B\} = \varphi(B \times [0,t]) \qquad (t \geq 0)$$

the *counting process* of B. It is a $(\mathcal{G}_t)$-Poisson process with parameter $\nu(B)$. Conversely, suppose that $e = (e(t), t \geq 0)$ is a stochastic process taking values in $E \cup \{\Upsilon\}$ such that, for every Borel subset B of E, the counting process $N_t^B = \text{Card}\{s \leq t : e(s) \in B\}$ is a Poisson process with intensity $\nu(B)$ in a given filtration $(\mathcal{G}_t)$. Then observe that counting processes associated to disjoint Borel sets never jump simultaneously and thus are independent according to Proposition 1. One then deduces that the associated random measure $\varphi = \sum_{t \geq 0} \delta_{(e(t),t)}$ is a Poisson measure with intensity μ.

We next present a useful probabilistic interpretation of the characteristic measure ν.

Proposition 2 *Let B be a Borel set with $0 < \nu(B) < \infty$. The first entrance time of e into B, $T_B = \inf\{t \geq 0 : e(t) \in B\}$, is a $(\mathcal{G}_t)$-stopping time and we have*

(i) *T_B has an exponential distribution with parameter $\nu(B)$.*

(ii) *The random variable $e(T_B)$ is independent of T_B and has the law $\nu(\cdot \mid B)$, that is for every Borel set A,*

$$\mathbb{P}(e(T_B) \in A) = \nu(A \cap B)/\nu(B) .$$

(iii) *The process e' given by $e'(t) = \Upsilon$ if $e(t) \in B$ and $e'(t) = e(t)$ otherwise $(t \geq 0)$ is a Poisson point process with characteristic measure $1_{B^c}\nu$, and is independent of $(T_B, e(T_B))$.*

The process $(e_t, 0 \leq t \leq T_B)$ is called stopped at the first point in B, its law is characterized by Proposition 2.

In practice, it is important to calculate certain expressions in terms of the characteristic measure. The following two formulas are the most useful:

Compensation formula *Let $H = (H_t, t \geq 0)$ be a predictable process taking values in the space of nonnegative measurable functions on $E \cup \{\Upsilon\}$, such that $H_t(\Upsilon) = 0$ for all $t \geq 0$. We have*

$$\mathbb{E}\left(\sum_{0 \leq t < \infty} H_t(e(t))\right) = \mathbb{E}\left(\int_0^\infty dt \int_E d\nu(\epsilon) H_t(\epsilon)\right) .$$

Exponential formula *Let f be a complex-valued Borel function on $E \cup \{\Upsilon\}$ with $f(\Upsilon) = 0$ and*

$$\int_E \nu(d\epsilon)|1 - e^{f(\epsilon)}| < \infty .$$

We have for every $t \geq 0$

$$\mathbb{E}\left(\exp\left\{\sum_{0\leq s\leq t} f(e(s))\right\}\right) = \exp\left\{-t\int_E \nu(d\epsilon)(1 - e^{f(\epsilon)})\right\} .$$

These two formulas are easy to prove when the space E is finite, using respectively the first and the third special martingale of section 4. The general case then follows from a monotone class theorem.

We conclude this section with a useful inequality which is a consequence of Doob's maximal inequality applied to the first special martingale of section 4.

Maximal inequality for compensated sums *Let f be a Borel function on $E \cup \{\Upsilon\}$ with $f(\Upsilon) = 0$. We have for every fixed $T > 0$*

$$\mathbb{E}\left(\sup\left\{\left|\sum_{0\leq s\leq t} f(e(s)) - t\int_E f(\epsilon)d\nu(\epsilon)\right|^2, 0 \leq t \leq T\right\}\right) \leq 4T\int_E f(\epsilon)^2 d\nu(\epsilon) .$$

6. Brownian motion

A real-valued stochastic process $B = (B_t, t \geq 0)$ is a (linear) *Brownian motion* if its sample paths are continuous a.s., its law at any fixed time $t > 0$ is the centred Gaussian distribution with variance t,

$$\mathbb{P}(B_t \in dx) = (2\pi t)^{-1/2}\exp\{-x^2/2t\}dx ,$$

and its increments are independent in the sense that for any $s, t > 0$, $B_{t+s} - B_t$ is independent of the σ-field generated by $(B_u, 0 \leq u \leq t)$. Note that this implies that $B_{t+s} - B_t$ has the centred Gaussian distribution with variance s, so that B is a Gaussian process with stationary (or, homogeneous) independent increments. Finally, a process $(B^1, \cdots, B^d)$ taking values in the d-dimensional Euclidean space is a Brownian motion if its coordinates $B^1, \cdots, B^d$ are independent linear Brownian motions.

There are several different constructions of Brownian motion; here is one of the simplest (see e.g. section I.1 in Revuz and Yor (1994)). First, a standard result guarantees the existence of a centred Gaussian process $\widetilde{B} = (\widetilde{B}_t, t \geq 0)$ with covariance $\mathbb{E}(\widetilde{B}_t\widetilde{B}_s) = s \wedge t$. Then one applies Kolmogorov's criterion to verify that there is a continuous version B of $\widetilde{B}$, that is $B_t = \widetilde{B}_t$ a.s. for every $t \geq 0$. Actually, Kolmogorov's criterion shows that the sample paths of B are a.s. Hölder-continuous with order $1/2-\varepsilon$ on every compact time interval and for every $\varepsilon > 0$. On the other hand, B is a.s. nowhere Hölder-continuous of order $1/2+\varepsilon$. In particular, B is nowhere differentiable and its total variation is infinite a.s. on any non-trivial time interval.

7. Regular variation and Tauberian theorems

In this section, we recall the basis of Karamata's theory, and refer to the first chapter of Bingham, Goldie and Teugels (1987) for details and proofs.

A measurable function $\ell : (0,\infty) \to (0,\infty)$ is *slowly varying* at 0+ (respectively, at ∞) if for every $\lambda > 0$, $\lim(\ell(\lambda x)/\ell(x)) = 1$ as x tends to 0+ (respectively, to ∞).

Representation of a slowly varying function *The function ℓ is slowly varying at 0+ (respectively, at ∞) if and only if it may be written in the form*

$$\ell(x) = \exp\left\{c(x) + \int_1^x \frac{\varepsilon(u)du}{u}\right\}$$

where $c, \varepsilon : (0,\infty) \to \mathbb{R}$ are two bounded measurable functions with $\lim c(x) = d \in \mathbb{R}$ and $\lim \varepsilon(x) = 0$ as $x \to 0+$ (respectively, as $x \to \infty$).

A measurable function $f : (0,\infty) \to (0,\infty)$ is *regularly varying* at 0+ (respectively, at ∞) if for every $\lambda > 0$, the ratio $f(\lambda x)/f(x)$ converges in $(0,\infty)$ as x tends to 0+ (respectively, to ∞).

Characterization of a regularly varying function *If the function f is regularly varying at 0+ (respectively, at ∞), then there exists a real number ρ, called the index, such that*

$$\lim(f(\lambda x)/f(x)) = \lambda^\rho \qquad (x \to 0+\ ,\ resp.\ \infty)$$

for every $\lambda > 0$. Moreover, $\ell(x) = f(x)x^{-\rho}$ is slowly varying at 0+ (respectively, at ∞).

Suppose now that $U : [0,\infty) \to [0,\infty)$ is an increasing right-continuous function, denote by $U(dx)$ the associated Stieltjes measure (by convention, this measure assigns a mass $U(0)$ at the origin) and by $\mathscr{L}U$ its Laplace transform,

$$\mathscr{L}U(\lambda) = \int_{[0,\infty)} e^{-\lambda x} U(dx) \in [0,\infty] \qquad (\lambda \geq 0).$$

Tauberian theorem *Let $\ell : (0,\infty) \to (0,\infty)$ be slowly varying at 0+ (respectively, at ∞) and $\rho \geq 0$. The following are equivalent:*

(i) $U(x) \sim x^{\rho}\ell(x)/\Gamma(1+\rho) \quad (x \to 0+\ , \text{resp.}\ \infty);$

(ii) $\mathscr{L}U(\lambda) \sim \lambda^{-\rho}\ell(1/\lambda) \quad (\lambda \to \infty,\ \text{resp.}\ 0+).$

It easy to see that if U is regularly varying with index $\rho > 0$, then its inverse function is again regularly varying, but with index $1/\rho$. Another useful property concerns the case when the Stieltjes measure $U(dx)$ is absolutely continuous with a monotone density.

Monotone density theorem *Suppose that $U(dx) = u(x)dx$, where $u : (0,\infty) \to (0,\infty)$ is monotone on some neighbourhood of 0+ (respectively, of ∞). If there exist a positive real number ρ and a function $\ell : (0,\infty) \to (0,\infty)$ that is slowly varying at 0+ (respectively, at ∞) such that*

$$U(x) \sim x^{\rho}\ell(x) \qquad (x \to 0+\ , \text{resp.}\ \infty),$$

then

$$u(x) \sim \rho x^{\rho-1}\ell(x) \qquad (x \to 0+\ , \text{resp.}\ \infty).$$

Conversely, it is easy to show that if $U(dx)$ is absolutely continuous with a density that is regularly varying with index $\rho - 1$ for some $\rho > 0$, then U is regularly varying with index ρ and $U(x) \sim \rho^{-1}xu(x)$.

Regularly varying functions often appear in probability in connection with weak limit theorems. Here is a classical example which has a particular interest for us. Let $(\xi_n, n \in \mathbb{N})$ be a sequence of independent identically distributed random variables taking values in $\mathbb{R}^d$. Suppose that for some sequence $(a_n, n \in \mathbb{N})$ of positive real numbers, the renormalized sum $a_n^{-1}(\xi_1 + \cdots + \xi_n)$ converges in distribution to some non-degenerate law ν (ν is not the point mass at 0), as n goes to ∞. Then ν is a strictly stable law of index $\alpha \in (0,2]$, that is an infinitely divisible law whose characteristic exponent Ψ fulfils $\Psi(\lambda) = |\lambda|^{\alpha}\Psi(\lambda/|\lambda|)$ for all $\lambda \in \mathbb{R}^d - \{0\}$. Moreover, there exists a function $a : (0,\infty) \to (0,\infty)$ which is regularly varying at ∞ with index $1/\alpha$ and such that $a_n = a(n)$ for all $n \in \mathbb{N}$.

I

Lévy Processes as Markov Processes

In this chapter, we introduce Lévy processes and specify their connections with infinitely divisible laws and the Lévy-Khintchine formula. Their first probabilistic properties (quasi-left-continuity, asymptotic behaviour at infinity, ...) are studied using fundamental analytic tools (semigroup, resolvent and infinitesimal generator) via the Markov property.

1. Lévy processes and the Lévy-Khintchine formula

To start with, we introduce the central notion of this text. Recall that we are using the canonical notation set down in section O.1.

Definition (Lévy process) *Let $\mathbb{P}$ be a probability measure on $(\Omega, \mathscr{F})$ with $\mathbb{P}(\zeta = \infty) = 1$. We say that X is a Lévy process for $(\Omega, \mathscr{F}, \mathbb{P})$ if for every $s, t \geq 0$, the increment $X_{t+s} - X_t$ is independent of the process $(X_v, 0 \leq v \leq t)$ and has the same law as X_s. In particular, $\mathbb{P}(X_0 = 0) = 1$.*

One often paraphrases the definition by saying that Lévy processes have stationary (or homogeneous) independent increments . They can be thought of as analogues of random walks in continuous time.

Consider an arbitrary Lévy process $(X, \mathbb{P})$. Using the decomposition

$$X_1 = X_{1/n} + (X_{2/n} - X_{1/n}) + \cdots + (X_{n/n} - X_{(n-1)/n}),$$

we observe that the distribution $\mathbb{P}(X_1 \in \cdot)$ is infinitely divisible, and we denote its characteristic exponent by Ψ,

$$\mathbb{E}(\exp\{i\langle\lambda, X_1\rangle\}) = \exp\{-\Psi(\lambda)\} \qquad (\lambda \in \mathbb{R}^d).$$

By a similar argument, we see that for any rational number $t \geq 0$, $\mathbb{P}(X_t \in \cdot)$ is infinitely divisible as well, and its characteristic function is given by

$$\mathbb{E}(\exp\{i\langle\lambda, X_t\rangle\}) = \exp\{-t\Psi(\lambda)\} \qquad (\lambda \in \mathbb{R}^d)\,. \tag{1}$$

Because X is right-continuous a.s., the mapping $t \to \mathbb{E}(\exp\{i\langle\lambda, X_t\rangle\})$ is right-continuous and (1) holds for all $t \geq 0$. The function $\Psi : \mathbb{R}^d \to \mathbb{C}$ is called the *characteristic exponent* of the Lévy process X. It characterizes the law $\mathbb{P}$ in the sense that two Lévy processes with the same characteristic exponent have the same law. More precisely, they have the same one-dimensional distributions, thus by the homogeneity and independence of the increments, they have the same finite-dimensional distributions, and finite-dimensional distributions determine laws on Ω, see e.g. Jacod and Shiryaev (1987), Lemma VI.3.19. One of our main concerns throughout this book will be to relate analytic properties of the characteristic exponent with the probabilistic behaviour of the Lévy process.

We have already considered two important examples of Lévy processes in sections O.4 and O.6: the Poisson process and Brownian motion for which we have $\Psi(\lambda) = c\left(1 - e^{i\lambda}\right)$ and $\Psi(\lambda) = \frac{1}{2}|\lambda|^2$, respectively. Here is another significant prototype. Let $\xi_1, \cdots, \xi_n, \cdots$ be independent random variables all having the same distribution ν on $\mathbb{R}^d - \{0\}$, and $S(n) = \xi_1 + \cdots + \xi_n$ be the corresponding random walk. Introduce $N = (N_t, t \geq 0)$, a Poisson process with parameter $c > 0$, independent of the ξ_n's. Then it is easy to check using the lack of memory of the exponential law that the process in continuous time,

$$e(t) = \begin{cases} \xi_n & \text{if } N_{t-} < n = N_t, \\ 0 & \text{otherwise,} \end{cases}$$

is a Poisson point process with characteristic measure $c\nu$, where 0 serves as isolated point. The time-changed random walk

$$S \circ N_t = \sum_{i=1}^{N_t} \xi_i = \sum_{0 \leq s \leq t} e(s) \qquad (t \geq 0)$$

is a Lévy process called a *compound Poisson process* with Lévy measure $c\nu$. Finally, it follows from the exponential formula of section O.5 that

$$\mathbb{E}(\exp\{i\langle\lambda, S \circ N_t\rangle\}) = \exp\{-t\psi(\lambda)\}\,, \qquad \lambda \in \mathbb{R}^d\,,$$

where

$$\psi(\lambda) = c\int_{\mathbb{R}^d} (1 - e^{i\langle\lambda, x\rangle})\, \nu(dx)\,,$$

a quantity that can be also written as

$$-ic\int_{\mathbb{R}^d}\langle\lambda,x\rangle\mathbf{1}_{\{|x|<1\}}\nu(dx)+\int_{\mathbb{R}^d}\left(1-e^{i\langle\lambda,x\rangle}+i\langle\lambda,x\rangle\mathbf{1}_{\{|x|<1\}}\right)c\nu(dx). \quad (2)$$

Equation (2) is a special case of the Lévy-Khintchine formula. The Lévy measure is $c\nu$, and it appears as the intensity of jumps of $S\circ N$. This also suggests the possibility of approaching a given infinitely divisible law using a sequence of compound Poisson processes. This idea is actually the key to Theorem 1 below.

Besides compound Poisson processes and Brownian motion, stable processes, which we now introduce, form another important sub-family of Lévy processes. For every $\alpha\in(0,2]$, a Lévy process with characteristic exponent Ψ is called a *stable process with index* α if $\Psi(k\lambda)=k^\alpha\Psi(\lambda)$ for every $k>0$ and $\lambda\in\mathbb{R}^d$. It is then straightforward to check the following *scaling property* : For every $k>0$, the rescaled process $(k^{-1/\alpha}X_{kt},t\geq 0)$ has the same law as X. For $\alpha\neq 2$, the Lévy measure of a stable process of index α can be expressed in polar coordinates $(r,\vartheta)\in[0,\infty)\times S_{d-1}$ (where S_{d-1} is the unit sphere of the Euclidean space $\mathbb{R}^d$) in the form

$$\Pi(dr,d\vartheta)=r^{-\alpha-d}dr\nu(d\vartheta),$$

where ν is some finite measure on S_{d-1}. Because Π is a Lévy measure, one must have $\int_0^\infty(1\wedge r^2)r^{-\alpha-1}dr<\infty$, which explains the restriction on the range of the index α. Stable processes appear in particular in limit theorems (cf. section O.7) and will be studied in depth in chapter VIII.

The main result of this section is that any infinitely divisible probability measure μ on $\mathbb{R}^d$ can be viewed as the distribution of a Lévy process evaluated at time 1 (the converse is obvious). The proof provides an explicit construction of this Lévy process, and sheds a probabilistic light on the Lévy-Khintchine formula.

Theorem 1 *Consider $a\in\mathbb{R}^d$, a positive semi-definite quadratic form Q on $\mathbb{R}^d$ and a measure Π on $\mathbb{R}^d-\{0\}$ such that $\int(1\wedge|x|^2)\Pi(dx)<\infty$. Put for every $\lambda\in\mathbb{R}^d$*

$$\Psi(\lambda)=i\langle a,\lambda\rangle+\frac{1}{2}Q(\lambda)+\int_{\mathbb{R}^d}\left(1-e^{i\langle\lambda,x\rangle}+i\langle\lambda,x\rangle\mathbf{1}_{\{|x|<1\}}\right)\Pi(dx)\,.$$

Then there exists a unique probability measure $\mathbb{P}$ on Ω under which X is a Lévy process with characteristic exponent Ψ. Moreover, the jump process of X, namely $\Delta X=(\Delta X_t,t\geq 0)$, is a Poisson point process with characteristic measure Π.

Proof Consider $B=(B_t,t\geq 0)$, a Brownian motion in $\mathbb{R}^d$, and $\Delta=(\Delta_t,t\geq 0)$, an independent Poisson point process with characteristic measure Π. Let $\sqrt{Q}$ be any matrix such that $\langle\sqrt{Q}\lambda, \sqrt{Q}\lambda\rangle = Q(\lambda)$, and put $X_t^{(1)} = \sqrt{Q}B_t - at$ $(t \geq 0)$. Using the Gaussian property of B, it is immediate that $X^{(1)}$ is a Lévy process with characteristic exponent

$$\Psi^{(1)}(\lambda) = \mathrm{i}\langle a, \lambda\rangle + \frac{1}{2}Q(\lambda)\,.$$

Let us next dwell on large values of Δ and introduce

$$\Delta_t^{(2)} = \begin{cases} \Delta_t & \text{if } |\Delta_t| \geq 1, \\ 0 & \text{otherwise}\end{cases}.$$

Observe that $\Delta^{(2)}$ is a Poisson point process with characteristic measure $\Pi^{(2)}(dx) = \mathbf{1}_{\{|x|\geq 1\}}\Pi(dx)$. The total mass of $\Pi^{(2)}$ is finite and $\Delta^{(2)}$ is discrete. We consider the partial sum, $X_t^{(2)} = \sum_{s\leq t}\Delta_s^{(2)}$ $(t \geq 0)$. This process has stationary independent increments and its sample paths are right-continuous with left limits. Hence $X^{(2)}$ is a Lévy process (actually, a compound Poisson process since the process that counts the jumps of $X^{(2)}$ is a non-degenerate Poisson process), and by the exponential formula of section O.5, its characteristic exponent is

$$\Psi^{(2)}(\lambda) = \int_{\mathbb{R}^d}(1 - \mathrm{e}^{\mathrm{i}\langle\lambda,x\rangle})\mathbf{1}_{\{|x|\geq 1\}}\,\Pi(dx)\,.$$

We then deal with the small values of Δ and introduce

$$\Delta_t^{(3)} = \begin{cases} \Delta_t & \text{if } |\Delta_t| < 1, \\ 0 & \text{otherwise.}\end{cases}$$

Note that $\Delta^{(3)}$ is a Poisson point process with characteristic measure $\Pi^{(3)}(dx) = \mathbf{1}_{\{|x|<1\}}\Pi(dx)$, and is independent of $\Delta^{(2)}$ (because $\Delta^{(2)}$ and $\Delta^{(3)}$ are two Poisson point processes in the same filtration which obviously never jump simultaneously, see Proposition O.1). Consider for every $\varepsilon > 0$ the process of compensated partial sums,

$$X_t^{(\varepsilon,3)} = \sum_{s\leq t}\mathbf{1}_{\{\varepsilon<|\Delta_s|<1\}}\Delta_s \;-\; t\int_{\mathbb{R}^d} x\mathbf{1}_{\{\varepsilon<|x|<1\}}\Pi(dx) \quad (t \geq 0).$$

On the one hand, by the same argument as above, $X^{(\varepsilon,3)}$ is a Lévy process with characteristic exponent

$$\Psi^{(\varepsilon,3)}(\lambda) = \int_{\mathbb{R}^d}(1 - \mathrm{e}^{\mathrm{i}\langle\lambda,x\rangle} + \mathrm{i}\langle\lambda, x\rangle)\mathbf{1}_{\{\varepsilon<|x|<1\}}\,\Pi(dx)\,.$$

On the other hand, the maximal inequality for compensated sums of section O.5 yields for every $t > 0$ and $\eta \in (0, \varepsilon)$,

$$\mathbb{E}(\sup_{s\leq t}|X_s^{(\eta,3)} - X_s^{(\varepsilon,3)}|^2) \leq 4t\int_{\mathbb{R}^d}|x|^2\mathbf{1}_{\{\eta<|x|<\varepsilon\}}\Pi(dx)\,.$$

Because the integral $\int(1 \wedge |x|^2)\Pi(dx)$ converges, this last quantity tends to 0 as ε goes to 0+. Hence $(X^{(\varepsilon,3)}, \varepsilon > 0)$ is a Cauchy family for the norm

$$\|Y\| = \mathbb{E}(\sup\{|Y_s|^2, 0 \leq s \leq t\})^{1/2}\,.$$

Its limit as ε goes to 0, denoted by $X^{(3)}$, has stationary independent increments and its sample path is right-continuous with left limits. It is a Lévy process with characteristic exponent

$$\Psi^{(3)}(\lambda) = \int_{\mathbb{R}^d} (1 - e^{i\langle \lambda, x\rangle} + i\langle \lambda, x\rangle)\mathbf{1}_{\{|x|<1\}}\, \Pi(dx)\,.$$

Moreover $X^{(3)}$ is measurable in the sigma-field generated by $\Delta^{(3)}$ and thus is independent of $X^{(2)}$.

Finally, $X^{(1)}+X^{(2)}+X^{(3)}$ is a Lévy process with characteristic exponent $\Psi = \Psi^{(1)} + \Psi^{(2)} + \Psi^{(3)}$ (because $X^{(1)}$, $X^{(2)}$ and $X^{(3)}$ are independent). By construction, its jump process is Δ. Its law $\mathbb{P}$ on Ω fulfils the requirements of the theorem. □

The proof of Theorem 1 actually says a bit more than was stated. We saw that one can express a Lévy process as the sum of three independent Lévy processes $X^{(1)}$, $X^{(2)}$ and $X^{(3)}$, where $X^{(1)}$ is a linear transform of a Brownian motion with drift (and in particular is continuous), $X^{(2)}$ is a compound Poisson process having only jumps of size at least 1, and finally $X^{(3)}$ is a pure-jump martingale having only jumps of size less than 1. This decomposition is clearly unique. It is interesting to note that the compound Poisson component $X^{(2)}$ does not contribute to the initial sample path behaviour of a Lévy process. This allows us to reduce some general studies to the case when the Lévy measure has compact support; in a picturesque style, one says that one throws away the big jumps.

The Lévy-Khintchine formula has a simpler expression when the sample paths of the Lévy process have bounded variation on every compact time interval a.s. For short, we will then say that the Lévy process has *bounded variation*. Specifically, it follows readily from the exponential formula of section O.5 that the series of the norm of the jumps, $\sum_{0\le s\le t} |\Delta X_s|$, converges for every $t > 0$ a.s. if and only if $\int (1 \wedge |x|)\Pi(dx) < \infty$. In that case, we see that X has bounded variation if and only if its Gaussian component has bounded variation as well, and since Brownian motion has infinite variation, the latter occurs if and only if the quadratic form Q is null. In conclusion, a Lévy process has bounded variation if and only if $Q = 0$ and $\int (1 \wedge |x|)\Pi(dx) < \infty$. In that case, the mapping

$$\lambda \to \int_{\mathbb{R}^d} \langle \lambda, x\rangle \mathbf{1}_{\{|x|<1\}} \Pi(dx)$$

is a well-defined linear function and the characteristic exponent can thus be re-expressed as

$$\Psi(\lambda) = -\mathrm{i}\langle \mathrm{d}, \lambda\rangle + \int_{\mathbb{R}^d} (1 - \mathrm{e}^{\mathrm{i}\langle\lambda,x\rangle})\Pi(dx) \qquad (\lambda \in \mathbb{R}^d),$$

for some $\mathrm{d} \in \mathbb{R}^d$ which is known as the *drift coefficient* . Moreover, it is easy to mimic the proof of Theorem 1 to check that if $\Delta = (\Delta_t, t \geq 0)$ is a Poisson point process with characteristic measure Π, then the process

$$X_t = \mathrm{d}t + \sum_{0\leq s\leq t} \Delta_s, \qquad t \geq 0,$$

is a Lévy process (recall that the series is absolutely convergent a.s. so X has bounded variation) with characteristic exponent Ψ. Of course, a compound Poisson process has bounded variation, and conversely, a Lévy process with bounded variation is a compound Poisson process if and only if its drift coefficient d is null and its Lévy measure Π has a finite mass.

We conclude this section by specifying the asymptotic behaviour at infinity of the characteristic exponent Ψ of a real-valued Lévy process. The multidimensional extension is straightforward, by considering the coordinates.

Proposition 2 *Suppose that the dimension $d = 1$.*

(i) *We have*

$$\lim_{|\lambda|\to\infty} \lambda^{-2}\Psi(\lambda) = Q/2,$$

where $Q \geq 0$ denotes the Gaussian coefficient.

(ii) *If X has bounded variation and drift coefficient d, then*

$$\lim_{|\lambda|\to\infty} \lambda^{-1}\Psi(\lambda) = -\mathrm{i}\mathrm{d}.$$

(iii) *If X is a compound Poisson process, then Ψ is bounded.*

Proof of Proposition 2 We only prove (i), the argument for (ii) is similar and (iii) is obvious. Plainly,

$$\lim_{|\lambda|\to\infty} \lambda^{-2}\left(1 - \mathrm{e}^{\mathrm{i}\lambda x} + \mathrm{i}\lambda x \mathbf{1}_{\{|x|<1\}}\right) = 0 \qquad \text{for every } x.$$

On the other hand, making use of the inequalities

$$|1 - \cos a| \leq 2(1 \wedge a^2) \quad \text{and} \quad |a - \sin a| \leq 2(|a| \wedge |a|^3),$$

it is easy to see that for every λ large enough

$$|\lambda^{-2}\left(1 - e^{i\lambda x} + i\lambda x \mathbf{1}_{\{|x|<1\}}\right)| \le 4(1 \wedge x^2).$$

So the theorem of dominated convergence applies and gives

$$\lim_{|\lambda|\to\infty} |\lambda|^{-2} \int_{\mathbb{R}-\{0\}} \left(1 - e^{i\lambda x} + i\lambda x \mathbf{1}_{\{|x|<1\}}\right) \Pi(dx) = 0.$$

We conclude the proof by the Lévy-Khintchine formula. □

It is interesting to note the following converse of (iii):

Corollary 3 *The characteristic exponent of a Lévy process is bounded only in the compound Poisson case.*

Proof Plainly, we need only consider the dimension $d = 1$. We deduce from Proposition 2(i) that Ψ cannot be bounded unless the Gaussian coefficient Q is null. In that case, the real part of Ψ has the simpler form

$$\Re\Psi(\lambda) = \int_{\mathbb{R}-\{0\}} (1 - \cos \lambda x)\Pi(dx).$$

Then recall that for every $x \in \mathbb{R}$ and $t > 0$

$$1 - e^{-tx^2/2} = \frac{1}{\sqrt{2\pi t}} \int_{-\infty}^{\infty} (1 - \cos \xi x) e^{-\xi^2/2t}\, d\xi.$$

By Fubini's theorem,

$$\begin{aligned}\int_{\mathbb{R}-\{0\}} \left(1 - e^{-tx^2/2}\right) \Pi(dx) &= \frac{1}{\sqrt{2\pi t}} \int_{-\infty}^{\infty} e^{-\xi^2/2t} \Re\Psi(\xi)\, d\xi \\ &\le \sup\{\Re\Psi(\xi) : \xi \in \mathbb{R}\}.\end{aligned}$$

Letting t go to ∞, we see that the Lévy measure is finite, and thus X has bounded variation. Then we apply Proposition 2(ii) and see that the drift coefficient is null. In conclusion, X is a compound Poisson process. □

However, the converse of (ii) is false; there exist Lévy processes with unbounded variation for which $\Psi(\lambda) = o(|\lambda|)$ at infinity. For instance, if we take $a = Q = 0$ and $\Pi(dx) = |x|^{-2}(|\log|x|| + 1)^{-1}dx$, then $\int_0^1 x\Pi(dx) = \infty$ and for every $\lambda > 0$

$$\Psi(\lambda) = 2\int_0^{\infty} \frac{1 - \cos \lambda x}{x^2(|\log x| + 1)} dx = 2\lambda \int_0^{\infty} \frac{1 - \cos y}{y^2(|\log \lambda - \log y| + 1)} dy.$$

It follows that $\lim_{\lambda\to\infty} \lambda^{-1}\Psi(\lambda) = 0$.

Henceforth, $\mathbb{P}$ will always be a probability measure for which X is a Lévy process with characteristic exponent Ψ, and the reference to $(\Omega, \mathcal{F}, \mathbb{P})$ will usually be omitted.

2. Markov property and related operators

Markov processes form one of the most important and best-known families of stochastic processes. Informally, the (simple) Markov property states that the future behaviour of the process after any fixed time depends on the past only through the current value of the process. The deep connections between Markov processes and potential theory lead to remarkable developments, in both the probabilistic and the analytic settings. The property of spatial homogeneity of Lévy processes yields significant simplifications of the theory, which make it even nicer. The purpose of this section is to introduce in the framework of Lévy processes some basic probabilistic and analytic notions related to the Markov property.

To give a rigorous definition for the past at a given time, we consider the filtration $(\mathscr{F}_t)$ associated with the coordinate process, that is $\mathscr{F}_t$ is the $\mathbb{P}$-completed sigma-field generated by $(X_s, s \leq t)$, and $\mathscr{F}_\infty$ stands for the $\mathbb{P}$-completion of $\mathscr{F}$. It follows immediately from the definition that for every $t \geq 0$, the process $X'_{\cdot} = X_{t+\cdot} - X_t$ is independent of $\mathscr{F}_t$. Moreover, its finite-dimensional distributions are the same as that of X, and since X' has right-continuous paths with left limits, X' again has the law $\mathbb{P}$ (because finite-dimensional distributions determine laws on Ω, see Jacod and Shiryaev (1987), Lemma VI.3.19). These two statements are usually referred to as the *simple Markov property* of Lévy processes.

We then observe that the Kolmogorov zero-one law entails a continuity property for the filtration $(\mathscr{F}_t)$.

Proposition 4 *The filtration $(\mathscr{F}_t)$ is right-continuous , that is $\mathscr{F}_t = \bigcap_{s>t} \mathscr{F}_s$ for every $t \geq 0$.*

Proof We first establish the assertion for $t = 0$. For every integer $n > 0$, introduce $\xi_n = \left(X_{t+2^{-n}} - X_{2^{-n}}, 0 \leq t \leq 2^{-n}\right)$. The simple Markov property shows that these random variables are independent, and because the path $\left(X_t, 0 \leq t \leq 2^{-n}\right)$ can be recovered from $\xi_{n+1}, \xi_{n+2}, \cdots$, the $\mathbb{P}$-completed sigma-field $\mathscr{G}_n$ generated by $\xi_{n+1}, \xi_{n+2}, \cdots$ coincides with $\mathscr{F}_{2^{-n}}$. According to Kolmogorov zero-one law, the tail sigma-field

$$\mathscr{G}_\infty = \bigcap_{n \in \mathbb{N}} \mathscr{G}_n = \bigcap_{s>0} \mathscr{F}_s$$

is $\mathbb{P}$-trivial, and the proposition is proven for $t = 0$.

For a general $t \geq 0$, we observe that for every $\varepsilon > 0$, $\mathscr{F}_{t+\varepsilon} = \mathscr{F}_t \vee \mathscr{F}'_\varepsilon$, where $\left(\mathscr{F}'_\varepsilon\right)$ is the $\mathbb{P}$-completed filtration generated by $X'_{\cdot} = X_{t+\cdot} - X_t$. The simple Markov property entails that $\mathscr{F}_t$ and $\mathscr{F}'_\varepsilon$ are independent, and we know from the foregoing that the tail sigma-field $\bigcap_{\varepsilon>0} \mathscr{F}'_\varepsilon$ is $\mathbb{P}$-trivial. The argument for Kolmogorov's zero-one law now applies under the conditional probability $\mathbb{P}(\cdot \mid \mathscr{F}_t)$ and shows that

$$\bigcap_{s>t} \mathscr{F}_s = \bigcap_{\varepsilon>0} (\mathscr{F}_t \vee \mathscr{F}'_\varepsilon) = \mathscr{F}_t .$$

We emphasize that the independence between $\mathscr{F}_t$ and $\mathscr{F}'_\varepsilon$ is crucial in this reasoning. □

In particular, the initial sigma-field $\mathscr{F}_0$ is $\mathbb{P}$-trivial. This property is often referred to as the *Blumenthal zero-one law* , because it was noted by Blumenthal for a wide class of Markov processes. This simple remark is very useful in practice, and here is a typical application. Suppose that X is real-valued, and consider the event $\Lambda = \{X_t > 0$ for some arbitrarily small $t > 0\}$. Plainly $\Lambda \in \bigcap_{s>0} \mathscr{F}_s = \mathscr{F}_0$ and the probability that X immediately enters the positive half-line is necessarily 0 or 1. Since obviously X leaves $\{0\}$ immediately a.s. whenever X is not a compound Poisson process, we deduce that in that case, either X immediately enters the positive half-line a.s. or X immediately enters the negative half-line a.s. (of course both events may occur simultaneously).

Next, we introduce some notation related to the simple Markov property. For every $x \in \mathbb{R}^d$, we denote by $\mathbb{P}_x$ the law of $X + x$ under $\mathbb{P}$, that is the law of the Lévy process started from x. In the sequel, we will write indifferently $\mathbb{P}_0$ or $\mathbb{P}$.

We consider the family of convolution operators on $L^\infty(\mathbb{R}^d)$ indexed by $t \geq 0$ and given for every $f \in L^\infty(\mathbb{R}^d)$ by

$$P_t f(x) = \mathbb{E}_x(f(X_t)) = \int_{\mathbb{R}^d} f(x+y)\mathbb{P}(X_t \in dy) .$$

We see that $(P_t, t \geq 0)$ is a Markov *semigroup* , in the sense that $P_0 =$Id, $P_t \circ P_s = P_{t+s}$ for every $t, s \geq 0$ and $0 \leq P_t f \leq 1$ whenever $0 \leq f \leq 1$. It will also be convenient to use the notation $P_t g(x) = \mathbb{E}_x(g(X_t)) \in [0, \infty]$ for any measurable function $g \geq 0$.

Consider now the sub-space $\mathscr{C}_0$ of $L^\infty(\mathbb{R}^d)$ consisting of continuous functions which tend to 0 at infinity, and equipped with the uniform norm.

Proposition 5 *The semigroup $(P_t, t \geq 0)$ has the Feller property , that is, for every $f \in \mathscr{C}_0$:*

(i) *$P_t f \in \mathscr{C}_0$ for every $t \geq 0$.*
(ii) *$\lim_{t\to 0} P_t f = f$ (uniformly).*

Proof (i) follows from the theorem of dominated convergence applied to $\mathbb{E}(f(X_t + x))$. The same argument using the right-continuity of the paths and the uniform continuity of f entails (ii). □

Next we reinforce the simple Markov property by showing that the fixed time t can be replaced by certain random times (viz., random variables with values in $[0, \infty]$). We say that a random time T is a *stopping time* if for every $t \geq 0$, the event $\{T \leq t\}$ is in the sigma-field $\mathcal{F}_t$. The filtration being right-continuous, we see that T is a stopping time if and only if $\{T < t\} \in \mathcal{F}_t$ for all $t \geq 0$. For an arbitrary random time R, we denote by $\mathcal{F}_R$ the $\mathbb{P}$-completion of the sigma-field generated by the process killed at time R, $X \circ \mathrm{k}_R$, and the variable $\mathbf{1}_{\{R<\infty\}} X_R$. It can be checked that when T is a stopping time, $\mathcal{F}_T$ consists of the events Λ such that $\{T \leq t\} \cap \Lambda \in \mathcal{F}_t$ for all $t \geq 0$.

We now assert the (strong) Markov property of Lévy processes.

Proposition 6 (Markov property) *Let T be a stopping time with $\mathbb{P}(T < \infty) > 0$. Then conditionally on $\{T < \infty\}$, the process $(X_{T+t} - X_T, t \geq 0)$ is independent of $\mathcal{F}_T$ and has the law $\mathbb{P}$.*

Proof It suffices to establish the statement when T is finite a.s. When T is deterministic, this is just the simple Markov property. It implies the result immediately when T is an elementary stopping time (i.e. T takes values in a discrete set). When T is an arbitrary stopping time, there exists a decreasing sequence of elementary stopping times T_n with limit T (for instance one can take $T_n = 2^{-n}[2^n T + 1]$, where $[\cdot]$ stands for the integer part). The result now follows from the right-continuity of the paths. □

The Markov property can be rephrased using the translation operators as follows. For every stopping time $T < \infty$ a.s., the conditional law of the shifted process $X \circ \theta_T = (X_{T+t}, t \geq 0)$ given $\mathcal{F}_T$ is $\mathbb{P}_x$, with $x = X_T$. Observe also that the same argument as in the proof of Proposition 4 shows that then

$$\mathcal{F}_T = \mathcal{F}_{T+} = \bigcap_{\varepsilon>0} \mathcal{F}_{T+\varepsilon}.$$

We know from the definition that the sample path of a Lévy process has left limits at each time, a.s. The Feller property of the semigroup yields an important reinforcement called *quasi-left-continuity* . See also Exercise 3 at the end of this chapter.

Proposition 7 *Let* $(T_n, n \in \mathbb{N})$ *be an increasing sequence of stopping times with* $\lim_{n\to\infty} T_n = T$ *a.s. Then* $\lim_{n\to\infty} X_{T_n} = X_T$ *a.s. on* $\{T < \infty\}$. *In particular, if* $T_n < T$ *a.s., then* X *is continuous at time* T *a.s. on* $\{T < \infty\}$.

Proof With no loss of generality, we may assume that $T < \infty$ and $T_n < T$ for all n a.s., and then $\lim_{n\to\infty} X_{T_n} = X_{T-}$ a.s. Let f and g be two functions in $\mathscr{C}_0$. For every $t > 0$, $T_n + t$ increases to $T + t$ a.s. and thus

$$\lim_{n\to\infty} \mathbb{E}(f(X_{T_n})g(X_{T_n+t})) = \mathbb{E}(f(X_{T-})g(X_{T+t-}))\,.$$

By the right-continuity of the paths,

$$\lim_{t\to 0+} \mathbb{E}(f(X_{T-})g(X_{T+t-})) = \mathbb{E}(f(X_{T-})g(X_T))\,.$$

On the other hand, the Markov property at time T_n yields

$$\mathbb{E}(f(X_{T_n})g(X_{T_n+t})) = \mathbb{E}(f(X_{T_n})P_t g(X_{T_n}))\,.$$

This quantity converges to $\mathbb{E}(f(X_{T-})P_t g(X_{T-}))$ as $n \to \infty$. Now we let t decrease to 0, so by the Feller property,

$$\lim_{t\to 0+} \mathbb{E}(f(X_{T-})P_t g(X_{T-})) = \mathbb{E}(f(X_{T-})g(X_{T-}))\,.$$

In conclusion we have

$$\mathbb{E}(f(X_{T-})g(X_{T-})) = \mathbb{E}(f(X_{T-})g(X_T))\,,$$

that is $X_{T-} = X_T$ a.s.

Finally, if $T_n < T < \infty$ a.s., the existence of a left limit at time T gives $X_{T-} = X_T$ a.s., that is X is continuous at T a.s. □

Propositions 6 and 7 emphasize the key rôle of stopping times in the study of Lévy processes. An important class of examples of stopping times is provided by the following random times. For every $B \subseteq \mathbb{R}^d$, we introduce

$$T_B = \inf\{t \geq 0 : X_t \in B\}\ ,\ T'_B = \inf\{t > 0 : X_t \in B\}.$$

One calls T_B the first passage (or entrance) time into B and T'_B the first hitting time of B. For simplicity, we will focus on the case when B is either open or closed. The study can be extended to the case when B is only analytic (in particular when B is a Borel set), but this requires a much deeper analysis relying on Choquet's theory of capacities, see for instance section I.10 in Blumenthal and Getoor (1968).

Corollary 8 (i) *If B is open, then T_B is a stopping time and $X_{T_B} \in \overline{B}$ a.s. on $\{T_B < \infty\}$, where $\overline{B}$ denotes the closure of B.*

(ii) *If B is closed and $(B_n, n \in \mathbb{N})$ is a decreasing sequence of open neighbourhoods of B with $\bigcap \overline{B_n} = B$, then T_{B_n} increases a.s. to T_B. This applies in particular when B_n is the set of points at distance from B less than $1/n$, so T_B is a stopping time with $X_{T_B} \in B$ a.s. on $\{T_B < \infty\}$.*

(iii) *If B is open or closed, then T'_B is a stopping time and $X_{T'_B} \in \overline{B}$ a.s. on $\{T'_B < \infty\}$.*

Proof (i) Using the right-continuity of the paths, we see that for B open and for each fixed $t > 0$, $T_B < t$ if and only if there exists a rational number $s < t$ such that $X_s \in B$. Hence $\{T_B < t\} \in \mathcal{F}_t$, that is T_B is a stopping time. On the event $\{T_B < \infty\}$, either $X_{T_B} \in B$ or X visits B immediately after time T_B. In the second case, the right-continuity implies that $X_{T_B} \in \overline{B}$.

(ii) Plainly $(T_{B_n}, n \in \mathbb{N})$ is an increasing sequence of stopping times, we denote its limit by T. We have $X_t \notin B_n$ for all $t < T_{B_n}$, so $T_B \geq T$. On the other hand, according to Proposition 7 and (i), $X_T \in B$ on the event $\{T < \infty\}$, and thus $T = T_B$. The last assertion is straightforward.

(iii) The (simple) Markov property and (i–ii) show that for every $\varepsilon > 0$, $T^{\varepsilon}_B := \inf\{t \geq \varepsilon : X_t \in B\}$ is a stopping time. Because T^{ε}_B decreases to T'_B as ε decreases to 0, this entails (iii). □

By the right-continuity of the paths, $T_B = T'_B$ $\mathbb{P}_x$-a.s. for every $x \in B^c \cup B^o$, where B^o is the interior of B and $B^c = \mathbb{R}^d - B$. In particular, $\mathbb{P}_\cdot(T_B = T'_B) = 1$ everywhere if B is open. Nonetheless, when B is closed, the preceding equality may fail for some points of the boundary ∂B which are then called *irregular* for B. Note that the Blumenthal zero-one law implies that $\mathbb{P}_x(T'_B > 0) = 1$ whenever x is irregular for B.

Next we introduce another family of linear operators U^q $(q > 0)$ associated with the Lévy process, called the *resolvent operators* . The resolvent operators correspond to the Laplace transform of the semigroup $(P_t, t \geq 0)$. They are given for every measurable function $f \geq 0$ by

$$U^q f(x) = \int_0^\infty e^{-qt} P_t f(x) dt = \mathbb{E}_x \left(\int_0^\infty e^{-qt} f(X_t) dt \right) .$$

From the probabilistic point of view, the resolvent operators describe the distribution of the Lévy process evaluated at independent exponential times. That is, if $\tau = \tau(q)$ has an exponential law with parameter $q > 0$ and is independent of X, then $\mathbb{E}_\cdot (f(X_\tau)) = qU^q f(\cdot)$. It is often more convenient to work with the resolvent operators than with the semigroup, thanks to the smoothing effect of the Laplace transform and to the lack of memory of exponential laws. The family of finite measures $(U^q(x,dy), x \in \mathbb{R}^d)$ associated to the resolvent operator by

$$U^q f(x) = \int_{\mathbb{R}^d} f(y) U^q(x,dy)$$

is called the *resolvent kernel* . It should be clear that U^q is a convolution operator, more precisely, if we introduce the finite measure

$$\Upsilon^q(\cdot) = \int_0^\infty e^{-qt} \mathbb{P}(-X_t \in \cdot)\, dt$$

then

$$U^q f = \Upsilon^q \star f. \tag{3}$$

One readily checks by Fubini's theorem that the semigroup property of $(P_t, t \geq 0)$ yields the so-called *resolvent equation*

$$U^q - U^r + (q-r) U^q U^r = 0 \quad (q,r > 0)$$

which can also be written

$$\Upsilon^q - \Upsilon^r + (q-r)\Upsilon^q \star \Upsilon^r = 0.$$

It is easy to deduce from the resolvent equation that the image of $\mathscr{C}_0$ under U^q does not depend on q, and we denote it by $\mathscr{D}$. Moreover the Feller property implies that for every function f in $\mathscr{C}_0$, $\lim_{q\to\infty} qU^q f = f$ in the sense of uniform convergence. Thus $\mathscr{D}$ is a dense subset of $\mathscr{C}_0$. We refer to section VII.1 in Revuz and Yor (1994) for a detailed argument and further properties.

Observe from the resolvent equation that if f and g are two functions in $\mathscr{C}_0$ such that $U^q f = U^q g$ for some $q > 0$, then $U^q f = U^q g$ for all $q > 0$, and thus, $f = \lim_{q\to\infty} qU^q f = \lim_{q\to\infty} qU^q g = g$. So the mapping $U^q : \mathscr{C}_0 \to \mathscr{D}$ is a bijection. We define the *infinitesimal generator* $\mathscr{A} : \mathscr{D} \to \mathscr{C}_0$ of the Lévy process by the relation

$$U^q(qI - \mathscr{A}) = I\ ,$$

where I denotes the identity operator. The resolvent equation guarantees that this definition is independent of $q > 0$. One calls $\mathscr{D}$ the *domain* of the infinitesimal generator $\mathscr{A}$.

The Fourier transform provides simple expressions for these operators in terms of the characteristic exponent Ψ.

Proposition 9 *Denote the Fourier transform of a function* g *in* $L^1(\mathbb{R}^d)$ *by*

$$\mathscr{F} g(\xi) = \int_{\mathbb{R}^d} e^{i\langle \xi, x\rangle} g(x) dx \quad (\xi \in \mathbb{R}^d).$$

For every $f \in L^1 \cap L^\infty$, we have for every $\xi \in \mathbb{R}^d$

$$\mathcal{F}(P_t f)(\xi) = \exp\{-t\Psi(-\xi)\}\mathcal{F}f(\xi) \quad (t \geq 0),$$

$$\mathcal{F}(U^q f)(\xi) = (q + \Psi(-\xi))^{-1}\mathcal{F}f(\xi) \quad (q > 0).$$

When moreover $f \in \mathcal{D}$ and $\mathcal{A}f \in L^1$,

$$\mathcal{F}(\mathcal{A}f)(\xi) = -\Psi(-\xi)\mathcal{F}f(\xi).$$

Proof We simply rewrite $\mathcal{F}(P_t f)(\xi)$ as

$$\mathbb{E}\left(\int e^{i\langle\xi,x\rangle} f(X_t + x)dx\right) = \mathbb{E}\left(\int e^{i\langle\xi,y-X_t\rangle} f(y)dy\right)$$
$$= \mathbb{E}(\exp\{i\langle -\xi, X_t\rangle\}) \int e^{i\langle\xi,y\rangle} f(y)dy\,.$$

The other formulas follow. □

Because the Fourier transform induces an isomorphism on the Schwartz space $\mathcal{S}$ of rapidly decreasing functions (see e.g. Gel'fand and Shilov (1964)) and $\Psi(\xi) = O(|\xi|^2)$ as $|\xi| \to \infty$ (according to Proposition 2), one deduces from Proposition 9 that $\mathcal{S} \subseteq \mathcal{D}$. For every $f \in \mathcal{S}$, one can use the Lévy-Khintchine formula to invert the Fourier transform $\mathcal{F}(\mathcal{A}f)$ and one gets

$$\mathcal{A}f(x) = -\langle a, f'(x)\rangle + \frac{1}{2}\sum_{i,j=1,\cdots,d} Q_{ij}f''_{ij}(x)$$
$$+ \int_{\mathbb{R}^d} \left(f(x+y) - f(x) - \mathbf{1}_{\{|y|<1\}}\langle y, f'(x)\rangle\right)\Pi(dy).$$

See section 4.1 in Skorohod (1991) for the details of the calculation. This formula extends to functions $f \in \mathcal{C}_b^2$, that is those which are twice continuously differentiable and are bounded together with their derivatives, and which are also in $\mathcal{C}_0$. As a consequence, such functions belong to the domain, see e.g. chapter II in Dynkin (1965). For instance, the infinitesimal generator of an isotropic stable process of index $\alpha \in (0, 2]$ is proportional to the so-called fractional power of the Laplacian, $\mathcal{A} = -c(-\Delta)^{\alpha/2}$. We refer to section V.1 in Stein (1970) for background on fractional powers of the Laplacian.

3. Absolutely continuous resolvents

To expand further the theory of Markov processes, it is often useful to introduce the condition that the resolvent kernels are absolutely continuous.

In particular, this is important for developing potential theory for Markov processes in duality, see Chapter VI in Blumenthal and Getoor (1968).

We have the following characterization.

Proposition 10 *The following assertions are equivalent:*

(i) *For some $q > 0$ and some $x \in \mathbb{R}^d$, the measure $U^q(x, dy)$ is absolutely continuous with respect to the Lebesgue measure.*

(ii) *For every $q > 0$ and every $x \in \mathbb{R}^d$, the measure $U^q(x, dy)$ is absolutely continuous with respect to the Lebesgue measure.*

(iii) *The resolvent operators have the strong Feller property, that is for every $q > 0$ and $f \in L^\infty(\mathbb{R}^d)$, the function $U^q f$ is continuous.*

Proof (i)⇔(ii) Recall that Υ^q is the finite measure which appears in (3). Suppose that (i) holds for $q = q_0 > 0$ and $x = x_0 \in \mathbb{R}^d$. Then Υ^{q_0} is absolutely continuous and it follows immediately from the resolvent equation that Υ^q is absolutely continuous as well for every $q > 0$. We then deduce (ii) from (3).

(ii)⇔(iii) Suppose that (ii) holds and let $\widehat{g^q} \in L^1(\mathbb{R}^d)$ be the density of Υ^q. Then for every bounded measurable function f, the convolution $\widehat{g^q} \star f$ is continuous and (iii) follows from (3). Conversely, suppose that (iii) holds, consider a Borel set N with zero Lebesgue measure, and take $f = \mathbf{1}_N$. Then by (3)

$$\int_{\mathbb{R}^d} U^q f(x)dx = \Upsilon^q(\mathbb{R}^d) \int_{\mathbb{R}^d} f(x)dx = 0$$

since the Lebesgue measure of N is null and the total mass of Υ^q equals $1/q$. So $U^q f(x) = 0$ for almost every $x \in \mathbb{R}^d$, and thus for all $x \in \mathbb{R}^d$ by continuity. That is to say that for every x, the measure $U^q(x, dy)$ gives no mass to sets of zero Lebesgue measure and thus is absolutely continuous. □

When the assertions of Proposition 10 hold, we say that the resolvent kernel is absolutely continuous. Then there exists for each $q > 0$ a measurable function $g^q : \mathbb{R}^d \to [0, \infty)$ such that for every measurable function $f \geq 0$ and $x \in \mathbb{R}^d$

$$U^q f(x) = \int_{\mathbb{R}^d} g^q(y - x) f(y) dy.$$

Any such function g^q is called a resolvent density (note that $g^q(x) = \widehat{g^q}(-x)$ for a.e. x in the notation of the proof of Proposition 10). Our next purpose is to show that one can then choose a special version of the resolvent density. First, we introduce an important family of functions.

Definition (Excessive function) *For every $q \geq 0$, a measurable function $f \geq 0$ is q-excessive if $rU^{r+q}f \leq f$ for all $r > 0$ and $\lim_{r\to\infty} rU^{r+q}f = f$ pointwise.*

This definition in terms of the resolvent operator U^r is equivalent to the usual one in terms of the semigroup P_t. Specifically, one can check that a measurable function $f \geq 0$ is q-excessive if and only if $e^{-qt}P_t f \leq f$ for all $t \geq 0$ and $\lim_{t\to 0+} e^{-qt}P_t f = f$ pointwise. See for instance Proposition 2.2 of chapter II in Blumenthal and Getoor (1968). Two classical prototypes of q-excessive functions are $U^q g$ where $g \geq 0$ is some measurable function (this follows from the resolvent equation), and $f(\cdot) = \mathbb{E}.(\exp\{-qT'\})$, where T' is the first hitting time of some set B, $T' = T'_B = \inf\{t > 0 : X_t \in B\}$. To check the second assertion, observe that for any $t > 0$, $T' \circ \theta_t \geq T' - t$, where θ is the translation operator. Therefore, we have

$$\mathbb{E}_x(f(X_t)) = \mathbb{E}_x(\exp\{-qT' \circ \theta_t\}) \leq e^{qt} f(x),$$

and hence

$$rU^{r+q}f(x) = \int_0^\infty re^{-(r+q)t}\mathbb{E}_x(f(X_t))dt \leq f(x).$$

Moreover, $T' \circ \theta_t$ tends to T' as t decreases to 0, and we immediately deduce from the foregoing by dominated convergence that $rU^{r+q}f$ converges pointwise to f as $r \to \infty$.

The next proposition points out that excessive functions are "smooth" when the resolvent operator has a density kernel with respect to the Lebesgue measure.

Proposition 11 *Assume that the resolvent kernel is absolutely continuous.*

(i) *The q-excessive functions are lower semi-continuous.*

(ii) *If f and h are two q-excessive functions with $f \geq h$ almost everywhere, then $f \geq h$ everywhere.*

Proof (i) Let f be q-excessive and $k, r > 0$. On the one hand, we have by monotone convergence $\sup_{k>0} U^{r+q}(f \wedge k) = U^{r+q}(f)$ and (since f is q-excessive) $\sup_{r>0} rU^{r+q}(f) = f$. On the other hand, recall from Proposition 10 that the resolvent operator has the strong Feller property, so $rU^{r+q}(f \wedge k)$ is continuous. Therefore the function $f = \sup_{r,k>0} rU^{r+q}(f \wedge k)$ is lower semi-continuous.

(ii) According to Proposition 10, the measure $U^r(x, dy)$ is absolutely continuous with respect to the Lebesgue measure for all $r > 0$ and $x \in \mathbb{R}^d$. It follows that $rU^{r+q}f \geq rU^{r+q}h$ everywhere. Since f and h are both q-excessive,

$$f = \lim_{r\to\infty} rU^{r+q}f \geq \lim_{r\to\infty} rU^{r+q}h = h\,.$$

□

We mention that conversely, if at least one of the conclusions (i-ii) holds, then the resolvent kernel is absolutely continuous; see Exercise 5. Next, we show that we can pick a nice version of the resolvent densities.

Proposition 12 *Assume that the resolvent kernel is absolutely continuous. Then there exists a unique measurable function $u^q : \mathbb{R}^d \to [0, \infty]$ such that*

(i) *For every measurable function $f \geq 0$ and $x \in \mathbb{R}^d$*

$$U^q f(x) = \int_{\mathbb{R}^d} f(y) u^q(y - x) dy\,.$$

(ii) *The function $\widehat{u^q} : x \to u^q(-x)$ is q-excessive.*

Proof Pick for each $r > 0$ an arbitrary version g^r of the resolvent density, that is

$$\int_0^\infty e^{-rt}\mathbb{P}(X_t \in A)dt = \int_A g^r(y)dy \qquad \text{for all Borel sets } A.$$

We see from the foregoing that $g^r \leq g^q$ almost everywhere if $q \leq r$, and that g^r converges in $L^1(\mathbb{R}^d)$ to 0 as $r \to \infty$. Observe also that for every measurable function $f \geq 0$, $U^r f = \widehat{g^r} \star f$, where $\widehat{h}(x) = h(-x)$ and $\star$ denotes the convolution.

First, we rewrite the resolvent equation as follows:

$$(r - q)\widehat{g^r} \star \widehat{g^q} = \widehat{g^q} - \widehat{g^r} \quad \text{a.e.} \tag{4}$$

In particular, note the inequality

$$(r - q)U^r\widehat{g^q} \leq \widehat{g^q} \qquad \text{a.e.} \tag{5}$$

Next, we check that the mapping $r \to (r - q)U^r\widehat{g^q}$ is increasing. Specifically, the resolvent equation $U^r f = U^{r+s}f + sU^{r+s}U^r f$ for any fixed $s > 0$ yields with $f = (r - q)\widehat{g^q}$

$$(r - q)U^r\widehat{g^q} = (r - q)U^{r+s}\widehat{g^q} + sU^{r+s}((r - q)U^r\widehat{g^q}).$$

It then follows from (5) and the absolute continuity of the resolvent operator that $U^{r+s}((r - q)U^r\widehat{g^q}) \leq U^{r+s}(\widehat{g^q})$ and thus

$$(r - q)U^r\widehat{g^q} \leq (r + s - q)U^{r+s}\widehat{g^q}.$$

Finally, denote by

$$\widehat{u^q} = \lim_{r\to\infty}(r-q)U^r\widehat{g^q} = \lim_{r\to\infty}(r-q)\widehat{g^r} \star \widehat{g^q}.$$

It follows from (4) and the convergence of g^r to 0 in $L^1(\mathbb{R}^d)$ that $\widehat{u^q} = \widehat{g^q}$ a.e., that is $u^q(x) = \widehat{u^q}(-x)$ is a version of g^q. In particular $\widehat{g^r} \star \widehat{u^q} = \widehat{g^r} \star \widehat{g^q} = U^r\widehat{u^q}$, and we see from the foregoing that $\widehat{u^q}$ is q-excessive. Uniqueness follows from Proposition 11. □

Next, we note a one-to-one correspondence between integrable excessive functions and finite measures in the case when the resolvent kernels are absolutely continuous. Here is the direct part.

Proposition 13 *Let $q > 0$ and assume that the resolvent kernel is absolutely continuous. Let u^q be the version of the density specified in Proposition 12 and ν a Radon measure on $\mathbb{R}^d$. Then the function $U^q\nu$ given by*

$$U^q\nu(x) = \widehat{u^q} \star \nu(x) = \int_{\mathbb{R}^d} u^q(y-x)\nu(dy), \qquad x \in \mathbb{R}^d,$$

is q-excessive. If moreover ν is finite, then $U^q\nu$ is integrable.

Proof Because the function $x \to u^q(y-x)$ is q-excessive for every $y \in \mathbb{R}^d$, it follows from Fubini's theorem that for every $r > 0$ and $x \in \mathbb{R}^d$

$$\begin{aligned}(r-q)U^r(U^q\nu)(x) &= (r-q)\int_{\mathbb{R}^d} \nu(dy)U^r(u^q(y-\cdot))(x)\\ &\le \int_{\mathbb{R}^d} \nu(dy)u^q(y-x) = U^q\nu(x),\end{aligned}$$

and by monotone convergence

$$\lim_{r\to\infty}(r-q)U^r(U^q\nu)(x) = U^q\nu(x).$$

This shows that $U^q\nu$ is q-excessive. Finally, another application of Fubini's theorem gives

$$\int_{\mathbb{R}^d} U^q\nu(x)\,dx = \nu(\mathbb{R}^d)\int_{\mathbb{R}^d} u^q(x)dx = \nu(\mathbb{R}^d)/q,$$

so $U^q\nu$ is integrable provided that $\nu(\mathbb{R}^d) < \infty$. □

Here is the converse part.

Proposition 14 *Let $q > 0$ and assume that the resolvent kernel is absolutely continuous. Let f be an integrable q-excessive function. Then there exists a unique finite measure ν such that $f = U^q\nu$.*

Proof Because f is q-excessive, $r(f - (r-q)U^r f)$ is a nonnegative measurable function. Using the identity

$$\int_{\mathbb{R}^d} U^r g(x)dx = \int_{\mathbb{R}^d}\int_{\mathbb{R}^d} g(y)u^r(y-x)\,dxdy = r^{-1}\int_{\mathbb{R}^d} g(y)dy$$

for any nonnegative measurable function g, we deduce from Fubini's theorem that

$$r\int_{\mathbb{R}^d} (f(x) - (r-q)U^r f(x))dx = q\int_{\mathbb{R}^d} f(x)dx.$$

Hence the family of finite measures $\{r(f(x) - (r-q)U^r f(x))dx : r > 0\}$ is relatively compact for the topology of vague convergence. Let v be the limit of a convergent sub-sequence as $r \to \infty$ and h a generic continuous function with compact support. In particular, the convolution $h \star u^q$ is a function in $\mathscr{C}_0$. On the one hand, we have by Fubini's theorem

$$\begin{aligned}\int_{\mathbb{R}^d} h(x)U^q v(x)dx &= \int_{\mathbb{R}^d} h \star u^q(x)v(dx)\\ &= \lim_{r\to\infty}\int_{\mathbb{R}^d} r(f(x) - (r-q)U^r f(x))(h \star u^q)(x)\,dx.\end{aligned}$$

On the other hand, the resolvent equation yields

$$\begin{aligned}&\int_{\mathbb{R}^d} r(f(x) - (r-q)U^r f(x))(h \star u^q)(x)\,dx\\ &\quad= \int_{\mathbb{R}^d} rU^q[f - (r-q)U^r f](x)h(x)dx\\ &\quad= \int_{\mathbb{R}^d} rU^r f(x)h(x)dx.\end{aligned}$$

Because f is q-excessive, we may apply the monotone convergence theorem to deduce that the foregoing quantity converges to $\int f(x)h(x)dx$ as $r \to \infty$. In conclusion, we have

$$\int_{\mathbb{R}^d} h(x)U^q v(x)dx = \int_{\mathbb{R}^d} f(x)h(x)dx.$$

Since h is arbitrary, we have $U^q v = f$ almost everywhere. Recall from Proposition 13 that $U^q v$ is q-excessive, so by Proposition 11, $U^q v = f$ everywhere.

Finally, the uniqueness follows from the identity

$$\int_{\mathbb{R}^d} U^q v(x)h(x)dx = \int_{\mathbb{R}^d} u^q \star h(x)v(dx)$$

and the fact that $\{u^q \star h, h \in \mathscr{C}_0\}$ is dense in $\mathscr{C}_0$ (Feller property). □

We conclude this section with the following simple and useful property of resolvent measures.

Proposition 15 *Suppose that X is not a compound Poisson process. Then the resolvent measures are diffuse, that is*

$$U^q(x,\{y\}) = 0 \qquad \text{for all } q > 0 \text{ and } x, y \in \mathbb{R}^d.$$

Proof With no loss of generality, we may restrict our attention to dimension $d = 1$.

Suppose first that $\mathbb{P}(T_{(0,\infty)} = 0) = 1$. The right-continuity of the paths then implies that $\mathbb{P}(T_{(0,\varepsilon]} = 0) = 1$ for every $\varepsilon > 0$. Since the first entrance time into $[\eta, \varepsilon]$ decreases to $T_{(0,\varepsilon]}$ as $\eta \to 0+$, we see that for every $\varepsilon > 0$ we can pick $\eta > 0$ small enough that $\mathbb{E}(\exp\{-qT_{[\eta,\varepsilon]}\}) > 1/2$. Hence there is a sequence $(\varepsilon_n, n \in \mathbb{N})$ which decreases to 0 with

$$\mathbb{E}(\exp\{-qT_{[\varepsilon_{n+1},\varepsilon_n]}\}) \geq 1/2 \qquad \text{for all } n \in \mathbb{N}.$$

Applying the Markov property at $T_{[\varepsilon_{n+1},\varepsilon_n]}$, we re-express $U^q(0,[\varepsilon_{n+1},\varepsilon_n])$ as

$$\begin{aligned}
&\mathbb{E}\left(\int_{T_{[\varepsilon_{n+1},\varepsilon_n]}}^{\infty} \mathbf{1}_{\{X_t \in [\varepsilon_{n+1},\varepsilon_n]\}} e^{-qt}\,dt\right)\\
&= \int_{[\varepsilon_{n+1},\varepsilon_n]} U^q(x,[\varepsilon_{n+1},\varepsilon_n])\mathbb{E}(\exp\{-qT_{[\varepsilon_{n+1},\varepsilon_n]}\}, X_{T_{[\varepsilon_{n+1},\varepsilon_n]}} \in dx)\\
&\geq \int_{[\varepsilon_{n+1},\varepsilon_n]} U^q(x,\{x\})\mathbb{E}(\exp\{-qT_{[\varepsilon_{n+1},\varepsilon_n]}\}, X_{T_{[\varepsilon_{n+1},\varepsilon_n]}} \in dx)\\
&\geq \frac{1}{2}U^q(0,\{0\}).
\end{aligned}$$

Since

$$\sum_{n\in\mathbb{N}} U^q(0,[\varepsilon_{n+1},\varepsilon_n]) \leq 2U^q(0,\mathbb{R}) = 2/q,$$

it follows that $U^q(0,\{0\}) = 0$.

Finally, observe that if $\mathbb{P}_x(T_{\{y\}} < \infty) = 0$, then *a fortiori* $U^q(x,\{y\}) = 0$. Applying the Markov property at the first passage time in y, we deduce that $U^q(x,\{y\}) = 0$ for every $x, y \in \mathbb{R}$. We conclude the proof by recalling that, except in the compound Poisson case, $\mathbb{P}(T_{(0,\infty)} = 0) = 1$ or $\mathbb{P}(T_{(-\infty,0)} = 0) = 1$, so that in the latter case the preceding argument applies to $-X$. □

Of course, resolvent measures are not diffuse in the compound Poisson case, more precisely $\mathbb{P}(X_t = 0) > 0$ for every $t \geq 0$. Proposition 15 also provides a simple argument for the well-known property (due to Hartman and Wintner) that an infinitely divisible distribution is diffuse, except when the Gaussian coefficient is zero and the Lévy measure finite. Specifically, suppose that for some $t > 0$, there is $a_t \in \mathbb{R}^d$ with $\mathbb{P}(X_t = a_t) > 0$. Because the convolution of a diffuse probability measure with an arbitrary probability measure is also diffuse, we see that for every $s \le t$, there is $a_s \in \mathbb{R}^d$ with $\mathbb{P}(X_s = a_s) > 0$. Consider now the symmetrized process $\widetilde{X} = X - X'$, where X' is an independent copy of X. For every $s \le t$, we have

$$\mathbb{P}\left(\widetilde{X}_s = 0\right) \ge \mathbb{P}(X_s = a_s)\,\mathbb{P}\left(X'_s = a_s\right) > 0,$$

so the resolvent measure of $\widetilde{X}$ has an atom at 0, and by Proposition 15, $\widetilde{X}$ is a compound Poisson process. This implies that X is a compound Poisson process with drift.

4. Transience and recurrence

We present here the first probabilistic applications of the material developed in the preceding sections. Specifically, our purpose is to point out a dichotomy for the asymptotic path behaviour of 'truly' d-dimensional Lévy processes. We will show that X either visits any given non-empty open set at arbitrarily large times $\mathbb{P}$-a.s., or goes to infinity a.s., and we will give an integral test in terms of the characteristic exponent to determine the asymptotic behaviour of the process. The results of this section are essentially due to Port and Stone (1971), and they provide the analogue in continuous time of the classification of random walks as being recurrent or transient (see Spitzer (1964) and Revuz (1984)).

To begin, we introduce the so-called *potential measures* $U(x, \cdot)$ ($x \in \mathbb{R}^d$) which correspond to the limit case $q = 0$ for the q-resolvent kernel. Specifically, we put for every $x \in \mathbb{R}^d$, $A \in \mathcal{B}(\mathbb{R}^d)$,

$$U(x, A) = \int_0^\infty \mathbb{P}_x(X_t \in A)dt = \mathbb{E}_x\left(\int_0^\infty \mathbf{1}_{\{X_t \in A\}}dt\right) \in [0, \infty]\,.$$

Recall that T_A stands for the first entrance time into A and observe that

$$\int_0^s \mathbf{1}_{\{X_t \in A\}}dt = 0 \qquad \text{for all } s \le T_A.$$

In the case when A is either open or closed, we will repeatedly use the Markov property as follows:

$$U(x, A) = \mathbb{E}_x\left(\int_{T_A}^\infty \mathbf{1}_{\{X_t \in A\}}dt\right) = \int_{\overline{A}} U(y, A)\mathbb{P}_x(X_{T_A} \in dy)\,,$$

where $\overline{A}$ stands for the closure of A. It will also be convenient in the sequel to use the notation

$$A + A' = \{x + x', x \in A \text{ and } x' \in A'\}\,,\ A - A' = \{x - x', x \in A \text{ and } x' \in A'\}$$

for all subsets A, A' of the Euclidean space. When A reduces to a single point $A = \{x\}$, we simply write $x + A'$ and $x - A'$, respectively.

Next we define transience and recurrence in terms of potential measures.

Definition (Transience and recurrence) (i) *We say that a Lévy process is transient if the potential measures are Radon measures, that is, for every compact set K*

$$U(x, K) < \infty, \qquad x \in \mathbb{R}^d.$$

(ii) *We say that a Lévy process is recurrent if $U(0, B) = \infty$ for every open ball B centred at the origin.*

Clearly, if $U(0, K)$ is finite for every compact set K, then $U(0, K - x) = U(x, K)$ is finite as well and the process is transient. In other words, it is enough to check Definition (i) for $x = 0$. It should be clear that results of sections 2–3 which have been established for $q > 0$ can be extended to $q = 0$ when the Lévy process is transient (beware however that in Proposition 10(iii), one needs to impose that f has compact support for $q = 0$). We will take this for granted in the following chapters.

A Lévy process cannot be both transient and recurrent. The next proposition shows that it is always either transient or recurrent.

Proposition 16 *Suppose that there exists $\varepsilon > 0$ such that $U(0, B) < \infty$, where B stands for the open ball centred at 0 with radius ε. Then X is transient.*

Proof Let B' stand for the closed ball centred at the origin with radius $\varepsilon/3$. Observe that $B' - B' \subseteq B$ and recall that the first entrance time into B', $T_{B'} = \inf\{t \geq 0 : X_t \in B'\}$, is a stopping time. It follows from the Markov property that for every $x \in \mathbb{R}^d$

$$U(x, B') = \mathbb{E}_x\left(\int_0^\infty \mathbf{1}_{\{X_t \in B'\}} dt\right) = \int_{B'} \mathbb{P}_x(X_{T_{B'}} \in dy) U(y, B') .$$

This last quantity is bounded from above by

$$\sup_{y \in B'} U(y, B') = \sup_{y \in B'} U(0, B' - y) \leq U(0, B' - B') \leq U(0, B) < \infty .$$

This implies that for every $y \in \mathbb{R}^d$, $U(x, y + B') = U(x - y, B') < \infty$. Since any compact set K can be covered by finitely many balls of the type $y + B'$, we deduce that $U(x, K) < \infty$. □

In dimension $d = 1$, a test of Chung and Fuchs (1951) enables us to decide whether a Lévy process with a finite first moment (i.e. $\mathbb{E}(|X_1|) < \infty$) is recurrent. Specifically, X is then recurrent if and only if $\mathbb{E}(X_1) = 0$. However, Fourier analysis yields the following completely general criterion (also due to Chung and Fuchs) for transience and recurrence in terms of the characteristic exponent, and we leave the proof of the test in the special case $d = 1$ as Exercise 10.

Theorem 17 *A Lévy process with characteristic exponent Ψ is transient if and only if for some $r > 0$ small enough,*

$$\limsup_{q \to 0+} \int_{B_r} \Re\left(\frac{1}{q + \Psi(\xi)}\right) d\xi < \infty ,$$

where $\Re z$ stands for the real part of the complex number z and B_r the ball with radius r centred at the origin.

It is natural to ask whether one can replace the condition on the characteristic exponent in Theorem 17 by the apparently weaker:

$$\int_{B_r} \Re\left(\frac{1}{\Psi(\xi)}\right) d\xi < \infty .$$

(To see that the latter condition is weaker than that in Theorem 17, just apply Fatou's lemma.) The answer is positive; but surprisingly, it seems quite difficult to establish this by purely analytic arguments. A probabilistic evidence in dimension $d = 1$ will be presented in Exercise V.6.; we refer to Port and Stone (1971) for the general case.

Proof For the sake of simplicity, we will assume that the dimension of the space is $d = 1$. The proof in higher dimensions is similar with heavier notation. Pick $r > 0$ arbitrarily small and consider the function $f = \mathbf{1}_{[-r,r]} \star \mathbf{1}_{[-r,r]}$, where $\star$ denotes the convolution operator. Clearly $f \geq 0$, and f is a continuous function with support $[-2r, 2r]$, and its Fourier transform is given by

$$\mathscr{F}f(\xi) = (2\xi^{-1}\sin(r\xi))^2 \quad (\xi \neq 0) \quad , \quad \mathscr{F}f(0) = 4r^2 \quad .$$

The latter is a nonnegative bounded and continuous function. Note also that $|q + \Psi|^{-1}$ is bounded since $\Re\Psi \geq 0$. Applying Fourier inversion and Proposition 9, we deduce that for every $q > 0$

$$\begin{aligned} U^q f(0) &= \frac{1}{2\pi} \int_{-\infty}^{\infty} (2\xi^{-1}\sin(r\xi))^2 (q + \Psi(-\xi))^{-1} d\xi \\ &= \frac{1}{2\pi} \int_{-\infty}^{\infty} (2\xi^{-1}\sin(r\xi))^2 \Re\left(\frac{1}{q + \Psi(-\xi)}\right) d\xi \quad , \end{aligned}$$

where we used the fact that $U^q f(0)$ is a real number.

Observe that $f \leq 2r\mathbf{1}_{[-2r,2r]}$, and deduce by monotone convergence that

$$2rU(0,[-2r,2r]) \geq \lim_{q\to 0+} U^q f(0) = \frac{1}{2\pi} \lim_{q\to 0+} \int_{-\infty}^{\infty} (2\xi^{-1}\sin(r\xi))^2 \Re\left(\frac{1}{q+\Psi(-\xi)}\right) d\xi .$$

The latter quantity is infinite whenener

$$\limsup_{q\to 0+} \int_{-r}^{r} \Re\left(\frac{1}{q+\Psi(\xi)}\right) d\xi = \infty ,$$

and then X is recurrent.

Conversely, assume that for some $r > 0$

$$\int_{-2r}^{2r} \Re\left(\frac{1}{q+\Psi(\xi)}\right) d\xi$$

remains bounded as $q \to 0+$. Then consider the function

$$g(x) = (2x^{-1}\sin(rx))^2 \quad (x \neq 0) \quad , \quad g(0) = 4r^2 .$$

By the preceding calculations, its Fourier transform is

$$\mathscr{F}g(\xi) = 2\pi \mathbf{1}_{[-r,r]} \star \mathbf{1}_{[-r,r]} .$$

Again by Fourier inversion and Proposition 9, we get

$$U^q g(0) = \frac{1}{2\pi} \int_{-\infty}^{\infty} \mathscr{F}g(\xi) \Re\left(\frac{1}{q+\Psi(-\xi)}\right) d\xi .$$

Using the inequality $g(x) \geq r^2$ whenever $|x| \leq \frac{\pi}{3r}$, we see by monotone convergence that

$$U(0,[-\frac{\pi}{3r},\frac{\pi}{3r}]) < \infty .$$

According to Proposition 16, X is transient. □

For instance, we see that a truly d-dimensional stable process of index $\alpha \in (0,2]$ is transient for $\alpha < d$ and recurrent otherwise.

We now turn our attention to the probabilistic interpretation of transience and recurrence. First, we denote the support of the potential measure $U(0,\cdot)$ by Σ and qualify the points in Σ as *possible* . By the right-continuity of the paths, a point $x \in \mathbb{R}^d$ is possible if and only if for every open ball B centred at x, there exists $t > 0$ such that $\mathbb{P}(X_t \in B) > 0$. Applying the Markov property at first passage times into balls, we also see that x is possible if and only if

$$\mathbb{P}(T_B < \infty) > 0 \quad \text{for every open ball } B \text{ centred at } x.$$

It is easily checked that Σ is a closed semigroup. Indeed, suppose that both x and y are possible, and denote the open ball centred at z with radius $\eta > 0$ by $B(z,\eta)$. Applying the Markov property at the first passage time into $B(x,\varepsilon)$, we see that $\mathbb{P}\left(T_{B(x+y,2\varepsilon)} < \infty\right) > 0$ provided that $\mathbb{P}_z\left(T_{B(x+y,2\varepsilon)} < \infty\right) > 0$ for every z with $|x-z| \le \varepsilon$, and the latter holds because

$$\mathbb{P}_z\left(T_{B(x+y,2\varepsilon)} < \infty\right) \ge \mathbb{P}\left(T_{B(y,\varepsilon)} < \infty\right) > 0.$$

To avoid some technical difficulties, we will focus on the case when

$$\text{(H)} \qquad \text{the group spanned by } \Sigma \text{ is } \mathbb{R}^d\,.$$

Loosely speaking, (H) means that the Lévy process does not live on any strict sub-group of $\mathbb{R}^d$. It should be clear that (H) holds whenever for some $t > 0$, the group spanned by the support of the distribution $\mathbb{P}(X_t \in \cdot)$ is $\mathbb{R}^d$.

Next, we introduce the set of points whose neighbourhoods are visited at arbitrarily large times a.s. Specifically, we denote by $\mathscr{R}$ the set of points $x \in \mathbb{R}^d$ such that for every open ball B centred at the origin

$$\mathbb{P}(\text{there exists } s \ge t \text{ such that } X_s \in x + B) = 1 \qquad \text{for all } t \ge 0.$$

Plainly, $\mathscr{R} \subseteq \Sigma$.

Lemma 18 *Suppose that (H) holds. Then $\mathscr{R} = \emptyset$ or $\mathscr{R} = \mathbb{R}^d$.*

Proof Suppose that $\mathscr{R}$ is not empty. First, we observe that

$$\text{if } x \in \Sigma \text{ and } y \in \mathscr{R}, \text{ then } y - x \in \mathscr{R}\,. \tag{6}$$

For this, let B and B' be two open balls, both centred at 0 and such that $\overline{B'} - \overline{B'} \subseteq B$. It is obvious that for any $y \in \mathscr{R}$, we have for every $t \ge 0$

$$\mathbb{P}\left(\exists s \ge t + T_{x+B'} : X_s \in y + B' \mid T_{x+B'} < \infty\right) = 1.$$

Applying the Markov property at $T_{x+B'}$, we can re-express the left-hand side as

$$\int_{\overline{x+B'}} \mathbb{P}_{x'}\left(\exists s \ge t : X_s \in y + B'\right) \mathbb{P}\left(X_{T_{x+B'}} \in dx' \mid T_{x+B'} < \infty\right).$$

Because $B' - \overline{B'} \subseteq B$, the foregoing quantity is bounded from above by

$$\mathbb{P}\left(\exists s \ge t : X_s \in y - x + B\right),$$

which entails (6).

Recall that $\mathscr{R} \subseteq \Sigma$. Pick $y \in \mathscr{R}$, so by (6), $0 = y - y \in \mathscr{R}$. Applying (6) again, we deduce that $-y = 0 - y \in \mathscr{R}$. Finally, for every $x, y \in \mathscr{R}$, x is possible and again by (6), $x - y \in \mathscr{R}$. Thus $\mathscr{R}$ is a group that contains Σ, and according to (H), $\mathscr{R} = \mathbb{R}^d$. □

We now give a probabilistic characterization of the analytic notion of transience.

Theorem 19 *Suppose that* (H) *holds. Then the following assertions are equivalent:*

(i) $\mathscr{R} = \emptyset$,
(ii) *the process is transient,*
(iii) $\lim_{t\to\infty} |X_t| = \infty$ *a.s.*

Proof (i)⇔(ii) Let B and B' be two closed balls centred at 0 with $B' + B' \subseteq B$, and put $A = B^c$, $A' = B'^c$. For every $y \in B'$, we have

$$\inf\{t \geq 0 : X_t + y \notin B\} \geq \inf\{t \geq 0 : X_t \notin B'\} .$$

Hence, recalling that T_A and $T_{A'}$ stand for the first exit time from B and B' respectively, we have

$$\inf_{y\in B'} \mathbb{E}_y(T_A) \geq \mathbb{E}_0(T_{A'}) > 0 .$$

Consider also the increasing sequence of stopping times $(T_1, T_1', T_2, T_2', \cdots)$, where $T_1 = T_A$ is the first exit time from B, $T_1' = \inf\{t > T_1 : X_t \in B'\}$ the first entrance time of B' after T_1, $T_2 = \inf\{t > T_1' : X_t \in A\}$ the first exit time from B after $T_1', \cdots$.

First, suppose that $\mathscr{R} = \mathbb{R}^d$, so the foregoing stopping times are finite a.s. We deduce using the Markov property at the second inequality below that

$$\begin{aligned} U(0, B) = \mathbb{E}\left(\int_0^\infty 1_{\{X_t\in B\}} dt\right) &\geq \mathbb{E}\left(\sum_{n=1}^\infty (T_{n+1} - T_n')\right) \\ &\geq \mathbb{E}\left(\sum_{n=1}^\infty \mathbb{E}_{X_{T_n'}}(T_A)\right) \\ &\geq \sum_{n=1}^\infty \inf_{y\in B'} \mathbb{E}_y(T_A) = \infty . \end{aligned}$$

Hence X is recurrent.

Conversely, suppose that $\mathscr{R} = \emptyset$. Then there exist $\varepsilon > 0$, a closed ball B centred at the origin and an instant $t_0 > 0$ such that

$$\mathbb{P}(\exists t \geq t_0, X_t \in B) < 1 - \varepsilon.$$

Let B' be the closed ball centred at the origin with radius half that of B, so $B' - B' = B$. Then for every $y \in B'$,

$$\mathbb{P}_y(\exists t \geq t_0, X_t \in B') \leq \mathbb{P}(\exists t \geq t_0, X_t \in B) < 1 - \varepsilon .$$

Now put $T_0 = 0$ and introduce by iteration $T_{n+1} = \inf\{t > T_n + t_0 : X_t \in B'\}$. Observe that the intervals $[T_n, T_n + t_0]$ $(T_n < \infty)$ cover the set of times when X visits B'. Therefore we have

$$U(0,B') = \mathbb{E}\left(\int_0^\infty \mathbf{1}_{\{X_s \in B'\}} ds\right) \leq \sum_{n=0}^{\infty} t_0 \mathbb{P}_0(T_n < \infty) .$$

On the other hand, the Markov property entails

$$\mathbb{P}_0(T_{n+1} < \infty \mid T_n < \infty, X_{T_n} = y) = \mathbb{P}_y(\exists t \geq t_0, X_t \in B') < 1 - \varepsilon .$$

We deduce that $\mathbb{P}_0(T_n < \infty) \leq (1-\varepsilon)^n$ and thus $U(0,B') < \infty$. Hence X is transient.

(i)$\Leftrightarrow$(iii) If $\mathscr{R} = \emptyset$, the preceding argument shows that $\lim_{n\to\infty} \mathbb{P}_0(T_n < \infty) = 0$. Let B'' be the closed ball centred at the origin with radius half of that of B', so $B'' - B'' = B'$. A standard argument based on the Markov property applied at the first entrance time into B'' shows that for every $x \in \mathbb{R}^d$

$$\lim_{t\to\infty} \mathbb{P}_x(\exists s \geq t, X_s \in B'') = 0 .$$

Covering a given compact set K with a finite number of closed balls of the type $y + B''$, we deduce that for every $x \in \mathbb{R}^d$

$$\lim_{t\to\infty} \mathbb{P}_x(\exists s \geq t, X_s \in K) = 0 .$$

The converse is obvious. □

We can also state the following probabilistic characterization of recurrence (see also Exercise 8).

Theorem 20 *Suppose that* (H) *holds. Then the following assertions are equivalent:*

(i) $\mathscr{R} = \mathbb{R}^d$,

(ii) *the process is recurrent,*

(iii) *for every Borel set K with positive Lebesgue measure,*

$$U(x,K) = \infty \qquad \text{for a.e. } x \in \mathbb{R}^d.$$

Proof By Lemma 18 and Theorem 19, all that is needed is to show that (iii) holds when X is recurrent (the converse implication is straightforward). To start with, we prove that

$$U(x,B) = \infty \qquad \text{for every } x \in \mathbb{R}^d \text{ and non-empty open ball } B . \quad (7)$$

With no loss of generality, we may suppose that B is centred at the origin. Let B' be another open ball centred at the origin, with radius less than half of that of B'. Then $\mathbb{P}(T_{x+B'} < \infty) = 1$ since $x \in \mathscr{R}$. Denote by $\overline{B'}$ the closure of B' and recall that $\overline{B'} - \overline{B'} \subseteq B$. As a consequence, we have that

$B' \subseteq x - y + B$ for every $y \in x + \overline{B'}$, and then $U(0, x - y + B) \geq U(0, B')$. We now deduce from the Markov property applied to the first entrance time into $x + B'$ that

$$\begin{aligned} U(0, x + B) &\geq \int_{\mathbb{R}^d} \mathbb{P}(X_{T_{x+B'}} \in dy) U(0, x - y + B) \\ &\geq \mathbb{P}(T_{x+B'} < \infty) \inf_{y \in x + \overline{B'}} U(0, x - y + B) \\ &\geq \mathbb{P}(T_{x+B'} < \infty)\ U(0, B') . \end{aligned}$$

Since $U(0, B') = \infty$, this shows that $U(0, x + B) = U(-x, B) = \infty$, and hence (7) holds.

Next, we suppose that C and K are both compact sets with positive Lebesgue measure. Fubini's theorem yields

$$\int_C U(x, K) dx = \int_{\mathbb{R}^d} U(0, dy) \int_C \mathbf{1}_K(x + y) dx.$$

The function $y \to \int_C \mathbf{1}_K(x + y) dx$ is continuous and not identically zero. It follows from (7) that

$$\int_C U(x, K) dx = \infty,$$

and since this holds for all compact sets C with positive Lebesgue measure, $U(x, K) = \infty$ for almost every x (by inner regularity of Lebesgue measure). This extends immediately to the general case when K is a Borel set with positive Lebesgue measure. □

We now end this section with an important result on the asymptotic behaviour of the potential measure of a real-valued transient Lévy process. It is known as the *renewal theorem* , and is the continuous-time analogue of the renewal theorem for real-valued transient random walks.

Theorem 21 (Renewal theorem) *Suppose that the dimension is $d = 1$ and that (H) holds. Assume also that $\mathbb{E}(|X_1|) < \infty$ and put $m = \mathbb{E}(X_1)$. If $m > 0$, then the measure $mU(x, \cdot)$ converges vaguely to the Lebesgue measure as $x \to -\infty$ and to 0 as $x \to \infty$.*

Proof The argument consists of reducing the result to the standard renewal theorem for random walks, for which we refer to section 5.3 of Revuz (1984). Consider the probability measure on $(-\infty, \infty)$,

$$\nu(dx) = \int_0^\infty \mathbb{P}(X_t \in dx) e^{-t} dt$$

which clearly has expectation m and Fourier transform

$$\mathscr{F}\nu(\lambda) = \int_{-\infty}^{\infty} e^{i\lambda x}\nu(dx) = \int_0^{\infty} e^{-t}e^{-t\Psi(\lambda)}dt = \frac{1}{1+\Psi(\lambda)}.$$

The potential measure $V(0,\cdot)$ at 0 of the random walk with step distribution ν has Fourier transform

$$\mathscr{F}V(0,\cdot)(\lambda) = \sum_{n=0}^{\infty}(1+\Psi(\lambda))^{-n} = 1 + \frac{1}{\Psi(\lambda)},$$

so that inverting Fourier transforms, we get $V(0,\cdot) = U(0,\cdot)+\delta_0$ where δ_0 stands for the Dirac mass at 0 and we simply apply the renewal theorem for the random walk. □

There is also a renewal theorem for Lévy processes that fulfil the conditions of Theorem 21 except that (H) fails (one says then that X is lattice-valued); it essentially reduces to the standard renewal theorem for lattice random walks.

5. Exercises

1. *(Lévy measure as a limit)* Prove that for every fixed $a > 0$, the measure $\frac{1}{\varepsilon}\mathbb{P}_0(X_\varepsilon \in dx)$ converges vaguely on $\{|x| > a\}$ as $\varepsilon \to 0+$ to $\Pi(dx)$. [Hint: use Proposition 9 and Fourier inversion.]
2. *(Independent components)* Let $X = (X^1,\cdots,X^d)$ be a d-dimensional Lévy process with Gaussian coefficient Q and Lévy measure Π. Show that the Lévy processes $X^1,\cdots,X^d$ are independent if and only if Q is a diagonal matrix and the support of Π is contained in the coordinate axes.
3. *(Another approach to quasi-left-continuity)* A stopping time $T < \infty$ is called *announceable* if there exists an increasing sequence of stopping times $(T_n, n \in \mathbb{N})$ such that $T_n < T$ for every n and $\lim T_n = T$. Show that if $(e(t), t \geq 0)$ is a Poisson point process with isolated point Υ and T an announceable stopping time, then $e(T) = \Upsilon$ a.s. Deduce that a Lévy process is a.s. continuous at every announceable stopping time.
4. *(Absolutely continuous semigroup)* Prove that the semigroup of a Lévy process has the strong Feller property (i.e. $P_t f$ is continuous for every measurable bounded function f and $t > 0$) if and only if the distribution $\mathbb{P}(X_s \in dx)$ is absolutely continuous with respect to Lebesgue measure for some (and then all) $s > 0$. Show that the

latter holds whenever the characteristic exponent Ψ is such that

$$\lim_{|\lambda|\to\infty} |\lambda|^{-\varepsilon}\Re\Psi(\lambda) = \infty$$

for some $\varepsilon > 0$. Check that a Poisson process with drift has a strong Feller resolvent but a singular semigroup.

5. *(Absolutely continuous resolvent)* (a) Prove that if all the q-excessive functions are lower semi-continuous, then the resolvent kernel is absolutely continuous. [Hint: check first that if N is a Borel set with zero Lebesgue measure, then the function $U^q(\cdot, N)$ is q-excessive and null almost everywhere.]
(b) Suppose that if f and h are two q-excessive functions with $f \geq h$ almost everywhere, then $f \geq h$ everywhere. Show that the resolvent kernel is absolutely continuous.

6. *(Exponential martingales)* Prove that for every $\lambda \in \mathbb{R}^d$, the process

$$\exp\{i\langle\lambda, X_t\rangle + t\Psi(\lambda)\} \qquad (t \geq 0)$$

is a complex-valued martingale.

7. *(Non-absolutely-continuous resolvent)* The purpose of this exercise is to provide an example of a Lévy process which is not a compound Poisson process and whose resolvent operators are not absolutely continuous. The dimension is $d = 1$. Put $x_k = 2^{2^k}$ for $k = 1, 2, \cdots$, and let Π be the Lévy measure obtained by putting atoms of mass $p_j = x_j^{1/2} j^{-1}$ at x_j^{-1} $(j = 1, 2, \cdots)$. Take $Q = 0$ and $a = -\int_0^1 x\Pi(dx)$, and let Ψ be the characteristic exponent given by the Lévy-Khintchine formula. Check that $\Psi(2\pi x_k)$ converges to 0 as $k \to \infty$. Deduce from the Riemann-Lebesgue theorem that the resolvent operators are not absolutely continuous with respect to the Lebesgue measure.

8. *(Recurrence with absolutely continuous resolvent)* (a) Suppose that X is recurrent and that its resolvent kernel is absolutely continuous. Check that every point is possible and then that for every $x \in \mathbb{R}^d$ and Borel set K with positive Lebesgue measure, $U(x, K) = \infty$. One says that X is *Harris-recurrent* .
(b) Give an example of a recurrent Lévy process such that every point is possible and $U(0, K) = 0$ for some compact set K with positive Lebesgue measure.

9. *(Strong law of large numbers)* The purpose of this exercise is to show that the strong law of large numbers (SLLN) holds for Lévy processes, that is

$$\lim_{t\to\infty} t^{-1}X_t = 0 \quad \text{a.s.}$$

whenever X is centred, viz. $\mathbb{E}(|X_1|) < \infty$ and $\mathbb{E}(X_1) = 0$. For simplicity, we assume that the dimension of the space is $d = 1$.
(a) Prove the SLLN in the compound Poisson case.
(b) Suppose that the Lévy measure has compact support and that X is centred. Check that X is a square-integrable martingale and use Doob's maximal inequality to show that there is a constant $c > 0$ such that

$$\mathbb{P}\left(\sup\{|X_s| : 0 \leq s \leq 1\} \geq x\right) \leq cx^{-2} \qquad (x > 0).$$

Finally, deduce from the standard SLLN for random walks and the Borel-Cantelli lemma that the SLLN holds for X.
(c) Conclude that the SLLN holds for any centred Lévy process.

10. *(Chung and Fuchs test in dimension $d = 1$)* The purpose of this exercise is to show that, in dimension $d = 1$, a centred Lévy process is recurrent. Note from the preceding exercise that if the first moment of X_1 exists and is not zero, then X is transient. In the sequel, we suppose that X is a transient Lévy process in dimension $d = 1$.
(a) Check for all $x \in \mathbb{R}$ and $h \geq 0$ the inequality $U(0, [x-h, x+h]) \leq U(0, [-2h, 2h])$ and deduce that $x^{-1}U(0, [-x, x])$ remains bounded as $x \to \infty$.
(b) Suppose that $\mathbb{E}(|X_1|) < \infty$ and put $\mathbb{E}(X_1) = \mu$. Using the weak law of large numbers, prove that $\mu \neq 0$.

6. Comments

Infinitely divisible laws were introduced by de Finetti (1929) and then studied by Lévy (1934) who established the Lévy-Khintchine formula; his proof was simplified by Khintchine. Exercise 1 is part of Lévy's original approach. See also Steutel (1970) for further references and developments. Lévy processes were previously called 'fonctions additives' by Lévy, 'decomposable processes' by Loève (1963), 'differentiable processes' by Itô and Mc Kean (1965), ... The denomination 'random processes with stationary independent increments' was widely used in the 60's and 70's. Theorem 1 is from Itô (1942), though it was probably known by Lévy, see the foot-note on page 186 in Lévy (1954). The present proof follows e.g. Dellacherie and Meyer (1987). We have discussed only a very small part of what is known on variation properties of Lévy processes. See Blumenthal and Getoor (1961), Greenwood (1969), Millar (1971),

Bretagnolle (1972), Fristedt and Greenwood (1972), Monroe (1972) and Fristedt and Taylor (1973) for much more on this topic.

Dynkin (1965), Blumenthal and Getoor (1968), Sharpe (1989) and the last two volumes of the treatise by Dellacherie and Meyer (1987; Dellacherie, Maisonneuve and Meyer, 1992) contain a complete account on Markov processes. The theory of Dirichlet forms provides a most efficient approach to investigating the special class of symmetric Markov processes. We refer to Fukushima, Oshima and Takeda (1994), where numerous applications to symmetric Lévy processes are discussed. The elementary treatment of entrance times that we have chosen here enables us to handle open or closed sets and countable unions thereof, which is sufficient for our purposes. Nonetheless it induces some loss of generality, and Choquet's theory of capacity is needed to consider general analytic sets; see Blumenthal and Getoor (1968), Port and Stone (1971) and Dellacherie and Meyer (1975). Sato (1972) and Berg and Forst (1975) contain a detailed exposition of the semigroup, resolvents and infinitesimal generator associated with a Lévy process.

The results of section 3 are from Blumenthal and Getoor (1968), who treat general Markov processes, and Hawkes (1979), who focuses on Lévy processes. The absolute continuity of the semigroup and its support were considered by Tucker (1965, 1975), Sharpe (1969) and Hawkes (1979). See also Yamazato (1978, 1982) and Sato (1990) for regularity properties of the semigroup.

Section 4 is essentially an excerpt from Port and Stone (1971), who treat Lévy processes on locally compact Abelian groups more generally; see also Berg and Forst (1975). We refer to Port and Stone for the case when hypothesis (H) fails.

Exercises 4 and 7 are from Hawkes (1979), 5 is from Zabczyk (1970), and 8 from Port and Stone (1971). Further developments related to the strong law of large numbers were made by Khintchine (1938, 1939). The approach to the Chung and Fuchs test proposed in Exercise 10 mimics that of Feller (1971).

II

Elements of Potential Theory

In this chapter, we develop some fundamental notions of potential theory for Lévy processes. Our main concern is to determine the sets of zero capacity, viz. the sets that are never visited by a given Lévy process. The solution involves the notion of energy, which is defined in terms of the characteristic exponent. Specifically, a set has zero capacity if and only if its energy is infinite. The results are then specialized to the case of a single point.

1. Duality and time reversal

We resume the introduction of analytic notions related to the Markov property of Lévy processes. This section is devoted to duality, which is a classical step to developing a substantial potential theory (see Chapter VI in Blumenthal and Getoor (1968)). For general Markov processes, one usually has to make some restrictive assumptions (such as the existence of densities for the resolvent kernel); however, the case of Lévy processes requires no special hypothesis.

We call $\widehat{X} = -X$ the *dual process* of X which is of course a Lévy process for $(\Omega, \mathcal{F}, \mathbb{P})$. We will often use the prefix *co-* for qualifiers (or a hat $\widehat{\ }$ in mathematical notation) related to the dual process. For instance, $\widehat{\mathbb{P}}$, $(\widehat{P}_t, t \geq 0)$ and $(\widehat{U}^q, q > 0)$ stand for its law, its semigroup and its resolvent operators respectively, and the last will be referred to as the co-resolvent operators. For every $x \in \mathbb{R}^d$, $\widehat{\mathbb{P}}_x$ denotes the law of $x + X$ under $\widehat{\mathbb{P}}$, that is the law of $\widehat{X}$ under $\mathbb{P}_{-x}$. Observe that the characteristic exponent of the dual process, $\widehat{\Psi}$, coincides with the complex conjugate of the characteristic exponent Ψ of the initial Lévy process, i.e. $\widehat{\Psi}(\lambda) = \Psi(-\lambda)$ for all $\lambda \in \mathbb{R}^d$. We say that the Lévy process is *symmetric* if its characteristic exponent is real-valued, which squares with $\mathbb{P} = \widehat{\mathbb{P}}$.

The terminology 'dual' stems from the following identity.

Proposition 1 *Let f and g be two nonnegative measurable functions. We have for every $t \geq 0$*

$$\text{(i)} \qquad \int_{\mathbb{R}^d} P_t f(x) g(x) dx = \int_{\mathbb{R}^d} f(x) \widehat{P}_t g(x) dx,$$

and for every $q > 0$

$$\text{(ii)} \qquad \int_{\mathbb{R}^d} U^q f(x) g(x) dx = \int_{\mathbb{R}^d} f(x) \widehat{U}^q g(x) dx.$$

Proof A change of variables in the second equality below yields

$$\begin{aligned}\int_{\mathbb{R}^d} P_t f(x) g(x) dx &= \mathbb{E}\left(\int_{\mathbb{R}^d} f(X_t + x) g(x) dx\right)\\ &= \mathbb{E}\left(\int_{\mathbb{R}^d} f(x) g(x - X_t) dx\right)\\ &= \widehat{\mathbb{E}}\left(\int_{\mathbb{R}^d} f(x) g(x + X_t) dx\right) = \int_{\mathbb{R}^d} f(x) \widehat{P}_t g(x) dx.\end{aligned}$$

The second assertion follows. Note also that alternatively, the result can be deduced by the Parseval identity from Proposition I.9. □

Applying Proposition 1 for $g \equiv 1$, we see in particular that the Lebesgue measure is *invariant* for P_t, that is

$$\int_{\mathbb{R}^d} f(x) dx = \int_{\mathbb{R}^d} P_t f(x) dx$$

for every measurable function $f \geq 0$.

For a general Markov process, the analytic notion of duality is related to the probabilistic notion of time reversal. This connection is one of the major (and demanding) achievements of the theory, see Chapter XVIII in Dellacherie, Maisonneuve and Meyer (1992), Nagasawa (1993) and the references therein. But once again, things are much simpler for Lévy processes, at least when we restrict our attention to time reversal at a fixed time, and the relation between a Lévy process and its dual via time reversal is straightforward. We make the convention that $X_{0-} = X_0$ in the sequel.

Lemma 2 (Duality lemma) *For each fixed $t > 0$, the reversed process $(X_{(t-s)-} - X_t, 0 \le s \le t)$ and the dual process $(\widehat{X}_s, 0 \le s \le t)$ have the same law under* $\mathbb{P}$.

Proof Put $Y = (Y_s, 0 \le s \le t)$, with $Y_s = X_{(t-s)-} - X_t$. Under $\mathbb{P}$, Y is a process started at 0, its paths are a.s. right-continuous and have left limits on $[0, t]$, and its increments are clearly independent and homogeneous. Finally, for every $0 \le s \le t$, $X_{(t-s)-} - X_t$ has the same distribution under $\mathbb{P}$ as $-X_s$, and hence the law of Y under $\mathbb{P}$ is identical to that of $(X_s, 0 \le s \le t)$ under $\widehat{\mathbb{P}}$. □

It is often convenient to rephrase the duality lemma in terms of conditional laws. Loosely speaking, the conditional version states that time reversal changes a bridge of a Lévy process into a bridge of the dual Lévy process. To make a formal statement, we fix $t > 0$ and consider $\mathbb{P}(\cdot \mid X_t = y)$ $(y \in \mathbb{R}^d)$, an arbitrary version of the conditional law $\mathbb{P}(\cdot \mid X_t)$ (see section O.1). For every $x \in \mathbb{R}^d$, we then denote by $\mathbb{P}_x(\cdot \mid X_t = y)$ the law of $x + X$ under $\mathbb{P}(\cdot \mid X_t = y - x)$, so that $\mathbb{P}_x(\cdot \mid X_t = y)$ $(y \in \mathbb{R}^d)$ is a version of $\mathbb{P}_x(\cdot \mid X_t)$.

Corollary 3 *For every $y \in \mathbb{R}^d$, the law of $(X_{(t-s)-}, 0 \le s \le t)$ under $\mathbb{P}_x(\cdot \mid X_t = y)$ $(x \in \mathbb{R}^d)$ is a version of the conditional law of $(X_s, 0 \le s \le t)$ under $\widehat{\mathbb{P}}_y(\cdot \mid X_t = x)$.*

Proof Consider $0 \le t_1 \le \cdots \le t_n \le t$ and a measurable function $f : (\mathbb{R}^d)^n \to [0, \infty)$. We have by definition

$$\begin{aligned}
&\int_{\mathbb{R}^d} \widehat{\mathbb{P}}_y(X_t \in dx)\mathbb{E}_x(f(X_{(t-t_1)-}, \cdots, X_{(t-t_n)-}) \mid X_t = y) \\
&= \int_{\mathbb{R}^d} \mathbb{P}(y - X_t \in dx)\mathbb{E}(f(X_{(t-t_1)-} + x, \cdots, X_{(t-t_n)-} + x) \mid X_t = y - x) \\
&= \mathbb{E}(f(X_{(t-t_1)-} + y - X_t, \cdots, X_{(t-t_n)-} + y - X_t)) \\
&= \widehat{\mathbb{E}}_y(f(X_{t_1}, \cdots, X_{t_n})),
\end{aligned}$$

where the last equality stems from the duality lemma. □

Combining Proposition 1 and Corollary 3, one obtains another interesting interpretation of the connection between duality and time reversal.

Loosely speaking, it says that on every finite time interval, when one time-reverses a Lévy process started according to the Lebesgue measure, one gets the dual Lévy process also started according to the Lebesgue measure. To give a rigorous statement, fix $t > 0$, put $\widetilde{X}_s = X_{(t-s)-}$ for $s \in [0,t]$, and consider an $\mathscr{F}_t$-measurable functional $F : \Omega \to [0,\infty)$. Applying Corollary 3, we see that

$$\begin{aligned}\int_{\mathbb{R}^d} \mathbb{E}_x(F(\widetilde{X}))dx &= \int_{\mathbb{R}^d} dx \int_{\mathbb{R}^d} \mathbb{P}_x(X_t \in dy)\mathbb{E}_x(F(\widetilde{X}) \mid X_t = y)\\ &= \int_{\mathbb{R}^d} dx \int_{\mathbb{R}^d} \mathbb{P}_x(X_t \in dy)\widehat{\mathbb{E}}_y(F(X) \mid X_t = x).\end{aligned}$$

According to Proposition 1, the measures $dx\mathbb{P}_x(X_t \in dy)$ and $dy\widehat{\mathbb{P}}_y(X_t \in dx)$ are the same, and we can re-express the last quantity as

$$\int_{\mathbb{R}^d} dy\widehat{\mathbb{E}}_y(F(X)).$$

In conclusion, we have shown that the processes $(\widetilde{X}_s, 0 \leq s \leq t)$ and $(\widehat{X}_s, 0 \leq s \leq t)$ have the same law under the measure $\int_{\mathbb{R}^d} \mathbb{P}_x(\cdot)dx$.

The main result of this section is that a relation of duality analogous to that of Proposition 1 holds as well for the Lévy process and its dual killed at the first passage into an open or a closed set B. First we develop some material about killed Lévy processes.

Recall that $T_B = \inf\{t \geq 0 : X_t \in B\}$ stands for the first passage time into B and k_t for the killing operator at time t. The sample path of $X \circ \mathrm{k}_{T_B}$ thus coincides with that of X on the time interval $[0, T_B)$ and stays at the cemetery point ∂ on the time interval $[T_B, \infty)$; $X \circ \mathrm{k}_{T_B}$ is called the *process killed at the first passage into B*. For every $x \in \mathbb{R}^d$, we denote its law under $\mathbb{P}_x$ by $\mathbb{P}_x^B$. Recall that we use canonical notation, in particular ζ stands for the lifetime and θ for the translation operator. It is easy to check that the killed process is a Markov process.

Proposition 4 *For every $t > 0$, under $\mathbb{P}_x^B$, conditionally on $t < \zeta, X_t = y$, the process $X \circ \theta_t = (X_{t+s}, s < \zeta - t)$ is independent of $\mathscr{F}_t$ and has the law $\mathbb{P}_y^B$.*

Proof To start with, observe that when $t < T_B$, first killing the path as it enters B and then translating it at time t is the same as first translating the path at time t and then killing it as it enters B. In other words, $\theta_t \circ \mathrm{k}_{T_B} = \mathrm{k}_{T_B} \circ \theta_t$ on the event $\{t < T_B\}$. Consider now a measurable functional $F \geq 0$ on Ω and $\Lambda \in \mathscr{F}_t$. The latter condition ensures that a.s.

on $\{t < T_B\}$, a path belongs to Λ if and only if the killed path belongs to Λ as well. We have then

$$\begin{aligned}\mathbb{E}_x^B(F(X \circ \theta_t), \Lambda, t < \zeta) &= \mathbb{E}_x(F(X \circ \theta_t \circ \mathrm{k}_{T_B}), \Lambda, t < T_B) \\ &= \mathbb{E}_x(F(X \circ \mathrm{k}_{T_B} \circ \theta_t), \Lambda, t < T_B)\,.\end{aligned}$$

Then one readily deduces from the Markov property that the foregoing quantity can be re-expressed as

$$\int \mathbb{P}_x^B(\Lambda, t < \zeta, X_t \in dy)\mathbb{E}_y^B(F(X)),$$

which establishes our claim. More generally, a similar argument yields the strong Markov property. □

Next we introduce the semigroup $(P_t^B, t \geq 0)$ and resolvent operators $(U_B^q, q > 0)$ of the killed process. For every measurable function $f \geq 0$ which vanishes on B,

$$P_t^B f(x) = \mathbb{E}_x^B(f(X_t), t < \zeta) = \mathbb{E}_x(f(X_t), t < T_B)$$

and

$$U_B^q f(x) = \int_0^\infty \mathrm{e}^{-qt}\mathbb{E}_x^B(f(X_t), t < \zeta)dt = \mathbb{E}_x\left(\int_0^{T_B} \mathrm{e}^{-qt} f(X_t)dt\right).$$

We also denote by $(\widehat{P}_t^B, t \geq 0)$ and $(\widehat{U}_B^q, q > 0)$ the semigroup and the resolvent operators of the dual process killed at its first passage into B. In particular

$$\widehat{U}_B^q f(x) = \widehat{\mathbb{E}}_x\left(\int_0^{T_B} \mathrm{e}^{-qt} f(X_t)dt\right).$$

Finally we state an identity of duality for killed Lévy processes, also known as *Hunt's switching identity* .

Theorem 5 *Suppose that B is either open or closed. For all nonnegative measurable functions f and g, we have for every $t \geq 0$*

$$\int_{\mathbb{R}^d} g(x)\, P_t^B f(x)dx = \int_{\mathbb{R}^d} f(x)\widehat{P}_t^B g(x)dx$$

and for every $q > 0$

$$\int_{\mathbb{R}^d} g(x)\, U_B^q f(x)dx = \int_{\mathbb{R}^d} f(x)\widehat{U}_B^q g(x)dx\,.$$

Proof The second statement in terms of the resolvent operators follows immediately from the first, so we focus on the semigroup of killed processes.

First assume that B is open. Using the right-continuity and the left continuity at time t, it is easy to see that the path $(X_s, 0 \le s \le t)$ enters B if and only if the reversed path $(X_{(t-s)-}, 0 \le s \le t)$ does. Therefore we have by Corollary 3

$$\begin{aligned}\int_{\mathbb{R}^d} g(x)\, P_t^B f(x) dx &= \int_{\mathbb{R}^d} dx g(x) \mathbb{E}_x(f(X_t), t < T_B) \\ &= \int_{\mathbb{R}^d} dx g(x) \int_{\mathbb{R}^d} \mathbb{P}_x(X_t \in dy) f(y) \mathbb{P}_x(t < T_B \mid X_t = y) \\ &= \int_{\mathbb{R}^d} dx g(x) \int_{\mathbb{R}^d} \mathbb{P}_x(X_t \in dy) f(y) \widehat{\mathbb{P}_y}(t < T_B \mid X_t = x).\end{aligned}$$

According to Proposition 1, the measures $dx\mathbb{P}_x(X_t \in dy)$ and $dy\widehat{\mathbb{P}_y}(X_t \in dx)$ are the same, so we can re-express the foregoing quantity as

$$\begin{aligned}&\int_{\mathbb{R}^d} dy f(y) \int_{\mathbb{R}^d} \widehat{\mathbb{P}}_y(X_t \in dx) g(x) \widehat{\mathbb{P}_y}(t < T_B \mid X_t = x) \\ &= \int_{\mathbb{R}^d} dy f(y) \widehat{\mathbb{E}}_y(g(X_t), t < T_B)) \\ &= \int_{\mathbb{R}^d} f(y) \widehat{P}_t^B g(y) dy .\end{aligned}$$

This proves the first assertion of the theorem when B is open.

Now assume that B is closed and consider a decreasing sequence $(B(n), n \in \mathbb{N})$ of open neighbourhoods of B with $\bigcap \overline{B(n)} = B$. According to Corollary I.8(ii), the sequence of stopping times $T_{B(n)}$ increases to T_B, $\mathbb{P}_x$-a.s. for every x. It follows that for every $x \in \mathbb{R}^d$, $P_t^{B(n)} f(x)$ increases to $P_t^B f(x)$; and we have a similar result in terms of the dual process. We now deduce by monotone convergence from the first part of the proof that the theorem holds when B is closed. □

2. Capacitary measure

The first step in the study of the probability that the Lévy process passes into a given set B consists in introducing the capacitary measures. Roughly, for every $q > 0$, the q-capacitary measure of B describes the law of the first point in B that is visited by the Lévy process, when the latter is initially distributed according to the Lebesgue measure and killed at rate q. Here is a formal definition; see also the end of this section for an analogous notion corresponding to the limit case $q = 0$, which is valid for transient Lévy processes.

Definition (Capacitary measure) *Let $q > 0$ and B be an open or a closed set. The measure μ_B^q given for every Borel set A by*

$$\mu_B^q(A) = q \int_{\mathbb{R}^d} \mathbb{E}_x(\exp\{-qT_B\}, X_{T_B} \in A)dx$$

is called the q-capacitary measure of B.

We point out that replacing the first passage time T_B by the first hitting time T'_B does not affect the definition. To see this, one only needs to verify that the set of irregular points for B,

$$\mathcal{I}_B := \{x \in B : \mathbb{P}_x(T'_B > 0) = 1\},$$

has zero Lebesgue measure. In this direction, it is easy to check that the set of times at which X visits $\mathcal{I}_B$ is at most countable, $\mathbb{P}_x$-a.s. for every $x \in \mathbb{R}^d$ (we refer to Blumenthal and Getoor (1968) on page 80 for an argument). As a consequence, $U^q(\cdot, \mathcal{I}_B) \equiv 0$, and this implies that the Lebesgue measure of $\mathcal{I}_B$ is null, because the Lebesgue measure is invariant.

Here is the first basic result on the q-capacitary measure.

Lemma 6 *The q-capacitary measure of B is a Radon measure, and its support is contained in the closure $\overline{B}$.*

Proof Let A and A' be the closed balls centred at 0 with radius R and $R/2$, respectively, where $R > 0$ is arbitrarily large. The observation that $A' + A' \subseteq A$ readily yields the inequality

$$U^q(y, A) \geq U^q(0, A') > 0 \qquad \text{for all } y \in A' .$$

Now the Markov property at the first passage time into A' entails that for every $x \in \mathbb{R}^d$

$$\begin{aligned} U^q(x, A) &\geq \mathbb{E}_x\left(\int_{T_{A'}}^{\infty} \mathbf{1}_{\{X_t \in A\}} e^{-qt} dt\right) \\ &\geq \int_{A'} \mathbb{E}_x(\exp\{-qT_{A'}\}, X_{T_{A'}} \in dy) U^q(y, A) \\ &\geq \mathbb{E}_x\left(\exp\{-qT_{A'}\}\right) \inf_{y \in A'} U^q(y, A) , \end{aligned}$$

with the usual convention that $\exp\{-q\infty\} = 0$. Hence, if we put $k = 1/U^q(0, A')$, we have

$$\mathbb{E}_x(\exp\{-qT_{A'}\}) \leq kU^q(x, A) \qquad \text{for all } x \in \mathbb{R}^d .$$

Then we deduce from the definition of the q-capacitary measure

$$\begin{aligned}\mu_B^q(A') &= q\int_{\mathbb{R}^d} dx \mathbb{E}_x(\exp\{-qT_B\}, X_{T_B} \in A') \\ &\le q\int_{\mathbb{R}^d} dx \mathbb{E}_x(\exp\{-qT_{B\cap A'}\}) \\ &\le q\int_{\mathbb{R}^d} dx \mathbb{E}_x(\exp\{-qT_{A'}\}) \\ &\le kq\int_{\mathbb{R}^d} U^q(x,A)dx = km(A) ,\end{aligned}$$

where $m(A)$ stands for the Lebesgue measure of A and the last equality follows from Proposition 1.

This shows that μ_B^q is a Radon measure. Finally, it is immediately deducible from Corollary I.8 that $\mu_B^q(\mathbb{R}^d - \overline{B}) = 0$. □

Except for a few specific examples, one does not know the q-capacitary measure of a set explicitly. However, Hunt's switching identity yields a simple formula which facilitates many important calculations. First, we need one more notation. For every Borel measure μ on $\mathbb{R}^d$ and $q > 0$, the measure given by

$$\mu U^q(A) = \int_{\mathbb{R}^d} U^q(x,A)\mu(dx), \qquad A \in \mathscr{B}(\mathbb{R}^d),$$

is called the *q-resolvent measure* of μ. The main result of this section is the following characterization of the q-capacitary measure. Recall the convention $e^{-\infty} = 0$.

Theorem 7 *Let $q > 0$ and suppose that B is either open or closed. Then the q-capacitary measure of B is the unique Radon measure μ such that*

$$\mu U^q(dx) = \widehat{\mathbb{E}}_x(\exp\{-qT_B\})dx \qquad (x \in \mathbb{R}^d).$$

Because the set of irregular points for B has zero Lebesgue measure, we may replace the first entrance time T_B by the first hitting time T_B' in Theorem 7.

Proof For every measurable function $f \ge 0$, we have by Proposition 1

$$\int_{\mathbb{R}^d} U^q f(x)dx = \frac{1}{q}\int_{\mathbb{R}^d} f(x)dx.$$

Applying the Markov property at the first passage time into B, T_B, we get the following expression for $U^q f(x)$:

$$\begin{aligned}
&\mathbb{E}_x\left(\int_0^\infty e^{-qt} f(X_t)dt\right) \\
&= \mathbb{E}_x\left(\int_0^{T_B} e^{-qt} f(X_t)dt\right) + \int_{\mathbb{R}^d} \mathbb{E}_x(\exp\{-qT_B\}, X_{T_B} \in dy) U^q f(y) \\
&= U_B^q f(x) + \int_{\mathbb{R}^d} \mathbb{E}_x(\exp\{-qT_B\}, X_{T_B} \in dy) U^q f(y) .
\end{aligned}$$

We now obtain the following identity by integration:

$$\frac{1}{q}\int_{\mathbb{R}^d} f(x)dx = \int_{\mathbb{R}^d} U_B^q f(x)dx + \frac{1}{q}\int_{\mathbb{R}^d} U^q f(x)\mu_B^q(dx) . \tag{1}$$

We then apply Hunt's switching identity (Theorem 5) to deduce

$$\begin{aligned}
\int_{\mathbb{R}^d} U_B^q f(x)dx &= \int_{\mathbb{R}^d} f(x)\widehat{U}_B^q \mathbf{1}(x)dx \\
&= \int_{\mathbb{R}^d} dx f(x) \int_0^\infty dt e^{-qt} \widehat{\mathbb{P}}_x(t < T_B) \\
&= \frac{1}{q}\int_{\mathbb{R}^d} dx f(x)\left(1 - \widehat{\mathbb{E}}_x(\exp\{-qT_B\})\right) .
\end{aligned} \tag{2}$$

Comparing (1) and (2), we conclude that

$$\int_{\mathbb{R}^d} U_B^q f(x)\mu_B^q(dx) = \int_{\mathbb{R}^d} dx f(x)\widehat{\mathbb{E}}_x(\exp\{-qT_B\}) ,$$

which establishes the first assertion.

Uniqueness follows from the fact that two Radon measures μ and μ' with the same q-resolvent measure are identical. Indeed, we have then

$$\int_{\mathbb{R}^d} U^q f(x)\mu(dx) = \int_{\mathbb{R}^d} U^q f(x)\mu'(dx)$$

for every function $f \in \mathscr{C}_0$. We know from section I.2 that the domain $\mathscr{D}$, that is the image of $\mathscr{C}_0$ under U^q, is dense in $\mathscr{C}_0$. Because μ and μ' are Radon measures, they are thus the same. □

It is instructive to relate Theorem 7 to the representation of excessive functions developed in section I.3. Specifically, suppose that the q-resolvent kernel is absolutely continuous and let u^q be the version of the density specified in Proposition I.12. The identity

$$\begin{aligned}
\int_{\mathbb{R}^d} \widehat{\mathbb{E}}_x(\exp\{-qT'_B\}) f(x)dx &= \int_{\mathbb{R}^d} U^q f(x)\mu_B^q(dx) \\
&= \int_{\mathbb{R}^d} \mu_B^q(dx) \int_{\mathbb{R}^d} u^q(y-x) f(y)dy
\end{aligned}$$

shows that for a.e. $x \in \mathbb{R}^d$,

$$\widehat{\mathbb{E}}_x\left(\exp\{-qT'_B\}\right) = u^q \star \mu_B^q(x).$$

Recalling from section I.3 that the function $x \to \widehat{\mathbb{E}}_x\left(\exp\{-qT'_B\}\right)$ is q-coexcessive, we deduce from Propositions I.11 and I.13 that the foregoing identity holds for every x.

When the Lévy process is transient and the set B bounded, one can still define an analogue of the q-capacitary measure of B for $q = 0$. For simplicity, we will only consider the case when $B = K$ is compact, though this restriction is unnecessary.

Corollary 8 *Assume that X is transient and K compact. Then μ_K^q converges weakly as $q \to 0+$. The limit measure μ_K^0 is the unique Radon measure μ given for every Borel set A by*

$$\mu U(A) := \int_{\mathbb{R}^d} U(x, A)\mu(dx) = \int_A \widehat{\mathbb{P}}_x(T_K < \infty)dx\,.$$

The measure μ_K^0 is called the *equilibrium measure* of K, and the exponent 0 is often omitted from the notation.

Proof Let B denote the closed unit ball centred at the origin and $K' = K + B$ the set of points which are at distance at most 1 from K. It is immediate that if $T = \inf\{t \geq 0 : |X_t| > 1\}$ stands for the first exit time from B, then for every $x \in K$

$$U^q(x, K') \geq \frac{1}{q}\mathbb{E}_0(1 - \exp\{-qT\}).$$

The right-hand side converges to $\mathbb{E}_0(T) > 0$ as q tends to $0+$, so that

$$\eta = \inf\{q^{-1}\mathbb{E}_0(1 - \exp\{-qT\}) : 0 < q \leq 1\} > 0.$$

Then, according to Lemma 6 and Theorem 7,

$$\begin{aligned}\eta\mu_K^q(\mathbb{R}^d) = \eta\mu_K^q(K) &\leq \int_K U^q(x, K')\mu_K^q(dx)\\ &\leq \int_{K'} \widehat{\mathbb{E}}_x(\exp\{-qT_K\})dx\\ &\leq m(K'),\end{aligned}$$

so that the total mass of μ_K^q is uniformly bounded for $q \in (0, 1]$. Hence the family of measures on the compact set K, $(\mu_K^q, 0 < q \leq 1)$, is weakly relatively compact.

Then consider a continuous function $f \geq 0$ with compact support. It follows from Dini's theorem that the continuous function $U^q f$ converges uniformly to Uf, and hence

$$\lim_{q \to 0+} \int_{\mathbb{R}^d} (Uf(x) - U^q f(x)) \mu_K^q(dx) = 0 .$$

On the other hand, $\widehat{\mathbb{E}}_x(\exp\{-qT_K\})$ increases to $\widehat{\mathbb{P}}_x(T_K < \infty)$ as q decreases to 0, so by monotone convergence and Theorem 7, we have

$$\lim_{q \to 0+} \int_{\mathbb{R}^d} Uf(x) \mu_K^q(dx) = \int_{\mathbb{R}^d} f(x) \widehat{\mathbb{P}}_x(T_K < \infty) dx .$$

But for a transient Lévy process, $U\mathscr{C}_0 = U^q \mathscr{C}_0$ is dense in $\mathscr{C}^0$, and this shows that there is only one possible limit for any weakly convergent sub-sequence of $(\mu_K^q, 0 < q \leq 1)$. Our assertion is now established. □

3. Essentially polar sets and capacity

Our interest in the problem of deciding whether a Lévy process ever visits a given set motivates the following definition.

Definition (Polar and essentially polar sets) *One says that a closed set B is polar if*

$$\mathbb{P}_x(X_t \in B \text{ for some } t > 0) = 0 \qquad \textit{for every } x,$$

and essentially polar if

$$\mathbb{P}_x(X_t \in B \text{ for some } t > 0) = 0 \qquad \textit{for almost every } x.$$

Obviously, a polar set is necessarily essentially polar. Conversely, it can be immediately seen that if the resolvent kernel is absolutely continuous, then an essentially polar set is in fact polar. Nonetheless there are Lévy processes for which the two classes differ. More precisely, Hawkes (1979) proved that every essentially polar set is polar if and only if the resolvent kernel is absolutely continuous.

We will always suppose in the sequel that B is either open or closed. The mass of the q-capacitary measure of B,

$$C^q(B) := \mu_B^q(\mathbb{R}^d) = q \int_{\mathbb{R}^d} \mathbb{E}_x(\exp\{-qT_B\}) dx,$$

is called the *q-capacity* of B. When X is transient and B compact, one can use Corollary 8 and define the 0-capacity of B by

$$C(B) = C^0(B) = \lim_{q \to 0+} C^q(B) .$$

The notion of capacity lies at the heart of the characterization of essentially polar sets. The terminology comes from the fact that the mapping $B \to C^q(B)$ can be extended to analytic sets and then defines a capacity in the sense of Choquet, see Port and Stone (1971) for a precise definition and details. Here is an easier result.

Lemma 9 *For every $q > 0$, the q-capacity has the following properties:*

(i) *If $B' \subseteq B$, then $C^q(B') \leq C^q(B)$.*

(ii) $C^q(A \cup B) + C^q(A \cap B) \leq C^q(A) + C^q(B)$.

(iii) *Suppose that $(B_n, n \in \mathbb{N})$ is an increasing sequence of open sets with $\bigcup B_n = B$. Then $C^q(B_n)$ converges to $C^q(B)$.*

(iv) *Suppose that B is closed. Then*

$$C^q(B) = \inf\{C^q(A) : A \text{ open and } B \subseteq A\}.$$

Proof (i) The obvious inequality $T_B \leq T_{B'}$ yields

$$\begin{aligned} C^q(B') = \mu^q_{B'}(\mathbb{R}^d) &= q \int_{\mathbb{R}^d} \mathbb{E}_x(\exp\{-qT_{B'}\})dx \\ &\leq q \int_{\mathbb{R}^d} \mathbb{E}_x(\exp\{-qT_B\})dx = \mu^q_B(\mathbb{R}^d) = C^q(B). \end{aligned}$$

(ii) From the inequality

$$\begin{aligned} \mathbb{P}_x(T_{A\cap B} \leq t) &\leq \mathbb{P}_x(T_A \leq t, T_B \leq t) \\ &\leq \mathbb{P}_x(T_A \leq t) + \mathbb{P}_x(T_B \leq t) - \mathbb{P}_x(T_{A\cup B} \leq t), \end{aligned}$$

we deduce that

$$\begin{aligned} \mathbb{E}_x(\exp\{-qT_{A\cup B}\}) + \mathbb{E}_x(\exp\{-qT_{A\cap B}\}) \leq & \mathbb{E}_x(\exp\{-qT_A\}) \\ & + \mathbb{E}_x(\exp\{-qT_B\}) \end{aligned}$$

and (ii) follows.

(iii) It can be immediately checked that the sequence of stopping times $(T_{B_n}, n \in \mathbb{N})$ decreases to T_B $\mathbb{P}_x$-a.s. for every $x \in \mathbb{R}^d$. The assertion then follows by monotone convergence.

(iv) Suppose first that B is compact and let B_n stand for the set of points which are at distance less than $1/n$ from B. According to Corollary I.8(ii), $(T_{B_n}, n \in \mathbb{N})$ increases to T_B, and it follows by dominated convergence that

$$\lim_{n\to\infty} \mu^q_{B_n}(\mathbb{R}^d) = \lim_{n\to\infty} \int_{B_1} \mathbb{E}_x(\exp\{-qT_{B_n}\})dx = \mu^q_B(B).$$

When B is closed, we may suppose that $C^q(B) < \infty$ since otherwise there is nothing to prove. We can express B as a countable union of compact sets, $B = \bigcup B_n$. Applying (iv) to the B_n's and (ii), it is easy to construct an open neighbourhood of B with finite q-capacity. The result then follows by an argument similar to the foregoing. □

We now identify essentially polar sets with sets of zero capacity.

Proposition 10 *If $C^q(B) = 0$ for some $q > 0$ ($q \geq 0$ in the transient case), then $C^q(B) = 0$ for all $q > 0$, and B is essentially polar. Conversely, if B is essentially polar, then $C^q(B) = 0$.*

Proof By definition, $C^q(B) = 0$ if and only if $\mathbb{E}_x(\exp\{-qT_B\}) = 0$ for almost every $x \in \mathbb{R}^d$. Because $T_B = T'_B$, $\mathbb{P}_x$-a.s. for almost every x, the latter assertion is equivalent to $\mathbb{P}.(T'_B < \infty) = 0$ almost everywhere. □

When every point is possible (in the sense of section I.4), one can reinforce Proposition 10 as follows. If the capacity of B is positive, then for almost every starting point, the probability that X visits B is positive. Here is the formal statement.

Proposition 11 *Assume that every $x \in \mathbb{R}^d$ is possible, i.e. the support of the potential measure $U(0,\cdot)$ is $\mathbb{R}^d$. Then $C^q(B) > 0$ if and only if $\mathbb{P}.(T'_B < \infty) > 0$ almost everywhere.*

Proof If $C^q(B) > 0$, then according to Proposition 8, the set of points x with $\mathbb{P}_x(T'_B < \infty) > 0$ has positive Lebesgue measure. Using inner regularity, we can therefore pick $\eta > 0$ and a compact set K with positive Lebesgue measure such that

$$\inf_{y \in K} \mathbb{P}_y(T'_B < \infty) \geq \eta.$$

Then for every $t > 0$ and $x \in \mathbb{R}^d$, we deduce from the simple Markov property that for every x

$$t\mathbb{P}_x(T'_B < \infty) \geq \int_K \int_0^t \mathbb{P}_x(X_s \in dy)\mathbb{P}_y(T'_B < \infty)ds$$
$$\geq \eta \int_0^t \mathbb{P}_x(X_s \in K)ds\,,$$

so that we need only check that

$$U^1(x,K) = \int_0^\infty e^{-s}\mathbb{P}_x(X_s \in K)ds > 0 \qquad \text{for almost every } x. \qquad (3)$$

Suppose that (3) fails, so that we can pick a Borel set A with positive Lebesgue measure such that

$$0 = \int_A U^q(x,K)dx = \int_{\mathbb{R}^d} U^q(0,dy) \int_A \mathbf{1}_K(x+y)dx .$$

But the function $y \to \int_A \mathbf{1}_K(x+y)dx$ is continuous and hence positive on some non-empty ball (because both A and K have positive Lebesgue measure). This easily yields a contradiction with the hypothesis that all the points are possible. □

4. Energy

Our aim in this section is to obtain an explicit characterization of essentially polar sets. This goal will be achieved using the notion of the *energy* of a probability measure, which we will define in terms of its Fourier transform and the characteristic exponent of the Lévy process. Then we obtain bounds for the capacity of a set in terms of its energy, that is the overall infimum of the energy of probability measures supported by the set. This enables us to identify essentially polar sets (which are sets with zero capacity according to Proposition 10) with sets with infinite energy. This section is essentially taken from Hawkes (1979).

In the classical potential theory, when the q-resolvent operator V^q of a Markov process possesses a density kernel $v^q(x,y)$, one defines the q-energy of a bounded measure ν by

$$\int_{\mathbb{R}^d} \nu(dx) \int_{\mathbb{R}^d} \nu(dy) v^q(x,y) .$$

For a Lévy process, one can get rid of the assumption of absolute continuity using the Parseval identity and Proposition I.9. Specifically, suppose for simplicity that ν is absolutely continuous with density $f \in L^1 \cap L^2$, so that the foregoing quantity can be re-expressed as

$$\int_{\mathbb{R}^d} f(x) U^q f(x) dx := e^q(\nu) .$$

Then recall that the real part of the characteristic exponent, $\Re\Psi$, is nonnegative and that $U^q f \in L^1 \cap L^2$. By the Parseval identity and Proposition I.9, we have

$$(2\pi)^d e^q(\nu) = \int_{\mathbb{R}^d} \frac{|\mathcal{F} f(\xi)|^2}{q+\Psi(\xi)} d\xi = \int_{\mathbb{R}^d} |\mathcal{F}\nu(\xi)|^2 \Re\left(\frac{1}{q+\Psi(\xi)}\right) d\xi ,$$

where $\mathcal{F}$ stands for the Fourier transform. We observe that the last expression is well defined (possibly infinite) even when the resolvent operators are not absolutely continuous, and this motivates the following definition.

Definition (Energy) *For every $q > 0$, the q-energy of a probability measure ν on $\mathbb{R}^d$ is given by*

$$e^q(\nu) = (2\pi)^{-d} \int_{\mathbb{R}^d} |\mathcal{F}\nu(\xi)|^2 \Re\left(\frac{1}{q+\Psi(\xi)}\right) d\xi \,.$$

For every Borel set $B \subseteq \mathbb{R}^d$, the q-energy of B is

$$e^q(B) = \inf\{e^q(\nu) : \nu \text{ probability measure with } \nu(B) = 1\} \,.$$

In the Brownian case, the lower bound in the definition can only be attained by a renormalization of the q-capacitary measure, and the energy always coincides with the inverse of the capacity. See Doob (1984) and Port and Stone (1978). More generally, this holds also for symmetric Markov processes, see e.g. Fukushima, Oshima and Takeda (1994). We will prove here a weaker result which is valid for the non-symmetric Lévy processes as well, and which will be sufficient for our purposes.

First, we establish a lower bound for the energy of open sets.

Lemma 12 *Let B be a relatively compact open set. Then $e^q(B) > 0$, and for every $\varepsilon > 0$ there exists a continuous function $h : \mathbb{R}^d \to [0, \infty)$ with $h(x) = 0$ for $x \in \mathbb{R}^d - B$ and $\int h(x)dx = 1$ (that is h is the density of a probability measure on B), such that*

$$\int_{\mathbb{R}^d} h(x) U^q h(x) dx \le e^q(B)(1+\varepsilon) \,.$$

Proof If the energy of B were zero, then there would be a sequence of probability measures $(\rho_n, n \in \mathbb{N})$ on B with $e^q(\rho_n) \le 1/n$. With no loss of generality, we may suppose that ρ_n converges weakly to some probability measure ρ on the compact set $\overline{B}$. An application of Fatou's lemma shows that then ρ would have zero energy,

$$\int_{\mathbb{R}^d} |\mathcal{F}\rho(\xi)|^2 \Re\left(\frac{1}{q+\Psi(\xi)}\right) d\xi = 0 \,.$$

Since $\Re(q+\Psi) \ge q > 0$, this would imply that $\mathcal{F}\rho(\xi) = 0$ for almost every ξ, i.e. $\rho = 0$, which is impossible, and thus B has positive energy.

Pick a probability measure ν on B such that $e^q(\nu) \le e^q(B) + \varepsilon$. For every $\eta > 0$, let B_η stand for the set of points at distance $> \eta$ from $\mathbb{R}^d - B$. So $(B_\eta, \eta > 0)$ is a decreasing family of open subsets of B, and $\bigcup B_\eta = B$. We may suppose that $\nu(B_\eta) \ge 1 - \varepsilon$ provided that η is small enough.

Then we consider a continuous function $a : \mathbb{R}^d \to [0,\infty)$ with $\int a(x)dx = 1$ and $a(x) = 0$ when $|x| > \eta$, and put

$$k = (\mathbf{1}_{B_\eta} \nu) \star a$$

where $\star$ denotes the convolution operator. By construction, k is a nonnegative function which vanishes away from B, and

$$\int_{\mathbb{R}^d} k(x)dx = \nu(B_\eta) \geq 1 - \varepsilon \, .$$

On the other hand, $|\mathcal{F}a| \leq 1$, and therefore

$$(2\pi)^{-d} \int_{\mathbb{R}^d} |\mathcal{F}(\nu \star a)(\xi)|^2 \Re\left(\frac{1}{q + \Psi(\xi)}\right) d\xi \leq e^q(B) + \varepsilon.$$

Note that $\mathcal{F}(\nu \star a)$ is a square-integrable function and recall that $|q+\Psi| \geq q$. We deduce from the Parseval identity and Proposition I.9 that

$$\int_{\mathbb{R}^d} (\nu \star a)(x) U^q(\nu \star a)(x)dx \leq e^q(B) + \varepsilon \, ,$$

and then *a fortiori* (because $k \leq \nu \star a$) that

$$\int_{\mathbb{R}^d} k(x) U^q k(x)dx \leq e^q(B) + \varepsilon \, .$$

Finally, we take $h = k/\nu(B_\eta)$, which is the continuous density of a probability measure on B such that

$$\int_{\mathbb{R}^d} h(x) U^q h(x)dx \leq (e^q(B) + \varepsilon)(1 - \varepsilon)^{-2} \, .$$

Since ε is arbitrary, the lemma is proven. □

Here is the main result of this section, which provides key bounds for the capacity of a set in terms of its energy.

Theorem 13 *Assume that B is bounded and either open or closed. Then*

$$e^q(B) \leq 1/C^q(B) \leq 4e^q(B).$$

To facilitate the proof, we first establish the theorem when B is open.

Proof of Theorem 13 for an open set Pick $\eta > 0$ and consider a continuous function $a_\eta : \mathbb{R}^d \to [0,\infty)$ which is the density of a probability measure on the ball centred at the origin with radius η, and is such that its Fourier transform $\mathcal{F}a_\eta$ is real-valued and nonnegative (such a function is easily constructed using convolution of smooth probability measures on the ball with radius $\eta/2$). For every $\varepsilon > \eta$, we denote by $B(\varepsilon)$ the set of points which are at distance $> \varepsilon$ from $\mathbb{R}^d - B$, and by μ^q_ε the q-capacitary measure of $B(\varepsilon)$. Next, we put $f = a_\eta \star \mu^q_\varepsilon$, which is a continuous function vanishing away from B (since $\varepsilon > \eta$). Using the fact that $\widehat{\mathbb{E}}_x(\exp\{-qT_B\}) = 1$ for all $x \in B$, we have by Theorem 7

$$\int_{\mathbb{R}^d} U^q f(x)\mu_B^q(dx) = \int_{\mathbb{R}^d} f(x)\widehat{\mathbb{E}}_x(\exp\{-qT_B\})dx$$
$$= \int_{\mathbb{R}^d} f(x)dx = \mu_\varepsilon^q(\mathbb{R}^d) = C^q(B(\varepsilon)).$$

On the other hand, it follows from Theorem 7 that

$$\int_{\mathbb{R}^d} U^q f(x)\mu_B^q(dx) = \int_{\mathbb{R}^d} f(x)\widehat{\mathbb{E}}_x(\exp\{-qT_B\})dx$$
$$\geq \int_{\mathbb{R}^d} f(x)\widehat{\mathbb{E}}_x(\exp\{-qT_{B(\varepsilon)}\})dx$$
$$= \int_{\mathbb{R}^d} U^q f(x)\mu_\varepsilon^q(dx).$$

Then, applying the Parseval identity and Proposition I.9, and recalling that $\mathscr{F}a_\eta(\xi) \geq 0$, we get

$$\int_{\mathbb{R}^d} U^q f(x)\mu_\varepsilon^q(dx) \geq (2\pi)^{-d} \int_{\mathbb{R}^d} |\mathscr{F}\mu_\varepsilon^q(\xi)|^2 \mathscr{F}a_\eta(\xi)\Re\left(\frac{1}{q+\Psi(\xi)}\right) d\xi.$$

When η tends to $0+$, $\mathscr{F}a_\eta$ converges to 1 and we get from Fatou's inequality

$$C^q(B(\varepsilon)) \geq (2\pi)^{-d} \int_{\mathbb{R}^d} |\mathscr{F}\mu_\varepsilon^q(\xi)|^2 \Re\left(\frac{1}{q+\Psi(\xi)}\right) d\xi$$
$$\geq C^q(B(\varepsilon))^2 e^q(B),$$

because $\mu_\varepsilon^q / C^q(B(\varepsilon))$ is a probability measure supported by B. Hence $1/C^q(B(\varepsilon)) \geq e^q(B)$, and since according to Lemma 9, $C^q(B(\varepsilon))$ increases to $C^q(B)$ as $\varepsilon \to 0+$, the first inequality of the theorem is proven.

To prove the second inequality, we pick $\varepsilon > 0$ and a function h which satisfies the conclusions of Lemma 12, so that

$$\int_{\mathbb{R}^d} h(x)U^q h(x)dx \leq e^q(B)(1+\varepsilon).$$

Then consider the set $B' = \{x \in B : U^q h(x) < 2e^q(B)\}$. Since $h \in \mathscr{C}_0$, $U^q h \in \mathscr{C}_0$ according to the Feller property of U^q, and B' is an open set. We have the inequalities

$$e^q(B)(1+\varepsilon) \geq \int_{\mathbb{R}^d} h(x)U^q h(x)dx \geq 2e^q(B) \int_{B-B'} h(x)dx,$$

and thus

$$\int_{B'} h(x)dx = \int_B h(x)dx - \int_{B-B'} h(x)dx$$
$$\geq 1 - \frac{1}{2}(1+\varepsilon) = \frac{1}{2}(1-\varepsilon). \tag{4}$$

Next, we consider the q-capacitary measure of B', $\mu^q_{B'}$. According to Theorem 7, we have

$$\int_{\mathbb{R}^d} h(x)\widehat{\mathbb{E}}_x(\exp\{-qT_{B'}\})dx = \int_{\mathbb{R}^d} U^q h(x)\mu^q_{B'}(dx) \,.$$

Plainly $\widehat{\mathbb{E}}_x(\exp\{-qT_{B'}\}) = 1$ for all $x \in B'$ and hence by (4)

$$\begin{aligned} \frac{1}{2}(1-\varepsilon) &\leq \int_{\mathbb{R}^d} U^q h(x)\mu^q_{B'}(dx) \\ &\leq 2e^q(B)\mu^q_{B'}(\overline{B'}) \qquad \text{(because } \mu^q_{B'} \text{ is supported by } \overline{B'}) \\ &= 2e^q(B)C^q(B') \,. \end{aligned}$$

Applying Lemma 9(i), we see that the foregoing quantity is less than or equal to $2e^q(B)C^q(B)$. This proves the second inequality of the theorem. □

Finally, the case when B is compact can be derived from the preceding by approximation.

Proof of Theorem 13 for a compact set The inequality $1/C^q(B) \leq 4e^q(B)$ follows immediately by approximation, using Lemma 9(iv) and the theorem for open sets. To prove the remaining inequality, we consider for every integer $n > 0$ the set B_n of points which are at distance $< 1/n$ from B, so B_n is open and relatively compact. Then we fix $\varepsilon > 0$ and we consider a probability measure ν_n on B_n such that $e^q(\nu_n) \leq e^q(B_n) + \varepsilon$. Taking a sub-sequence if necessary, we may suppose that ν_n converges vaguely to a probability measure ν on B, so that $\mathscr{F}\nu_n$ converges pointwise to $\mathscr{F}\nu$. We now deduce from Fatou's lemma, the theorem for open sets and Lemma 9(iv) that

$$e^q(\nu) \leq \liminf_{n\to\infty} e^q(\nu_n) \leq \liminf_{n\to\infty} e^q(B_n) + \varepsilon \leq 1/C^q(B) + \varepsilon \,.$$

Since ε is arbitrary, the desired inequality is established, and the proof of Theorem 13 is now complete. □

We finish this section by presenting two important consequences of Theorem 13. The first provides the desired characterization of essentially polar sets in terms of the characteristic exponent.

Corollary 14 *Suppose that B is a closed set. Then B is essentially polar if and only if its energy is infinite, that is if and only if*

$$\int_{\mathbb{R}^d} |\mathscr{F}\nu(\xi)|^2 \Re\left(\frac{1}{q+\Psi(\xi)}\right) d\xi = \infty$$

for every probability measure ν on B.

Proof By Theorem 13, $e^q(B) = \infty$ is equivalent to $C^q(B) = 0$. Proposition 10 shows that the latter holds if and only if B is essentially polar. □

The second result compares essentially polar sets for two different Lévy processes.

Corollary 15 *Let $\widetilde{\mathbb{P}}$ be the law of a second Lévy process with characteristic exponent $\widetilde{\Psi}$. If there exist $M > 0$ such that for all $\xi \in \mathbb{R}^d$*

$$\Re\left(\frac{1}{q+\widetilde{\Psi}(\xi)}\right) \geq M\Re\left(\frac{1}{q+\Psi(\xi)}\right),$$

then for any open or closed set B,

$$C^q(B) \geq \frac{1}{4}M\widetilde{C}^q(B).$$

In particular, any essentially polar set with respect to $\mathbb{P}$ is also essentially polar with respect to $\widetilde{\mathbb{P}}$.

Proof This is immediate from Theorem 13 and the definition of the q-energy. The last assertion stems from Proposition 10. □

Corollary 15 was proved first by Orey (1967) and Kanda (1976) under some restrictive conditions. In particular, it was used by Kanda to show that all the stable processes of the same index have the same class of polar sets, solving a conjecture of Taylor (1973).

5. The case of a single point

We now specialize to the case when B is a single point. For the sake of simplicity, we will focus on the case $B = \{0\}$, the extension to an arbitrary point being immediate. We denote the q-capacity of $\{0\}$ by $C^q = C^q(\{0\})$ for every $q > 0$ ($q \geq 0$ when the Lévy process is transient) and state the following criterion to decide whether $\{0\}$ is essentially polar for the Lévy process.

Theorem 16 *The following assertions are equivalent for every $q > 0$ ($q \geq 0$ in the transient case):*

(i) *Single points are not essentially polar, that is $C^q > 0$.*

(ii) $$\int_{\mathbb{R}^d} \Re\left(\frac{1}{q+\Psi(\xi)}\right) d\xi < \infty.$$

(iii) *The measure $U^q(0, dx)$ is absolutely continuous with respect to the Lebesgue measure and has a bounded density.*

(iv) *Let $B(\varepsilon)$ be the closed ball with radius ε centred at the origin. Then*

$$\liminf_{\varepsilon \to 0+} \varepsilon^{-d} U^q(0, B(\varepsilon)) < \infty .$$

Theorem 16 is essentially due to Kesten (1969), see also Bretagnolle (1971). Actually Kesten and Bretagnolle have a sharper result. They proved that in dimension $d = 1$, except when X is a compound Poisson process, $\{0\}$ is polar if and only if it is essentially polar. This refinement will be established for increasing Lévy processes in section III.2 below.

Proof We will focus on the case $q > 0$, the arguments for $q = 0$ in the transient case are similar, using Corollary 8.

(i)$\Leftrightarrow$(ii) The equivalence is obvious from Corollary 14 since the Dirac mass δ_0 is the unique probability measure on $\{0\}$.

(i)$\Rightarrow$(iii) The q-capacitary measure of $\{0\}$ is $C^q \delta_0$, and we deduce from Theorem 7 that

$$C^q U^q(0, A) = \int_A \widehat{\mathbb{E}}_x(\exp\{-qT_{\{0\}}\}) dx \qquad \text{for every Borel set } A .$$

Thus, if (i) holds, then $U^q(0, dx)$ is absolutely continuous with respect to the Lebesgue measure and has density $(1/C^q)\widehat{\mathbb{E}}_x(\exp\{-qT_{\{0\}}\})$ which is obviously bounded.

(iii)$\Rightarrow$(i) Suppose now that $U^q(0, dx)$ has a bounded density. Then we know from Propositions I.11 and I.12 that its canonical density u^q is lower semi-continuous. Pick any x such that $u^q(-x) > 0$, consider for $\varepsilon > 0$ the closed ball $B(\varepsilon)$ centred at the origin with radius ε, and denote its Lebesgue measure by $m(B(\varepsilon))$. Using the lower semi-continuity, we see that, provided that ε is small enough,

$$U^q(x, B(\varepsilon)) = \int_{|y|<\varepsilon} u^q(y - x) dy \geq \frac{1}{2} u^q(-x) m(B(\varepsilon)) .$$

On the other hand, the Markov property applied at the first passage time into $B(\varepsilon)$ yields

$$\begin{aligned} U^q(x, B(\varepsilon)) &= \mathbb{E}_x \left(\int_0^\infty e^{-qt} \mathbf{1}_{\{X_t \in B(\varepsilon)\}} dt \right) \\ &= \mathbb{E}_x \left(\exp\{-qT_{B(\varepsilon)}\} U^q(X_{T_{B(\varepsilon)}}, B(\varepsilon)) \right) \\ &\leq \mathbb{E}_x \left(\exp\{-qT_{B(\varepsilon)}\} \right) \| u^q \|_\infty m(B(\varepsilon)) , \end{aligned}$$

where in the last inequality, we used the fact that for all y

$$U^q(y, B(\varepsilon)) = \int_{|z|<\varepsilon} u^q(z - y) dz \leq \| u^q \|_\infty m(B(\varepsilon)) .$$

It follows that

$$\mathbb{E}_x\left(\exp\{-qT_{B(\varepsilon)}\}\right) \geq \frac{1}{2}u^q(-x)/\parallel u^q \parallel_\infty > 0\,.$$

But according to Corollary I.8, when ε decreases to 0+, $T_{B(\varepsilon)}$ increases $\mathbb{P}_x$-a.s. to the first passage time into $\{0\}$, so that $\mathbb{P}_x(T_{\{0\}} < \infty) > 0$. This holds for all $x \neq 0$ with $u^q(-x) > 0$, which form a set of positive Lebesgue measure. Thus according to Proposition 10, $C^q > 0$.

(iii)$\Leftrightarrow$(iv) Assume that there is a sequence of positive real numbers $(\varepsilon_n, n \in \mathbb{N})$ that tends to 0, such that $\varepsilon_n^{-d}U^q(0, B(\varepsilon_n)) < M$, where M is some fixed real number. Then fix $x \in \mathbb{R}^d$ and apply the Markov property at the first passage time into $x + B(\varepsilon_n/2)$ to get

$$\begin{aligned}
U^q(0, x + B(\varepsilon_n/2)) &= \mathbb{E}_0\left(\int_0^\infty e^{-qt}\mathbf{1}_{\{X_t \in x + B(\varepsilon_n/2)\}}dt\right)\\
&\leq \int_{x+B(\varepsilon_n/2)} \mathbb{P}_0(X_{T_{x+B(\varepsilon_n/2)}} \in dy)U^q(y, x + B(\varepsilon_n/2))\\
&\leq \mathbb{P}(T_{x+B(\varepsilon_n/2)} < \infty)U^q(0, B(\varepsilon_n))\\
&\leq \varepsilon_n^d M,
\end{aligned}$$

where we used the obvious inequality

$$U^q(y, x + B(\varepsilon_n/2)) \leq U^q(0, B(\varepsilon_n)) \quad \text{for every } y \in x + B(\varepsilon_n/2).$$

This shows that the potential measure has a bounded density. The converse assertion is obvious. □

For instance, in dimension $d = 1$, points are polar for a stable process of index α for $\alpha < 1$ but not for $\alpha > 1$; and for $\alpha = 1$, points are polar for the symmetric Cauchy process (because the resolvent kernel of a stable process is absolutely continuous, there is no need to distinguish between essentially polar and polar). Another interesting application of Theorem 16 is that single points are always essentially polar in dimension $d \geq 2$.

Corollary 17 *One has $C^q = 0$ in dimension $d \geq 2$.*

Proof Suppose first that the Gaussian coefficient Q is zero and recall from Proposition I.2(i) that $|\Psi(\xi)| = o(|\xi|^2)$ as $|\xi| \to \infty$. Consequently, for every $\lambda \in \mathbb{R}^d$,

$$\mathbb{E}\left(\exp\{i\langle\lambda t^{-1/2}, X_t\rangle\}\right) = \exp\{-t\Psi(\lambda t^{-1/2}\}) \to 1 \quad \text{as } t \to 0+\,.$$

That is $t^{-1/2}X_t$ converges in probability to 0 as t tends to 0+. Then let B_η stand for the ball with radius η centred at the origin. For every fixed

$\varepsilon > 0$, there exists $t_\varepsilon > 0$ such that

$$\mathbb{P}_0(|X_t| < \sqrt{\varepsilon t}) \geq 1/2 \qquad \text{for all } 0 \leq t \leq t_\varepsilon.$$

Then for every $\eta \in (0, \varepsilon t_\varepsilon)$ small enough, we have

$$\begin{aligned}\eta^{-d/2} U^1(0, B_{\sqrt{\eta}}) &= \eta^{-d/2} \int_0^\infty dt e^{-t} \mathbb{P}_0(|X_t| < \sqrt{\eta}) \\ &\geq \eta^{-d/2} \int_0^{\eta/\varepsilon} dt e^{-t} \mathbb{P}_0(|X_t| < \sqrt{\varepsilon t}) \\ &\geq \varepsilon^{-1}/4.\end{aligned}$$

This shows that the assertion (iv) in Theorem 16 fails and hence $C^q = 0$.

Suppose now that $Q \neq 0$ and recall that $\Psi(\xi) = O(|\xi|^2)$ as $|\xi| \to \infty$. On the other hand, there is an open wedge $\mathscr{W}$ oriented in the direction of an eigenvector of Q such that

$$\liminf_{\xi \in \mathscr{W}, |\xi| \to \infty} |\xi|^{-2} Q(\xi) > 0,$$

and then, again by the Lévy-Khintchine formula,

$$\liminf_{\xi \in \mathscr{W}, |\xi| \to \infty} |\xi|^{-2} \Re\Psi(\xi) > 0.$$

Using polar coordinates, this entails that for some constants $c, c' > 0$,

$$\begin{aligned}\int_{\mathbb{R}^d} \Re\left(\frac{1}{1+\Psi(\xi)}\right) d\xi &\geq c \int_{\mathscr{W}} \frac{\Re\Psi(\xi)}{1+|\xi|^4} d\xi \\ &\geq c' \int^\infty r^{d-1} \frac{r^2}{1+r^4} dr = \infty,\end{aligned}$$

that is the assertion (ii) of Theorem 16 fails. □

We now assume that $C^q > 0$, so we are implicitly working on the real line. Our next purpose is to get precise information on the hitting time of single points. We will henceforth denote by u^q the q-coexcessive version of the density of the measure $U^q(0, \cdot)$, see Proposition I.12. The following result specifies the distribution of the first hitting time on $\{0\}$. For simplicity, we will write $T' = T'_{\{0\}} = \inf\{t > 0 : X_t = 0\}$ in the sequel.

Corollary 18 *Suppose that the assertions of Theorem 16 hold, that is the resolvent kernel is absolutely continuous with a bounded density. Then*

$$\mathbb{E}_x(\exp\{-qT'\}) = C^q u^q(-x), \qquad \textit{for all } q > 0, x \in \mathbb{R}.$$

Moreover, in the transient case, $\mathbb{P}_x(T' < \infty) = Cu(-x)$.

Proof We already observed in section I.3 that the function

$$f : x \to \mathbb{E}_x(\exp\{-qT'\})$$

is q-excessive. On the other hand, because $\mu^q_{\{0\}} = C^q \delta_0$, we deduce from Theorem 7 that

$$(1/C^q)\hat{\mathbb{E}}_\cdot(\exp\{-qT'\})$$

is a version of the density of $U^q(0,\cdot)$. We conclude the proof by Proposition I.11(ii). The case $q = 0$ when X is transient is similar. □

Next, we are interested in the question of whether the Lévy process started at the origin returns to the origin at arbitrarily small times. More precisely, recall that according to the Blumenthal zero-one law $\mathbb{P}_0(T' = 0) = 0$ or 1. We say that 0 is *irregular for itself* if this probability is zero, and *regular for itself* otherwise.

Theorem 19 *Take any $q > 0$ ($q \geq 0$ in the transient case) and assume that the assertions of Theorem 16 hold.*

(i) *If 0 is regular for itself, then u^q is continuous and $u^q > 0$ everywhere.*
(ii) *Conversely, if there exists a continuous version of the density of $U^q(0,\cdot)$, then 0 is regular for itself (and thus u^q is continuous).*
(iii) *If u^q is continuous, then*

$$u^q(0) = \frac{1}{2\pi}\int_{-\infty}^{\infty} \Re\left(\frac{1}{q+\Psi(\xi)}\right) d\xi$$

and for every real number $x \neq 0$

$$2u^q(0) - (u^q(x) + u^q(-x)) = \frac{1}{\pi}\int_{-\infty}^{\infty} (1-\cos\xi x)\Re\left(\frac{1}{q+\Psi(\xi)}\right) d\xi\,.$$

For instance, one gets that in the stable case with index $\alpha \in (1,2]$, $u^q(0) = q^{-1+1/\alpha}u^1(0)$.

Proof (i) We will focus on the case $q > 0$, the case when the Lévy process is transient and $q = 0$ is similar. Assume that 0 is regular for itself. According to Corollary 18, we have

$$\mathbb{E}_0(\exp\{-qT'\}) = 1 = C^q u^q(0)\,,$$

and then for all $x \in \mathbb{R}$,

$$\mathbb{E}_x(\exp\{-qT'\}) = u^q(-x)/u^q(0) \leq 1.$$

By the lower semi-continuity of u^q, this forces $\lim_{x\to 0} u^q(x) = u^q(0)$. Next, for every $x, y \neq 0$, the Markov property implies

$$\mathbb{E}_x(\exp\{-qT'\}) \geq \mathbb{E}_{x+y}(\exp\{-qT'\})\mathbb{E}_{-y}(\exp\{-qT'\})$$

and thus we have $u^q(-x) \geq u^q(-x-y)u^q(y)C^q$. Then $u^q(y)$ converges to $u^q(0) = 1/C^q$ as $y \to 0$, and hence

$$u^q(-x) \geq \limsup_{y \to 0} u^q(-x-y).$$

Using again the lower semi-continuity of u^q, we deduce that u^q is continuous at $-x$.

Finally, because u^q is continuous and $u^q(0) > 0$, we have that $u^q > 0$ on some open ball centred at the origin. On the other hand, the Markov property at T'_x entails the inequality

$$\mathbb{E}(\exp\{-qT'_{2x}\}) \geq \mathbb{E}(\exp\{-qT'_x\})^2,$$

and by induction

$$\mathbb{E}(\exp\{-qT'_{nx}\}) \geq \mathbb{E}(\exp\{-qT'_x\})^n \quad \text{for every integer } n > 0.$$

According to the foregoing, the right-hand side is positive provided that $|x|$ is small enough, and by Corollary 18, this shows that $u^q > 0$ eveywhere.

(ii) Suppose that there exists a continuous version v of the density of $U^q(0,\cdot)$. For every $\varepsilon > 0$, the Markov property at the first passage time into $(-\varepsilon,\varepsilon)$ yields that for all $x \neq 0$

$$\begin{aligned} \varepsilon^{-1}\int_{-\varepsilon}^{\varepsilon} dy v(y-x) &= \varepsilon^{-1}U^q(x,(-\varepsilon,\varepsilon)) \\ &= \int_{[-\varepsilon,\varepsilon]} \mathbb{E}_x(\exp\{-qT_{(-\varepsilon,\varepsilon)}\}, X_{T_{(-\varepsilon,\varepsilon)}} \in dz)\varepsilon^{-1}\int_{-\varepsilon}^{\varepsilon} dy v(y-z). \end{aligned}$$

On the one hand, it follows from Corollary I.8 that the family of sub-probability measures

$$\mathbb{E}_x(\exp\{-qT_{(-\varepsilon,\varepsilon)}\}, X_{T_{(-\varepsilon,\varepsilon)}} \in \cdot)$$

converges vaguely as ε tends to 0+ to $\mathbb{E}_x(\exp\{-qT'\})\delta_0$, where δ_0 stands for the Dirac mass at the origin. Making use of the continuity of v, we deduce that

$$\mathbb{E}_x(\exp\{-qT'\}) = v(-x)/v(0) \quad (x \neq 0). \tag{5}$$

On the other hand, because $\mathbb{E}_\cdot(\exp\{-qT'\})$ is q-excessive,

$$\mathbb{E}_0(\exp\{-qT'\}) \geq \int_{\mathbb{R}} rU^{r+q}(0,dx)\mathbb{E}_x(\exp\{-qT'\})$$

for every $r > 0$. According to Proposition I.10, the probability measure $(r+q)U^{r+q}(0,\cdot)$ gives no mass to $\{0\}$ and converges to the Dirac mass at 0 as $r \to \infty$. We deduce from (5) and the continuity of v that $\mathbb{E}_0(\exp\{-qT'\}) = 1$, that is 0 is regular for itself.

(iii) This is an easy matter of Fourier inversion. For every $\varepsilon > 0$, let $g_\varepsilon(y) = (2\pi\varepsilon)^{-1/2}\exp(-y^2/2\varepsilon)$ denote the Gaussian density. Because u^q is continuous,

$$\lim_{\varepsilon\to 0+} U^q g_\varepsilon(0) = \lim_{\varepsilon\to 0+}\int_{-\infty}^{\infty} g_\varepsilon(y)u^q(y)dy = u^q(0).$$

On the other hand, $U^q g_\varepsilon$ is continuous, and by Proposition I.9

$$\mathscr{F}U^q g_\varepsilon(\xi) = \frac{\mathscr{F}g_\varepsilon(\xi)}{q+\Psi(-\xi)} = \frac{\exp(-\varepsilon\xi^2/2)}{q+\Psi(-\xi)}, \qquad \xi\in\mathbb{R}.$$

The latter function is integrable and by Fourier inversion,

$$\begin{aligned} U^q g_\varepsilon(0) &= \frac{1}{2\pi}\int_{-\infty}^{\infty}\frac{\exp(-\varepsilon\xi^2/2)}{q+\Psi(\xi)}d\xi \\ &= \frac{1}{2\pi}\int_{-\infty}^{\infty}\exp(-\varepsilon\xi^2/2)\Re\left(\frac{1}{q+\Psi(\xi)}\right)d\xi \end{aligned}$$

because $U^q g_\varepsilon(0)$ is real. The preceding quantity converges monotonely as $\varepsilon\to 0+$ and the first identity of the lemma is established. The second can be proven similarly, considering now the function $2g_\varepsilon(y) - (g_\varepsilon(y+x)+g_\varepsilon(y-x))$. □

We now conclude this section with a simple criterion (in dimension $d = 1$) for the regularity of single points and essential polarity, respectively, in terms of the characteristic exponent Ψ. Recall that when X has bounded variation, then

$$\Psi(\lambda) = -\mathrm{i}\mathrm{d}\lambda + \int_{\mathbb{R}}(1-e^{\mathrm{i}x\lambda})\Pi(dx),$$

where $\mathrm{d}\in\mathbb{R}$ is the so-called drift coefficient.

Corollary 20 (i) *Suppose that*

$$\int_{\mathbb{R}}\frac{1}{|1+\Psi(\xi)|}d\xi < \infty;$$

then $C^q > 0$ and $\{0\}$ is regular for itself.

(ii) *Suppose that X has bounded variation and drift coefficient d. Then $C^q > 0$ if and only if $\mathrm{d}\neq 0$, and in that case $\{0\}$ is irregular for itself.*

Proof (i) By Fourier inversion and Proposition I.9, we see that the measure $U^1(0,\cdot)$ is absolutely continuous with respect to the Lebesgue measure and has a continuous density. The assertion now follows from Theorem 19. Observe also that the condition (i) is plainly stronger than that in Theorem 16(ii).

(ii) The proof relies on the fact that when X has bounded variation and drift coefficient d,

$$\lim_{|\lambda|\to\infty} \lambda^{-1}\Psi(\lambda) = -\mathrm{id}, \tag{6}$$

see Proposition I.2 (ii). As a consequence, the quantity

$$\mathbb{E}\left(\exp\{i\lambda t^{-1}X_t\}\right) = \exp\{-t\Psi(\lambda t^{-1})\}$$

converges to $\exp\{-\lambda \mathrm{d}\}$ as $t \to 0+$ for every $\lambda \in \mathbb{R}$, that is

$$\lim_{t\to 0+} t^{-1}X_t = \mathrm{d} \qquad \text{in probability,} \tag{7}$$

First, assume that $\mathrm{d} = 0$. Then for every fixed $\varepsilon > 0$, there exists $t_\varepsilon > 0$ such that

$$\mathbb{P}_0(|X_t| < \varepsilon t) \geq 1/2 \qquad \text{for all } 0 \leq t \leq t_\varepsilon.$$

Then for every $\eta \in (0, \varepsilon t_\varepsilon)$ small enough, we have

$$\begin{aligned} \eta^{-1}U^1(0,[-\eta,\eta]) &= \eta^{-1}\int_0^\infty dte^{-t}\mathbb{P}_0(|X_t| < \eta) \\ &\geq \eta^{-1}\int_0^{\eta/\varepsilon} dte^{-t}\mathbb{P}_0(|X_t| < t\varepsilon) \geq \varepsilon^{-1}/4. \end{aligned}$$

This shows that the assertion (iv) in Theorem 16 fails and hence $C^q = 0$.

Next, assume that $\mathrm{d} \neq 0$. We deduce from (6) that the hypothesis (ii) of Theorem 16 holds whenever

$$\int_{\mathbb{R}} \frac{\Re\Psi(\lambda)}{1+\lambda^2}d\lambda = \int_{\mathbb{R}} \Pi(dx)\int_{\mathbb{R}} \frac{1-\cos\lambda x}{1+\lambda^2}d\lambda < \infty. \tag{8}$$

On the other hand, the inequality $1 - \cos\vartheta \leq 2 \wedge \vartheta^2$ readily yields

$$\int_{\mathbb{R}} \frac{1-\cos\lambda x}{1+\lambda^2}d\lambda \leq 10(1\wedge|x|).$$

Since the Lévy process has bounded variation, $\int(1\wedge|x|)\Pi(dx) < \infty$, and (8) follows. Finally, Proposition VI.11(ii) below shows that (7) holds a.s. and hence 0 is necessarily irregular for itself when $\mathrm{d} \neq 0$. □

We mention for completeness that Bretagnolle (1971) established that $\{0\}$ is always regular for itself in the case when $C^q > 0$ and X has unbounded variation.

6. Exercises

1. *(Co-capacity)* Denote by $\widehat{C}^q(B)$ the q-co-capacity of B, that is the q-capacity of B for the dual process. Check that $\widehat{C}^q(B) = C^q(-B) = C^q(B)$.

2. *(Sausage and capacity)* For every $t > 0$ and closed set B, consider the so-called *sausage*

$$S(t, B) = \{X_s + B : 0 \leq s \leq t\}.$$

Denote the volume (i.e. the Lebesgue measure) of $S(t, B)$ by $V(t, B)$. Check the following formula for the q-capacity of B:

$$C^q(B) = q \int_0^\infty \mathbb{E}(V(t, B)) e^{-qt} dt.$$

3. *(Intersection of two independent Lévy processes)* Let $\widetilde{X}$ be a Lévy process taking values in $\mathbb{R}^d$ and with characteristic exponent $\widetilde{\Psi}$, which we assume independent of the Lévy process X. Consider the random probability measure

$$\nu(dx) = \int_0^\infty e^{-t} \mathbf{1}_{\{\widetilde{X}_t \in dx\}} dt,$$

which is obviously supported by the closure of the range of $\widetilde{X}$. Check that

$$\mathbb{E}(|\mathscr{F}\nu(\xi)|^2) = \Re\left(\frac{1}{1 + \widetilde{\Psi}(\xi)}\right).$$

Assume now that

$$\int_{\mathbb{R}^d} \Re\left(\frac{1}{1 + \widetilde{\Psi}(\xi)}\right) \Re\left(\frac{1}{1 + \Psi(\xi)}\right) d\xi < \infty.$$

Deduce that the closure of the range of $\widetilde{X}$ has a positive capacity (with respect to X). Check that this entails that there exists a Borel set $B \subseteq \mathbb{R}^d$ with positive Lebesgue measure such that for every $x \in B$, the $\mathbb{P}_x$-probability that X visits the range of $\widetilde{X}$ is positive. See also Exercise V.5 for an alternative approach.

4. *(Point recurrence)* Suppose that a real-valued Lévy process is recurrent and that its resolvent kernel is absolutely continuous with a bounded density. That is to say in terms of the characteristic exponent that

$$\int_{\mathbb{R}} \Re\left(\frac{1}{1 + \Psi(\xi)}\right) d\xi < \infty \quad \text{and} \quad \int_{-1}^{1} \Re\left(\frac{1}{\Psi(\xi)}\right) d\xi = \infty .$$

Prove that X visits the origin at arbitrarily large times, $\mathbb{P}_x$-a.s. for every $x \in \mathbb{R}$.

5. *(Regularity of points)* Check that the condition of Corollary 20(i) is fulfilled whenever the Gaussian coeffient of the Lévy process is positive. Show that for any symmetric Lévy process, either single points are essentially polar, or 0 is regular for itself.

7. Comments

The material developed in this chapter partially extends the classical potential theory for Brownian motion, see Doob (1984) and Port and Stone (1978). Some aspects however are peculiar to Lévy processes, we refer to Hawkes (1984) for a discussion and further references.

Blumenthal and Getoor (1968) and Dellacherie, Maisonneuve and Meyer (1992) contain a complete account of potential theory for general Markov processes, which stems from Hunt's pioneer work. Sections 2-3 and 4 are based on Port and Stone (1971) and Hawkes (1979), respectively. The link between sausages and capacity (cf. Exercise 2) originates from results of Spitzer in the Brownian case, see Port and Stone (1978). The case of a Lévy process was considered first by Getoor (1965) and then by Hawkes (1984), see also Port (1990) for further developments in the stable case and Evans (1994). Polar sets for Lévy processes were studied first by Orey (1967) and Kanda (1976, 1978, 1983). Frostman's lemma and the comparison test of Corollary 15 yield handy criteria for polarity of a set in terms of its Hausdorff dimension, see Orey (1967) and Hawkes (1970, 1971-a).

The major open problem in this field was raised by Getoor, who asked for which Lévy processes *semipolar* sets are always polar. Here, a set is called semipolar if it can be expressed as the countable union of thin sets, and a set B thin if there are no points in B that are regular for B. The condition that a semipolar set is polar appears as Hypothesis (H) in Hunt (1958) and entails numerous potential theoretic properties. In particular, Blumenthal and Getoor (1970) proved that (H) is a necessary and sufficient condition for a maximum principle. Kanda (1976, 1978) obtained sufficient conditions for (H) to hold; see also Zabczyk (1975) and Rao (1977, 1987, 1988). We refer to Fitzsimmons and Kanda (1992) for recent developments.

Section 5 presents the main part of the results of Kesten (1969), our approach follows Port and Stone (1971). The arguments of Kesten were simplified by Bretagnolle (1971). Previously, Bretagnolle and Dacunha-Castelle (1968) had considered the problem of pointwise recurrence (cf. Exercise 4).

III

Subordinators

Subordinators form the sub-class of increasing Lévy processes. Distributions related to their first passage time above a fixed level are specified, which yields an important family of limit theorems known as the arcsine laws. The rate of growth of their sample paths is studied, in particular, laws of the iterated logarithm are established for a wide class of subordinators. Finally, the Hausdorff dimension of the range of a subordinator is determined in terms of its Laplace exponent.

1. Definitions and first properties

A *subordinator* is a Lévy process taking values in $[0, \infty)$, which implies that its sample paths are increasing. Subordinators not only form a significant sub-class of Lévy processes, but also play a key rôle in the study of Markov processes as will be emphasized in the next chapter. The terminology comes from the fact that when one time-changes a Markov process M by an independent subordinator T, the resulting process $M \circ T$ is again a Markov process. This transformation was introduced by Bochner (1955), who called it *subordination* . We refer to section X.7 in Feller (1971) for a detailed account; see also Exercise 1 and the survey by Lee and Whitmore (1990).

Some authors also use the term subordinator to designate a slightly more general class of processes, which will here be referred to as *killed subordinators* . More precisely, if X is a subordinator and $\tau = \tau(q)$ an independent exponential time with parameter $q > 0$, the process $X^{(q)}$ taking values in $[0, \infty]$ and given by

$$X_t^{(q)} = X_t \text{ if } t \in [0, \tau), \; X_t^{(q)} = \infty \text{ if } t \in [\tau, \infty)$$

is called a subordinator killed at rate q. It can be immediately checked that a right-continuous process $Y = (Y_t, t \geq 0)$ taking values in $[0, \infty]$ is a subordinator killed at rate $q > 0$ if and only if $\mathbb{P}(Y_t < \infty) = e^{-qt}$ and conditionally on $Y_t < \infty$, the increment $Y_{t+s} - Y_t$ is independent of $(Y_v, 0 \leq v \leq t)$ and has the same law as Y_s. Properties of a killed subordinator are easy to derive from those of the corresponding (conservative) subordinator, and therefore we will concentrate our attention on the latter.

We will assume throughout this chapter that X is a subordinator.

An important feature is that one can now work with the Laplace transform (instead of the Fourier transform) to analyse functions and measures related to subordinators. The infinite divisibility of the law of X implies that its Laplace transform can be expressed in the form

$$\mathbb{E}(\exp -\lambda X_t) = \exp -t\Phi(\lambda) \qquad (\lambda \geq 0), \tag{1}$$

where $\Phi : [0, \infty) \to [0, \infty)$ is called the *Laplace exponent* , or also the *cumulant* . Specifically, X obviously has bounded variation and no negative jumps, so that the discussion of section I.1 shows that the Lévy measure Π has support in $[0, \infty)$ and fulfils the extra condition

$$\int_{(0,\infty)} (1 \wedge x)\Pi(dx) < \infty. \tag{2}$$

The characteristic exponent Ψ can be expressed as

$$\Psi(\lambda) = -\mathrm{i}\mathrm{d}\lambda + \int_{(0,\infty)} (1 - e^{\mathrm{i}\lambda x})\Pi(dx) \qquad (\lambda \in \mathbb{R}),$$

where d is the drift coefficient. We then see that the functions $\lambda \to \Psi(\lambda)$ and $\lambda \to \mathbb{E}(\exp\{\mathrm{i}\lambda X_t\})$ can be extended analytically on the complex upper half-plane so that $\mathbb{E}(\exp\{\mathrm{i}\lambda X_t\}) = \exp\{-t\Psi(\lambda)\}$ whenever $\Im\lambda \geq 0$. This yields (1) with

$$\Phi(\lambda) = \Psi(\mathrm{i}\lambda) = \mathrm{d}\lambda + \int_{(0,\infty)} (1 - e^{-\lambda x})\Pi(dx) \qquad (\lambda \geq 0). \tag{3}$$

Alternatively, one can also rewrite the Lévy-Khintchine formula in terms of the Laplace transform of the *tail of the Lévy measure* , $\overline{\Pi}(x) = \Pi((x, \infty))$:

$$\Phi(\lambda)/\lambda = \mathrm{d} + \int_0^\infty e^{-\lambda t}\overline{\Pi}(t)dt.$$

Note also that condition (2) then squares with $\int_0^1 \overline{\Pi}(t)dt < \infty$. Moreover, recall from Proposition I.2(ii) that $\lim_{|\lambda|\to\infty} \lambda^{-1}\Psi(\lambda) = -\text{id}$, which implies that X_t/t converges in probability to d as t tends to $0+$. This forces $\mathrm{d} \geq 0$. Conversely, it is obvious that a Lévy process whose characteristic exponent Ψ has the expression (3) with $\int(1 \wedge x)\Pi(dx) < \infty$ and $\mathrm{d} \geq 0$ is in fact a subordinator with Laplace exponent Φ. We will henceforth refer to (3) as the Lévy-Khintchine formula for subordinators. It readily entails that the Laplace exponent Φ is concave, and more precisely (by integration by parts) that its derivative Φ' is completely monotone.

More generally, the Laplace exponent $\Phi^{(q)}$ of a killed subordinator $X^{(q)}$ obtained from X by killing at rate $q > 0$ is given by $\Phi^{(q)} = \Phi + q$ and fulfils the relation

$$\mathbb{E}(\exp\{-\lambda X_t^{(q)}\}) = \exp\{-t\Phi^{(q)}(\lambda)\}, \qquad \lambda > 0,$$

with the convention $\mathrm{e}^{-\infty} = 0$.

We have already encountered two important families of subordinators. The first consists of the Poisson processes with intensity $c > 0$, and the second of stable subordinators. Specifically, a subordinator is stable with index $\alpha \in (0,1)$ if its Laplace exponent is proportional to

$$\Phi(\lambda) = \lambda^\alpha = \frac{\alpha}{\Gamma(1-\alpha)} \int_0^\infty (1 - \mathrm{e}^{-\lambda x})x^{-1-\alpha}dx.$$

The restriction on the range of the index in comparison with general stable Lévy processes (recall that α varies in $(0,2]$ in the latter case) is due to the requirement (2). The boundary case $\alpha = 1$ is degenerate since it corresponds to the deterministic process $X_t \equiv t$, and is usually implicitly excluded. A third family of examples is provided by the *Gamma processes* with parameters $a, b > 0$, for which the Laplace exponent is

$$\Phi^{(a,b)}(\lambda) = a\log(1 + \lambda/b) = \int_0^\infty (1 - \mathrm{e}^{-\lambda x})ax^{-1}\mathrm{e}^{-bx}dx\,,$$

where the second equality is known as *the Frullani integral* . We see that the Lévy measure is $\Pi^{(a,b)}(dx) = ax^{-1}\mathrm{e}^{-bx}dx$ and the drift coefficient is zero. Recall that the distribution of the Gamma(a,b) process evaluated at time $t > 0$, $X_t^{(a,b)}$, is

$$\mathbb{P}(X_t^{(a,b)} \in dx) = \frac{b^{at}}{\Gamma(at)}x^{ta-1}\mathrm{e}^{-bx}dx \qquad (x \geq 0).$$

For $at = 1$, this is the standard exponential law with parameter b.

It is clear that $\lim_{t\to\infty} X_t = \infty$ a.s., and a subordinator is a transient

Lévy process. Its potential measure

$$U(0,A) = \mathbb{E}\left(\int_0^\infty \mathbf{1}_{\{X_t\in A\}}dt\right)$$

is a Radon measure. For simplicity, in the sequel we will denote the potential measure by U (and to avoid possible confusion, the potential operator will be denoted by U^0). Observe that its Laplace transform is given by

$$\mathscr{L}U(\lambda) = \mathbb{E}_0\left(\int_0^\infty \exp\{-\lambda X_t\}dt\right) = \frac{1}{\Phi(\lambda)}.$$

The distribution function $\mathscr{U}$ of the potential measure is called the *renewal function* of the subordinator. Note that if we denote the first passage time above x by $T(x) = T_{(x,\infty)}$, then, since X is increasing, we have

$$\mathscr{U}(x) = U([0,x]) = \mathbb{E}(T(x)).$$

The Markov property then entails the following handy inequality for every $x, y \geq 0$:

$$\mathscr{U}(x+y) \leq \mathscr{U}(x) + \mathscr{U}(y).$$

We now end this section with a useful comparison between the renewal function and the Laplace exponent. For every pair of functions $f, g > 0$, we will write $f \asymp g$ if there exists a positive real number c such that $cf(x) \leq g(x) \leq f(x)/c$ for all x.

Proposition 1 *We have*

$$\mathscr{U}(x) \asymp 1/\Phi(1/x) \quad \text{and} \quad \Phi(x)/x \asymp I(1/x) + \mathrm{d}$$

where I denotes the integrated tail of the Lévy measure, viz. $I(x) = \int_0^x \overline{\Pi}(t)dt$.

Proof We first deduce from the Lévy-Khintchine formula that for every $\lambda > 0$,

$$\frac{1}{\Phi(\lambda)} = \int_0^\infty \mathrm{e}^{-y}\mathscr{U}(y/\lambda)dy \tag{4}$$

and since $\mathscr{U}$ increases, we deduce that

$$\mathrm{e}^k/\Phi(\lambda) \geq \mathscr{U}(k/\lambda) \qquad \text{for all } \lambda, k > 0. \tag{5}$$

To prove a converse bound, we recall that the Laplace exponent is concave and nonnegative, the inequality

$$\Phi(\lambda) \leq k\Phi(\lambda/k) \qquad \text{for all } \lambda > 0 \text{ and } k > 1 \tag{6}$$

follows. Then we have for all $x > 0$

$$\frac{1}{\Phi(\lambda)} \le \mathcal{U}(x/\lambda)\int_0^x \mathrm{e}^{-y}dy + \int_x^\infty \mathrm{e}^{-y}\mathcal{U}(y/\lambda)dy \qquad \text{[by (4)]}$$

$$\le \mathcal{U}(x/\lambda) + \frac{1}{\Phi(\lambda/2)}\int_x^\infty \mathrm{e}^{-y}\mathrm{e}^{y/2}dy \qquad \text{[by (5)]}$$

$$\le \mathcal{U}(x/\lambda) + \frac{4}{\Phi(\lambda)}\mathrm{e}^{-x/2} \qquad \text{[by (6)]}.$$

We pick $x = 2\log 8$, so that $4\mathrm{e}^{-x/2} = 1/2$, and we finally get the inequality

$$1/\Phi(\lambda) \le 2\mathcal{U}(x/\lambda) \qquad \text{for all } \lambda > 0\,.$$

The inequality (6) completes the proof of the first estimate. The proof of the second is similar, using the identity

$$\int_0^\infty \mathrm{e}^{-y}\,\big(I(y/\lambda) + \mathrm{d}\big)\,dy = \Phi(\lambda)/\lambda\,, \qquad \lambda > 0$$

which stems from the Lévy-Khintchine formula. □

In the case when the Laplace exponent Φ is regularly varying, say with index $\alpha \in [0,1]$ at $0+$ (respectively, at ∞), the Tauberian theorem yields the following reinforcement of Proposition 1:

$$\Gamma(1+\alpha)\mathcal{U}(x) \sim 1/\Phi(1/x) \quad \text{as } x \to \infty \text{ (respectively, as } x \to 0+).$$

By a further application of the monotone density theorem, we see that if $\alpha < 1$, then the asymptotic behaviour of the tail of the Lévy measure is given by

$$\Gamma(1-\alpha)\overline{\Pi}(x) \sim \Phi(1/x) \quad \text{as } x \to \infty \text{ (respectively, as } x \to 0+).$$

2. Passage across a level

This section concerns the passage of a subordinator strictly above a fixed level; the main problem consists in specifying the probability that the passage occurs continuously, that is that the passage time is not an instant when the process jumps. Except in the compound Poisson case, the sample paths of X are strictly increasing, and then the first passage time in (x,∞) occurs continuously if and only X reaches $\{x\}$. We know from Corollary II.20 that the probability of this event is zero for almost every x if and only if the drift coefficient is zero. But this first result is not completely satisfactory, because we should like to determine this probability for all x.

Introduce for every $x \ge 0$ the first passage time strictly above x,

$$T(x) = \inf\{t \ge 0 : X_t > x\} = T_{(x,\infty)}\,.$$

The law of the jump at this time is specified by the following.

Proposition 2 (i) *For each fixed* $x \geq 0$ *and every* $0 \leq y \leq x < z$, *we have*

$$\mathbb{P}\left(X_{T(x)-} \in dy, X_{T(x)} \in dz\right) = U(dy)\Pi(dz - y)\,.$$

(ii) *For every* $x > 0$, *we have* $\mathbb{P}(X_{T(x)-} < x = X_{T(x)}) = 0$.

Proof (i) Let $f, g \geq 0$ be two Borel functions with $g(x) = 0$. Applying the compensation formula of section O.5 to the jump process of X, ΔX, we obtain

$$\begin{aligned}
&\mathbb{E}(f(X_{T(x)-})g(X_{T(x)})) \\
&= \mathbb{E}\left(\sum_{t\geq 0} f(X_{t-})g(X_{t-} + \Delta X_t)\mathbf{1}_{\{X_{t-}\leq x, \Delta X_t > x - X_{t-}\}}\right) \\
&= \int_0^\infty dt \mathbb{E}\left(f(X_{t-})\mathbf{1}_{\{X_{t-}\leq x\}}\int_0^\infty \Pi(ds)g(X_{t-} + s)\mathbf{1}_{\{s > x - X_{t-}\}}\right)\,.
\end{aligned}$$

Therefore we have

$$\begin{aligned}
&\int_{0\leq y\leq x<z} f(y)g(z)\mathbb{P}\left(X_{T(x)-} \in dy, X_{T(x)} \in dz\right) \\
&= \int_0^\infty dt \int_{0\leq y\leq x, s>x-y} f(y)g(y+s)\mathbb{P}(X_t \in dy)\Pi(ds) \\
&= \int_{0\leq y\leq x<z} f(y)g(z)U(dy)\Pi(dz - y)\,,
\end{aligned}$$

which establishes our assertion.

(ii) In the compound Poisson process case, the identity $\mathbb{P}(X_{T(x)} = x) = 0$ is plain from the definition of $T(x)$. We suppose henceforth that X is not a compound Poisson process. Then the same argument as for (i) based on the compensation formula gives

$$\mathbb{P}(X_{T(x)-} < x = X_{T(x)}) = \int_{[0,x)} U(dy)\Pi(\{x - y\}).$$

Recall from Proposition I.15 that the potential measure is diffuse (that is assigns no mass to single points), so the right-hand side is zero since there are at most countably many $y \in [0, x)$ with $\Pi(\{x - y\}) > 0$. □

Proposition 2(ii) means that if a subordinator jumps at the first instant when it exceeds a given level, then it necessarily jumps strictly above this level. The next step of our analysis is the following.

Lemma 3 *Assume that for some* $x > 0$, $\mathbb{P}(X_{T(x)} = x) > 0$. *Then for every* $\varepsilon > 0$, *there exists* $y \in (0, \varepsilon)$ *with* $\mathbb{P}(X_{T(y)} = y) > 1 - \varepsilon$.

Proof With no loss of generality, we may exclude the compound Poisson case. We deduce from Proposition 2(ii) that X is a.s. continuous at time $T(x)$ on the event $\{X_{T(x)} = x\}$, and because X increases, this implies

$$\{X_{T(x)} = x\} \subseteq \bigcap_{n=1,\cdots} \{X_{T(x-1/n)} < x\} \qquad \text{a.s.}$$

Conversely, on the event $\liminf_{n\to\infty}\{X_{T(x-1/n)} < x\}$, we have by quasi-left-continuity $X_{T(x-)} = x$, where $T_{(x-)} = \lim_{n\to\infty} T(x-1/n)$. Then by the Markov property at time $T(x-)$, we see that $X_t > x$ for all $t > T(x-)$ and thus $T(x) = T(x-)$. In conclusion,

$$\{X_{T(x)} = x\} = \bigcap_{n=1,\cdots} \{X_{T(x-1/n)} < x\} \qquad \text{a.s.}$$

Next, we apply the Markov property at time $T(x-1/n)$ and deduce from Proposition 2(ii)

$$\begin{aligned}\mathbb{P}(X_{T(x)} = x) &= \int_{[x-1/n,x)} \mathbb{P}(X_{T(x-1/n)} \in dz)\mathbb{P}(X_{T(x-z)} = x - z) \\ &\leq \mathbb{P}(X_{T(x-1/n)} < x) \sup_{y\in(0,1/n]} \mathbb{P}(X_{T(y)} = y)\,.\end{aligned}$$

We conclude using the foregoing and Fatou's lemma. □

We are now able to prove that the passage above any fixed level is a.s. realized by a jump when the subordinator has zero drift.

Theorem 4 *Assume that the drift coefficient is zero. Then* $\mathbb{P}(X_{T(x)} > x) = 1$ *for every* $x > 0$.

Proof We will repeatedly invoke the following inequality which stems from the Markov property applied at time $T(a)$: For every $0 < a < b$, we have

$$\mathbb{P}(X_{T(b)} = b) \leq \mathbb{P}(X_{T(a)} = a)\mathbb{P}(X_{T(b-a)} = b-a) + 1 - \mathbb{P}(X_{T(a)} = a)\,. \quad (7)$$

Assume that $\mathbb{P}(X_{T(x)} = x) > 0$ for some $x > 0$. Then, according to Lemma 3, we may suppose that $\mathbb{P}(X_{T(x)} = x) > 0.7$, and there is a decreasing sequence of positive real numbers $y_n < x$ $(n \in \mathbb{N})$ which tends to 0, with

$$\mathbb{P}(X_{T(y_n)} = y_n) > 0.9\,.$$

Applying (7) for $b = x$ and $a = y_n$, we get that

$$\mathbb{P}(X_{T(x-y_n)} = x - y_n) > 0.6\,.$$

Next, we introduce the integer

$$k_n = \inf\{k \in \mathbb{N} : \mathbb{P}(X_{T(x-ky_n)} = x - ky_n) < 0.5\} .$$

The key step consists in proving that

$$\limsup_{n\to\infty} k_n y_n > 0 . \tag{8}$$

Suppose that (8) fails, that is that $y_n k_n$ converges to 0. Applying (7) for $b = x - (k_n - 1)y_n$ and $a = y_n$, we obtain

$$0.4 < \mathbb{P}(X_{T(x-k_n y_n)} = x - k_n y_n) < 0.5 . \tag{9}$$

Similarly, applying (7) for $b = x$ and $a = x - k_n y_n$ and recalling that $\mathbb{P}(X_{T(x)} = x) > 0.7$, we get from (9)

$$\mathbb{P}(X_{T(k_n y_n)} = k_n y_n) > 0.1 . \tag{10}$$

Then consider the sets

$$A(n) = \{k_0 y_0, \cdots, k_n y_n\}, \; A = \bigcup A(n) = \{k_0 y_0, \cdots, k_n y_n, \cdots\}.$$

The first entrance time $T_{A(n)}$ in $A(n)$, is a stopping time (because $A(n)$ is finite) which decreases to the first entrance time T_A in A when $n \to \infty$. Since the sample paths of X increase, we must have $T_A \le T(k_n y_n)$ on the event $\{X_{T(k_n y_n)} = k_n y_n\}$, and we see from (10) that $\mathbb{P}(T_A \le T(k_n y_n)) \ge 0.1$ for every n. On the one hand, because $k_n y_n$ tends to 0,

$$\lim_{n\to\infty} \mathbb{P}(\eta < T(k_n y_n) < \infty) = 0 \qquad \text{for every } \eta > 0.$$

It follows now from Fatou's lemma that $\mathbb{P}(T_A = 0) \ge 0.1$. On the other hand, T_A is a stopping time (T_A is a limit of stopping times) and the Blumenthal zero-one law entails $T_A = 0$ a.s. We can thus pick n such that $\mathbb{P}(T_{A(n)} < \infty) > 0.9$. Finally, we deduce from the Markov property at $T_{A(n)}$ that $\mathbb{P}(X_{T(x)} = x)$ is bounded from above by

$$\begin{aligned} &\mathbb{P}(T_{A(n)} = \infty) + \sum_{p=0}^{n} \mathbb{P}(X_{T_{A(n)}} = k_p y_p)\mathbb{P}(X_{T(x-k_p y_p)} = x - k_p y_p) \\ &< \; 0.1 + 0.5 \qquad [\text{by } (9)]. \end{aligned}$$

In conclusion, if (8) failed, then we would have $\mathbb{P}(X_{T(x)} = x) < 0.6$, which is in contradiction with our initial hypothesis.

Now we obtain from the Markov property applied at time $T(x-(p+1)y_n)$:

$$\begin{aligned} &U([x-(k_n-1)y_n, x)) \\ &= \sum_{p=0}^{k_n-2} U([x-(p+1)y_n, x-py_n)) \\ &\geq \sum_{p=0}^{k_n-2} \mathbb{P}(X_{T(x-(p+1)y_n)} = x-(p+1)y_n)U([0,y_n)) \\ &\geq \frac{1}{2}(k_n-1)U([0,y_n)), \end{aligned}$$

where the last inequality comes from the fact that

$$\mathbb{P}(X_{T(x-(p+1)y_n)} = x-(p+1)y_n) \geq 1/2, \quad p=0,\cdots,k_n-2,$$

by the very definition of k_n. In particular, we have

$$\liminf_{n\to\infty} \frac{U([0,y_n))}{y_n} \leq 2\liminf_{n\to\infty} \frac{U([0,x))}{(k_n-1)y_n}$$

and the right-hand side is finite according to (8). It follows now from Theorem II.16 that the capacity of single points is positive, which implies by Corollary II.20 that the drift coefficient is positive. □

Finally we specify the probability of continuous passage across x when the drift coefficient is positive. Recall from Corollary II.20, Theorem II.16 and Proposition I.12 that the potential measure then is absolutely continuous with a co-excessive density, and that the capacity of single points is positive.

Theorem 5 *Assume that the drift coefficient* d *is positive, and let* $u : (-\infty,\infty) \to [0,\infty)$ *be the co-excessive version of the density of the potential measure. Then* u *is continuous and positive on* $(0,\infty)$, $u(0+) = 1/\mathrm{d}$, *and for every* $x > 0$,

$$\mathbb{P}(X_{T(x)} = x) = \mathrm{d}u(x).$$

We stress that the density u is *not* continuous on $\mathbb{R}$ (clearly, $0 = u(0) < u(0+) = 1/\mathrm{d}$), and this agrees with Corollary II.20.

Proof. Recall that for every $x > 0$, the events $\{X_{T(x)} = x\}$ and $\{T_{\{x\}} < \infty\}$ coincide, and hence, by Corollaries II.18 and II.20, that

$$\mathbb{P}(X_{T(x)} = x) = Cu(x),$$

where $C > 0$ denotes the capacity of single points. Then consider a sequence $(x_n, n \in \mathbb{N})$ of positive real numbers which decrease (respectively, increase) to $x > 0$. The stopping times $T(x_n)$ decrease (respectively, increase) to $T(x)$ a.s. It follows from the right-continuity of the paths (respectively, from the quasi-left-continuity property) and Fatou's lemma that

$$\limsup_{n\to\infty} \mathbb{P}(X_{T(x_n)} = x_n) \leq \mathbb{P}(X_{T(x)} = x),$$

that is $\limsup_{n\to\infty} u(x_n) \leq u(x)$. Because u is lower semi-continuous, this shows that u is actually continuous at x.

Next, recall that $\overline{\Pi}$ stands for the tail of the Lévy measure; we deduce from Proposition 2 by integration that

$$\mathbb{P}(X_{T(x)} = x) = 1 - \mathbb{P}(X_{T(x)} > x) = 1 - \int_0^x u(y)\overline{\Pi}(x-y)dy.$$

Applying the Laplace transform $\mathcal{L}$ and recalling that $\mathcal{L}u = 1/\Phi$, we obtain

$$C\Phi(\lambda)^{-1} = \lambda^{-1} - \Phi(\lambda)^{-1}\int_0^\infty e^{-\lambda t}\overline{\Pi}(t)dt.$$

We deduce from the Lévy-Khintchine formula that $C = \mathrm{d}$.

Pick $x > 0$ with $u(x) > 0$, and suppose that $y = \inf\{x' > x : u(x') = 0\} < \infty$. Then $y > x$, $u(y) = 0$ and $u > 0$ on $[x, y)$ since u is continuous. By Lemma 3, there exists $z \in (0, y-x)$ with $u(z) > 0$, and we see from the Markov property applied at the first passage time in z that

$$\begin{aligned} Cu(y) = \mathbb{P}(X_{T(y)} = y) &\geq \mathbb{P}(X_{T(z)} = z)\mathbb{P}(X_{T(y-z)} = y-z) \\ &= C^2u(z)u(y-z) \end{aligned}$$

and the ultimate quantity is positive since $x \leq y-z < y$. This contradicts the fact that $u(y) = 0$, and thus $u > 0$ on $[x, \infty)$. But according to Lemma 3, x can be chosen arbitrarily small, and this proves that u is positive on $(0, \infty)$. Finally an easy modification of this argument (using then the full strength of Lemma 3) shows that $\lim_{x\to 0+} \mathbb{P}(X_{T(x)} = x) = 1$, that is $u(0+) = 1/\mathrm{d}$. □

Without additional work, we can also determine the distribution of the passage time at a point when the drift coefficient d is positive. It is easy to deduce from the resolvent equation that the q-coexcessive version of the density of the q-resolvent density is given by

$$u^q = u - q\widehat{U}^q u,$$

where $\widehat{U}^q$ stands for the q-resolvent operator of the dual process. So Corollary II.18 gives that for all $x > 0$

$$\mathbb{E}(\exp\{-qT_{\{x\}}\}) = C^q u^q(x).$$

This implies first that $u^q(x) > 0$ (because $\mathbb{P}(T_{\{x\}} < \infty) > 0$), and second that $1/C^q = \lim_{x\to 0+} u^q(x)$ (because $T_{\{x\}}$ converges in probability to 0 according to Theorem 5). Finally, it follows from the strong Feller property that $\widehat{U}^q u$ is continuous, so u^q is continuous on $(0,\infty)$, identically zero on $(-\infty, 0]$ and $C^q = \mathrm{d}$.

3. The arcsine laws

In one of his most celebrated papers on Brownian motion, Lévy (1939) proved that the total time spent in $[0,\infty)$ during the time interval $[0,1]$ by a real-valued Brownian motion B started from 0 has the so-called arcsine law. That is

$$\mathbb{P}\left(\int_0^1 \mathbf{1}_{[0,\infty)}(B_s)ds \in dt\right) = \left(\pi\sqrt{t(1-t)}\right)^{-1} dt \qquad (t \in [0,1]).$$

He later showed that the last zero of B before time 1, $\sup\{s < 1 : B_s = 0\}$, is also distributed according to the arcsine law, see Lévy (1965). We will see in Section VI.3 that these two results can be extended to a wide class of Lévy processes. These extensions rely essentially on a limit theorem concerning the distribution of a subordinator immediately before it crosses some level, which is the main result of this section. It can be viewed as the analogue in continuous times of the Dynkin-Lamperti theorem, see e.g. Bingham, Goldie and Teugels (1987) on page 361, and also has interesting applications to Markov processes, see for instance Exercise IV.5. At the heart of the matter lies the notion of regular variation; see section O.7

Theorem 6 *The following assertions are equivalent:*

(i) *The random variables $x^{-1}X_{T(x)-}$ converge in distribution as $x \to \infty$ (respectively as $x \to 0+$).*

(ii) $\lim x^{-1}\mathbb{E}(X_{T(x)-}) = \alpha \in [0,1]$ *as $x \to \infty$ (respectively, as $x \to 0+$).*

(iii) *The Laplace exponent Φ is regularly varying at $0+$ (respectively, at ∞) with index $\alpha \in [0,1]$.*

In this case, the limit distribution in (i) *is specified by the following. For $\alpha = 0$ (respectively, $\alpha = 1$), it is the Dirac point mass at 0 (respectively, at 1). For $\alpha \in (0,1)$, it is the generalized arcsine law with parameter α, that is the measure on* $[0,1]$ *given by*

$$\frac{s^{\alpha-1}(1-s)^{-\alpha}}{\Gamma(\alpha)\Gamma(1-\alpha)}ds = \frac{\sin\alpha\pi}{\pi}s^{\alpha-1}(1-s)^{-\alpha}ds \qquad (0 < s < 1).$$

Before proving the arcsine laws, we point out that Φ is regularly varying at 0+ with index $\alpha \in (0,1)$ if and only if the tail of the Lévy measure $\overline{\Pi}$ is regularly varying at ∞ with index $-\alpha$ (this follows readily from the Lévy-Khintchine formula, the Tauberian theorem and the monotone density theorem). This holds if and only if X_t belongs to the domain of attraction of the stable law with index α. The condition that Φ is regularly varying at ∞ with index $\alpha \in (0,1)$ is equivalent to the condition that the drift coefficient is zero and the tail of the Lévy measure $\overline{\Pi}$ is regularly varying at 0+ with index $-\alpha$. This holds if and only if $\varphi(1/t)X_t$ converges in distribution as t tends to 0+ to the stable law with index α, where φ denotes the inverse function of Φ. Finally, it is a necessary and sufficient condition for the existence of an increasing function g such that $g(1/t)X_t$ converges in distribution to some non-degenerate law as t tends to 0+. Observe also that if X is a stable subordinator with index $\alpha \in (0,1)$, then Theorem 6 and the scaling property imply that for every $x > 0$, the distribution of $x^{-1}X_{T(x)-}$ is the generalized arcsine law with parameter α.

The first step for the proof of Theorem 6 is the following.

Lemma 7 *Fix $x > 0$ and put $A_t(x) = x^{-1}X_{T(tx)-}$ for every $t > 0$. Then*

$$\int_0^\infty e^{-qt}\mathbb{E}(\exp\{-\lambda A_t(x)\})dt = \frac{\Phi(q/x)}{q\Phi((\lambda+q)/x)}$$

for every $q, \lambda > 0$.

Proof According to Proposition 2, we have

$$\begin{aligned}
&\int_0^\infty e^{-qt}\mathbb{E}(\exp\{-\lambda A_t(x)\})dt \\
&= \int_0^\infty \int_{0\le y<tx} e^{-\lambda y/x}\overline{\Pi}(tx-y)e^{-qt}U(dy)dt \\
&+ \int_0^\infty \int_{0\le y\le tx} e^{-\lambda y/x}e^{-qt}\mathbb{P}(X_{T(tx)-}\in dy, X_{T(tx)} = tx)dt.
\end{aligned}$$

Applying Theorems 4 and 5, we see that the second term of the sum in the right-hand side is simply

$$\int_0^\infty e^{-(\lambda+q)t}\mathrm{d}u(tx)dt = \mathrm{d}\left[x\Phi((\lambda+q)/x)\right]^{-1},$$

where d is the drift coefficient and u the density of the potential measure when $\mathrm{d} > 0$, $\mathrm{d}u \equiv 0$ otherwise.

On the other hand, straightforward calculations based on the Lévy-Khintchine formula show that the first term in the sum equals

$$\int_0^\infty \int_{0\leq y<tx} \mathrm{e}^{-(\lambda+q)y/x} U(dy)\, \mathrm{e}^{-q(tx-y)/x} \overline{\Pi}(tx-y)dt$$
$$= \Phi((\lambda+q)/x)^{-1} \left(\int_0^\infty \mathrm{e}^{-qs/x} \overline{\Pi}(s) x^{-1} ds \right)$$
$$= \Phi((\lambda+q)/x)^{-1} \left(\frac{1}{q}\Phi(q/x) - \frac{\mathrm{d}}{x} \right) .$$

Putting the pieces together, we deduce the formula of Lemma 7. □

We are now able to prove Theorem 6.

Proof of Theorem 6 We will only present the proof as x goes to ∞, the argument when x tends to $0+$ is similar.

(i)⇒(iii) Plainly, (i) holds if and only if for every $t > 0$, the random variable $A_t(x)$ appearing in Lemma 7 converges in law as x goes to ∞. It then follows from Lemma 7 that the latter implies that for every $\lambda, q > 0$, $\Phi(q/x)\Phi((\lambda+q)/x)^{-1}$ converges as x goes to ∞, or equivalently that Φ is regularly varying at $0+$ with index, say, α. Finally, it is immediately deducible from the Lévy-Khintchine formula that necessarily $\alpha \in [0, 1]$.

(iii)⇒(i) When Φ is regularly varying at $0+$ with index $\alpha \in (0, 1)$, we have according to Lemma 7 for every $q, \lambda > 0$

$$\lim_{x\to\infty} \int_0^\infty \mathrm{e}^{-qt} \mathbb{E}(\exp\{-\lambda A_t(x)\})dt = q^{\alpha-1}(\lambda+q)^{-\alpha}$$
$$= \int_0^\infty dt\, \mathrm{e}^{-qt} \int_0^t \mathrm{e}^{-\lambda s} \frac{s^{\alpha-1}(t-s)^{-\alpha}}{\Gamma(\alpha)\Gamma(1-\alpha)} ds.$$

Note that for each $x > 0$, the process $t \to A_t(x)$ increases, so the function $t \to \mathbb{E}(\exp\{-\lambda A_t(x)\})$ decreases. We then obtain by integration by parts

$$\lim_{x\to\infty} \int_0^\infty \mathrm{e}^{-qt} d\left(1 - \mathbb{E}(\exp\{-\lambda A_t(x)\})\right)$$
$$= \int_0^\infty dt\, \mathrm{e}^{-qt}\, d\left(1 - \int_0^t \mathrm{e}^{-\lambda s} \frac{s^{\alpha-1}(t-s)^{-\alpha}}{\Gamma(\alpha)\Gamma(1-\alpha)}\right) .$$

This implies that for almost every $t > 0$

$$\lim_{x\to\infty} \mathbb{E}(\exp\{-\lambda A_t(x)\}) = \int_0^t \mathrm{e}^{-\lambda s} \frac{s^{\alpha-1}(t-s)^{-\alpha}}{\Gamma(\alpha)\Gamma(1-\alpha)} ds ,$$

which yields (i). The case when $\alpha = 0$ or 1 is similar (with simpler calculations).

(iii)⇔(ii) The variables $x^{-1}X_{T(x)-}$ take values in $[0, 1]$, so (ii) follows from (i). Conversely, if (ii) holds, then for every $t > 0$, $\lim_{x\to\infty} \mathbb{E}(A_t(x)) = \alpha t$, and we deduce from Lemma 7 that

$$\lim_{t\to 0+} t\Phi'(t)\Phi(t)^{-1} = \alpha\,.$$

Hence the logarithmic derivative of the function $t \to t^{-\alpha}\Phi(t)$ can be expressed in the form $t \to \varepsilon(t)/t$, where $\lim_{t\to 0+} \varepsilon(t) = 0$. It follows from the representation theorem of slowly varying functions (see section O.7) that $t^{-\alpha}\Phi(t)$ is slowly varying at 0+, i.e.

$$\lim_{t\to 0+} \frac{(at)^{-\alpha}\Phi(at)}{t^{-\alpha}\Phi(t)} = 1$$

for every $a > 0$. This implies that Φ is regularly varying at 0+ with index α. □

4. Rates of growth

We turn our attention to the study of the asymptotic rate of growth of the sample paths of subordinators, at the origin and at infinity. The very same arguments work for small times as well as for large times, so we will only present a detailed account for small times and merely state without proofs the corresponding results for large times. We also point out that results on the behaviour at infinity follow from corresponding statements for increasing random walks using an immediate argument of discretization of times and monotonicity, but this argument is not available for small times.

We begin with a simple and very useful property.

Proposition 8 *Recall that* $\mathrm{d} \geq 0$ *denotes the drift coefficient of* X. *We have* $\mathbb{P}$*-a.s.*

$$\lim_{t\to 0+} t^{-1}X_t = \mathrm{d}\,.$$

Proof It is plainly sufficient to prove the result for $\mathrm{d} = 0$. Throwing away the large jumps if necessary, we may also assume that the Lévy measure Π has bounded support. Recall then from Proposition I.2(ii) that

$$\lim_{t\to 0+} t\Phi(\lambda/t) = 0 \qquad \text{for all } \lambda \geq 0\,,$$

so that X_t/t converges in probability to 0. To check that the convergence holds a.s., we will show that the process $(X_t/t, t > 0)$ is a reversed martingale, so that we will then be able to invoke the theorem of a.s. convergence for martingales. By the Markov property, all that is needed is to verify that for every $\lambda \geq 0$ and $s < t$

$$\mathbb{E}\left((X_s/s - X_t/t)\exp\{-\lambda X_t/t\}\right) = 0 .$$

The quantity in the left-hand side can be rewritten as

$$\begin{aligned}
&\mathbb{E}\left[\left((X_s/s - X_s/t) + (X_s/t - X_t/t)\right)\exp\{-\lambda X_s/t\}\exp\{-\lambda(X_t - X_s)/t\}\right] \\
&= (s^{-1} - t^{-1})s\Phi'(\lambda/t)\exp\{-s\Phi(\lambda/t)\}\exp\{-(t-s)\Phi(\lambda/t)\} \\
&\quad - t^{-1}\exp\{-s\Phi(\lambda/t)\}\exp\{-(t-s)\Phi(\lambda/t)\}(t-s)\Phi'(\lambda/t) .
\end{aligned}$$

After cancellation, we see that the foregoing quantity is zero, which establishes the martingale property. □

Proposition 8 reduces the study of the asymptotic behaviour for small times of general subordinators to that of subordinators with zero drift. First, we have the following integral test for the upper functions.

Theorem 9 *Let X be a subordinator with zero drift and Lévy measure Π. Suppose that $h : [0,\infty) \to [0,\infty)$ is an increasing function such that the function $t \to h(t)/t$ increases as well. Then the following assertions are equivalent:*

(i) $$\limsup_{t\to 0+} \left(X_t/h(t)\right) = \infty \qquad a.s.;$$

(ii) $$\int_0^1 \overline{\Pi}(h(t))dt = \infty ;$$

(iii) $$\int_0^1 \{\Phi(1/h(t)) - (1/h(t))\Phi'(1/h(t))\}dt = \infty.$$

Finally, if these assertions fail to be true, then

$$\lim_{t\to 0+} \left(X_t/h(t)\right) = 0 \qquad a.s.$$

For instance, we see that when X is a stable subordinator with index $\alpha \in (0,1)$,

$$\limsup_{t\to 0+} \left(X_t/h(t)\right) = \infty \text{ or } 0 \qquad \text{a.s.}$$

according as the integral $\int_{(0,1)} h(t)^{-\alpha}dt$ diverges or converges.

Proof (i) ⇔ (ii) Assume that (ii) holds, so that for every $c > 1$ and $\varepsilon > 0$

$$\int_0^\varepsilon \overline{\Pi}(ch(t))dt \geq \int_0^\varepsilon \overline{\Pi}(h(ct))dt = \infty.$$

Recall that the jump process of X, ΔX, is a Poisson point process with characteristic measure Π, so that for every $\eta \in (0, \varepsilon)$, the random variable

$$\mathrm{Card}\{t \in [\eta, \varepsilon] : \Delta X_t > ch(t)\}$$

has a Poisson distribution with intensity $\int_\eta^\varepsilon \overline{\Pi}(ch(t))dt$. We deduce that a.s., there exist infinitely many instants $s \in (0, \varepsilon)$ with $\Delta X_s > ch(s)$. *A fortiori*, $X_s > ch(s)$ for such s, and hence $\limsup_{t\to 0+} (X_t/h(t)) \geq c$.

Conversely, assume that (ii) fails to be true and denote the inverse function of h by $\tilde{h}$. The process $(\tilde{h}(\Delta X_t), t \geq 0)$ is a Poisson point process with characteristic measure $\tilde{\Pi}$ which is specified by the relation $\tilde{\Pi}[t, \infty) = \Pi[h(t), \infty)$.

Integrating by parts, we have

$$\int_{(0,\infty)} (1 \wedge x)\tilde{\Pi}(dx) = \int_0^1 \tilde{\Pi}[t, \infty)dt < \infty,$$

so that $\tilde{\Pi}$ can be thought of as the Lévy measure of a subordinator. Specifically, the process

$$\tilde{X}_t = \sum_{0\leq s\leq t} \tilde{h}(\Delta X_s) \qquad (t \geq 0)$$

is a subordinator with zero drift and Lévy measure $\tilde{\Pi}$. The hypothesis that the function $t \to h(t)/t$ increases easily yields the inequality $h(a+b) \geq h(a) + h(b)$ for every $a, b > 0$, from which we deduce that $X_t \leq h(\tilde{X}_t)$. But according to Proposition 8, $\lim_{t\to 0+} t^{-1}\tilde{X}_t = 0$ a.s., and hence, for every $\varepsilon > 0$, we have $X_t \leq h(\varepsilon t)$ whenever $t > 0$ is small enough, a.s. Recalling that $t \to h(t)/t$ increases, we deduce that $h(\varepsilon t) \leq \varepsilon h(t)$ provided that $\varepsilon \leq 1$, and in conclusion $\lim_{t\to 0+} (X_t/h(t)) = 0$ a.s.

(ii) $\Leftrightarrow$ (iii) Since the drift coefficient is zero, $\Phi(\lambda)/\lambda$ is the Laplace transform of the tail of the Lévy measure and then it follows that

$$\int_0^1 \{\Phi(1/h(t)) - (1/h(t))\Phi'(1/h(t))\}dt = \int_0^1 dt h(t)^{-2} \int_0^\infty x e^{-x/h(t)}\overline{\Pi}(x)dx$$

$$= \int_0^1 dt \int_0^\infty x e^{-x}\overline{\Pi}(xh(t))dx.$$

The inequality

$$\overline{\Pi}(h(t)) \leq \overline{\Pi}(xh(t)) \leq \overline{\Pi}(h(xt)), \quad \text{for every } 0 < x < 1,\, t > 0,$$

now readily implies that the integrals in (ii) and (iii) converge or diverge simultaneously. □

Theorem 9 is not completely satisfactory, because the condition that the mapping $x \to h(x)/x$ increases may be quite restrictive. For instance, it often fails when h is a typical sample path of an increasing process. Nonetheless, it may be dropped when the integrated tail $I(t) = \int_0^t \overline{\Pi}(x)dx$ has *positive increase* , in the sense that

$$\liminf_{x\to 0+} \left(I(2x)/I(x)\right) > 1$$

(see Exercise 7 for alternative formulations of this condition).

Proposition 10 *Suppose that I has positive increase and let $h : [0,\infty) \to [0,\infty)$ be an increasing function. Then the following assertions are equivalent:*

(i) $$\limsup_{t\to 0+} \left(X_t/h(t)\right) = \infty \quad a.s.;$$

(ii) $$\int_0^1 \overline{\Pi}(h(t))dt = \infty;$$

(iii) $$\int_0^1 \Phi(1/h(t))dt = \infty.$$

Finally, if these assertions fail to be true, then $\lim_{t\to 0+} \left(X_t/h(t)\right) = 0$ *a.s.*

Proof On the one hand, since $\overline{\Pi}$ is non-increasing, we have $I(x) \geq x\overline{\Pi}(x)$. On the other hand, the hypothesis that I has positive increase and the inequality

$$I(2x) = I(x) + \int_x^{2x} \overline{\Pi}(t)dt \leq I(x) + x\overline{\Pi}(x)$$

entail $\liminf_{x\to 0+} \left(x\overline{\Pi}(x)/I(x)\right) > 0$. It then follows from Proposition 1 that (ii) and (iii) are equivalent.

Then suppose that (ii) holds. The same argument as in the proof of Theorem 9 shows that $\limsup_{t\to 0+} \left(X_t/h(t)\right) \geq 1$ a.s. But, since Φ is concave, (iii) holds as well when h is replaced by kh for any $k > 1$. As a consequence, $\limsup_{t\to 0+}(X_t/h(t)) = \infty$ a.s.

On the other hand, the obvious inequality

$$\begin{aligned}\mathbb{P}(X_t \geq a) &\leq (1 - 1/\mathrm{e})^{-1}\mathbb{E}(1 - \exp\{-a^{-1}X_t\}) \\ &= (1 - 1/\mathrm{e})^{-1} \left(1 - \exp\{-t\Phi(1/a)\}\right)\end{aligned}$$

applied for $t = 2^{-n+1}$ and $a = h(2^{-n})$ entails that

$$\mathbb{P}(X_{2^{-n+1}} \geq h(2^{-n})) \leq 2(1 - 1/\mathrm{e})^{-1}2^{-n}\Phi(1/h(2^{-n})).$$

Since $t \to \Phi(1/h(t))$ decreases, we deduce that the series $\sum 2^{-n}\Phi(1/h(2^{-n}))$ converges whenever (iii) fails. Then, by the Borel-Cantelli lemma,

$$X_{2^{-n+1}} < h(2^{-n}) \quad \text{for all integers } n \text{ large enough, a.s.}$$

An immediate monotonicity argument shows that the latter implies that

$X_t < h(t)$ for all real numbers t small enough, a.s. Since (iii) still fails when one replaces h by εh for any $\varepsilon \in (0,1)$ (because Φ is concave), this proves that $\lim_{t\to 0+} \left(X_t/h(t)\right) = 0$ a.s. □

Theorem 9 and Proposition 10 show that the upper envelope of a subordinator with no drift is quite irregular, in the sense that for any increasing function h, either the ratio $X_t/h(t)$ converges to 0 or its limsup is infinite a.s. It is remarkable that the lower envelope is much smoother, at least for a wide class of subordinators. Specifically, let φ be the inverse function of Φ, and consider

$$f(t) = \frac{\log|\log t|}{\varphi(t^{-1}\log|\log t|)} \qquad (0 < t < 1/e).$$

Then, under a mild condition on Φ, there is a positive finite constant c, such that $\liminf \left(X_t/f(t)\right) = c$ a.s. Recall that regularly varying functions have been defined in Section O.7.

Theorem 11 (Law of the iterated logarithm) *Suppose that the Laplace exponent Φ is regularly varying at ∞ with index $\alpha \in (0,1)$, Then we have $\mathbb{P}$-a.s.*

$$\liminf_{t\to 0+} \left(X_t/f(t)\right) = \alpha(1-\alpha)^{(1-\alpha)/\alpha}.$$

We refer to the preceding section for the discussion of the condition that the Laplace exponent Φ is regularly varying. The proof of Theorem 11 relies on the following sharp estimate for the distribution of X_t.

Lemma 12 *Assume that the hypotheses of Theorem 11 are fulfilled. Then for every $c > 0$, we have*

$$-\log \mathbb{P}(X_t \le cf(t)) \sim (1-\alpha)(\alpha/c)^{\alpha/(1-\alpha)} \log|\log t| \qquad (t \to 0+).$$

First, we establish the lower bound, which is the easiest part.

Proof of the lower bound Applying Chebyshev's inequality, we have for every $\lambda > 0$

$$\mathbb{P}(X_t \le cf(t)) \le \exp\{\lambda cf(t)\}\mathbb{E}(\exp -\lambda X_t)\,,$$

and thus

$$-\log \mathbb{P}(X_t \le cf(t)) \ge t\Phi(\lambda) - \lambda cf(t)\,.$$

Next we choose $\lambda = \lambda(t)$ such that $t\Phi(\lambda) = k\log|\log t|$ for some real number $k > 0$ that will be specified later on, that is $\lambda = \varphi(kt^{-1}\log|\log t|)$. Since Φ is regularly varying at ∞ with index α, its inverse φ is regularly varying at ∞ with index $1/\alpha$ (see section O.7) and

$$\lambda = \lambda(t) \sim k^{1/\alpha}\varphi(t^{-1}\log|\log t|).$$

This implies

$$t\Phi(\lambda) - \lambda c f(t) \sim (k - ck^{1/\alpha}) \log|\log t| \qquad (t \to 0+).$$

We now choose k in such way that $k - ck^{1/\alpha}$ is maximal, that is

$$k = (\alpha/c)^{\alpha/(1-\alpha)} \text{ and } k - ck^{1/\alpha} = (1-\alpha)(\alpha/c)^{\alpha/(1-\alpha)}.$$

In conclusion, we have established that

$$\liminf_{t\to 0+} \left(-\log \mathbb{P}(X_t \le cf(t)) / \log|\log t|\right) \ge (1-\alpha)(\alpha/c)^{\alpha/(1-\alpha)}.$$

□

The upper bound requires a more delicate analysis which is reminiscent of standard arguments of the theory of large deviations. The key idea consists in applying an exponential tilting. That is, we use an exponential martingale to define a new probability measure on the sigma-field $\mathscr{F}_1$, which is absolutely continuous with respect to $\mathbb{P}$, and under which $(X_t, 0 \le t \le 1)$ is still a subordinator (on the unit time interval). Specifically, for each $v > 0$, we consider the probability measure $\mathbb{P}^{(v)}$ on $(\Omega, \mathscr{F}_1)$ given by

$$\mathbb{P}^{(v)}(\Lambda) = \mathbb{E}(\exp\{-vX_1 + \Phi(v)\}, \Lambda), \quad \Lambda \in \mathscr{F}_1.$$

Because the process $\left(\exp\{-vX_t + \Phi(v)t\},\, t \ge 0\right)$ is a positive martingale (see e.g. Exercise I.6), we deduce that for every $t \in [0,1]$ and $\Lambda \in \mathscr{F}_t$, we also have

$$\mathbb{P}^{(v)}(\Lambda) = \mathbb{E}(\exp\{-vX_t + \Phi(v)t\}, \Lambda).$$

Observe that for $s, t \ge 0$ with $s + t \le 1$ and $\Lambda \in \mathscr{F}_t$,

$$\begin{aligned} &\mathbb{E}^{(v)}(\exp\{-\lambda(X_{t+s} - X_t)\}, \Lambda) \\ &= \mathbb{E}(\exp\{-vX_{t+s} + \Phi(v)(t+s)\} \exp\{-\lambda(X_{t+s} - X_t)\}, \Lambda) \\ &= \exp\{-s(\Phi(v+\lambda) - \Phi(v))\} \mathbb{E}(\exp\{-vX_t + t\Phi(v)\}, \Lambda) \\ &= \exp\{-s(\Phi(v+\lambda) - \Phi(v))\} \mathbb{P}^{(v)}(\Lambda). \end{aligned}$$

This shows that under $\mathbb{P}^{(v)}$, X has independent homogeneous increments; plainly it is an increasing right-continuous process and hence it is a subordinator (on the unit time interval). Finally its Laplace exponent is

$$\Phi^{(v)}(\lambda) = \Phi(\lambda + v) - \Phi(v).$$

Note that its mean and variance are

$$\mathbb{E}^{(v)}(X_t) = t\Phi'(v) \quad \text{and} \quad \mathbb{E}^{(v)}((X_t - t\Phi'(v))^2) = -t\Phi''(v). \tag{11}$$

Proof of the upper bound We can now express $\mathbb{P}(X_t \le cf(t))$ in terms of $\mathbb{P}^{(v)}$ as

$$\begin{aligned} &\mathbb{E}^{(v)}\left(\exp\{vX_t - \Phi(v)t\}, X_t \le cf(t)\right) \\ &\ge \exp\{vc(1-2\varepsilon)f(t) - \Phi(v)t\} \mathbb{P}^{(v)}\left(c(1-2\varepsilon)f(t) \le X_t \le cf(t)\right), \end{aligned}$$

where $\varepsilon > 0$ is arbitrarily small. Next, we choose $v = v(t)$ in such way that $t\Phi'(v) = c(1-\varepsilon)f(t)$. Recall that Φ is concave, so according to the monotone density theorem (see section O.7), $\Phi'(x) \sim \alpha\Phi(x)/x$ $(x \to \infty)$. This entails that as t goes to $0+$

$$v = v(t) \sim k^{1/\alpha}\varphi(t^{-1}\log|\log t|) \quad \text{with} \quad k = \left(\frac{\alpha}{c(1-\varepsilon)}\right)^{\alpha/(1-\alpha)}.$$

On the one hand, the choice for v allows us to write

$$\mathbb{P}^{(v)}\left(c(1-2\varepsilon)f(t) \le X_t \le cf(t)\right) = \mathbb{P}^{(v)}\left(|X_t - t\Phi'(v)| \le c\varepsilon f(t)\right).$$

Chebyshev's inequality and (11) yield

$$\mathbb{P}^{(v)}\left(|X_t - t\Phi'(v)| > c\varepsilon f(t)\right) \le -t\Phi''(v)(c\varepsilon f(t))^{-2}.$$

Another application of the monotone density theorem shows that

$$\Phi''(v) \sim (\alpha-1)\Phi'(v)/v,$$

and a straightforward calculation shows that the right-hand side of the foregoing inequality tends to 0 as t goes to $0+$. In conclusion

$$\lim_{t\to 0+} \mathbb{P}^{(v)}\left(c(1-2\varepsilon)f(t) \le X_t \le cf(t)\right) = 1.$$

On the other hand, we have as t tends to $0+$

$$\begin{aligned} vc(1-2\varepsilon)f(t) - \Phi(v)t &\sim \frac{1-2\varepsilon}{1-\varepsilon}vt\Phi'(v) - \Phi(v)t \\ &\sim \left(\frac{1-2\varepsilon}{1-\varepsilon}\alpha - 1\right)\Phi(v)t \\ &\sim \left(\frac{1-2\varepsilon}{1-\varepsilon}\alpha - 1\right)\left(\frac{\alpha}{c(1-\varepsilon)}\right)^{\alpha/(1-\alpha)} \log|\log t|. \end{aligned}$$

Putting the pieces together, we finally obtain

$$\limsup_{t\to 0+} -\frac{\log \mathbb{P}(X_t \le cf(t))}{\log|\log t|} \le \left(1 - \alpha\frac{1-2\varepsilon}{1-\varepsilon}\right)\left(\frac{\alpha}{c(1-\varepsilon)}\right)^{\alpha/(1-\alpha)},$$

and since ε can be chosen arbitrarily small, the lemma is proven. □

Theorem 11 follows from the estimate of Lemma 12 by standard arguments.

Proof of Theorem 11 It will be convenient to use the notation $X_t = X(t)$ in the sequel. First we prove the lower bound. Take $r < 1$ and $0 < c < c' < \alpha(1-\alpha)^{(1-\alpha)/\alpha}$. Using the fact that φ is regularly varying at ∞ with index $1/\alpha$, we see that for n large enough and r close enough to 1

$$\mathbb{P}(X(r^n) \le cf(r^{n-1})) \le \mathbb{P}(X(r^n) \le c'f(r^n)).$$

Observe that

$$(1-\alpha)(\alpha/c')^{\alpha/(1-\alpha)} \log|\log r^n| \ge (1+\eta)\log n$$

for some $\eta > 0$, so we deduce from Lemma 12 that the series

$$\sum_n \mathbb{P}(X(r^n) \leq cf(r^{n-1}))$$

converges. According to the Borel-Cantelli lemma, we have $X(r^n) \leq cf(r^{n-1})$ for all n large enough, a.s. By an immediate monotonicity argument, $X(t) \geq X(r^n)$ and $f(t) \leq f(r^{n-1})$ whenever $r^n \leq t \leq r^{n-1}$ and we have proven that

$$\liminf_{t\to 0+} \left(X(t)/f(t)\right) \geq c \text{ a.s.}$$

Next we prove the converse upper bound. Take $c'' > \alpha(1-\alpha)^{(1-\alpha)/\alpha}$ and $r > 1$, and put $t_n = \exp\{-n^r\}$. Observe that

$$(1-\alpha)(\alpha/c'')^{\alpha/(1-\alpha)} \log(n^r) \leq (1-\eta)\log n$$

for some $\eta > 0$, provided that r has been chosen close enough to 1. Using the obvious inequality

$$\mathbb{P}(X(t_n) - X(t_{n+1}) \leq x) \geq \mathbb{P}(X(t_n) \leq x)$$

and Lemma 12, we see that the series

$$\sum_n \mathbb{P}(X(t_n) - X(t_{n+1}) \leq c'' f(t_n))$$

diverges. According to the Borel-Cantelli lemma for independent events, we have

$$\liminf \frac{X(t_n) - X(t_{n+1})}{f(t_n)} \leq c'' \quad \text{a.s.}$$

We will prove below that

$$\lim \left(X(t_{n+1})/f(t_n)\right) = 0 \quad \text{a.s.} \tag{12}$$

We can then deduce from the foregoing that $\liminf_{t\to 0+} \left(X(t)/f(t)\right) \leq c''$ a.s., so all that is needed now is to check (12).

We start from the following elementary inequality for every $x, t > 0$:

$$\begin{aligned}\mathbb{P}(X(t) > x) &\leq (1-1/\mathrm{e})^{-1}\mathbb{E}\left(1-\exp\{-x^{-1}X(t)\}\right)\\ &= (1-1/\mathrm{e})^{-1}\left(1-\exp\{-t\Phi(1/x)\}\right).\end{aligned}$$

Aiming to apply this to $t = t_{n+1}$ and $x = \varepsilon f(t_n)$ for some $\varepsilon > 0$, we are led to consider the quantity

$$\begin{aligned}t_{n+1}\Phi\left(\varepsilon^{-1}/f(t_n)\right) &= \exp\{-(n+1)^r\}\Phi\left(\frac{\varphi(\exp(n^r) r \log n)}{\varepsilon r \log n}\right)\\ &\leq \exp\{-(n+1)^r\}\exp\{n^r\} r \log n\,,\end{aligned}$$

where the last inequality holds provided that $\varepsilon r \log n > 1$. Since $(n+1)^r - n^r \geq rn^{r-1}$, the foregoing quantity is bounded from above by $\exp\{-rn^{r-1}\} r \log n$. In conclusion, we have for n large enough

$$\mathbb{P}(X(t_{n+1}) > \varepsilon f(t_n)) \leq (1-\mathrm{e}^{-1})^{-1}\exp\{-rn^{r-1}\} r \log n,$$

and the series $\sum \mathbb{P}(X(t_{n+1}) > \varepsilon f(t_n))$ converges. According to the Borel-Cantelli lemma,

$$\limsup \frac{X(t_{n+1})}{f(t_n)} \leq \varepsilon \qquad \text{a.s.,}$$

which establishes (12) since $\varepsilon > 0$ can be chosen arbitrarily small. The proof of Theorem 11 is now complete. □

We now conclude this section by describing the rate of growth for large times. First, the standard strong law of large numbers in discrete times readily extends to continuous times by a monotonicity argument (see also Exercise I.9).

Strong law of large numbers *We have a.s.*

$$\lim_{t\to\infty} t^{-1} X_t = \mathbb{E}(X_1) .$$

In the case when $\mathbb{E}(X_1)$ is infinite, one has the following analogue of Theorem 9.

Theorem 13 *Let X be a subordinator with infinite mean. Suppose that $h : [0,\infty) \to [0,\infty)$ is an increasing function such that $t \to h(t)/t$ increases. Then the following assertions are equivalent:*

(i) $$\limsup_{t\to\infty} \left(X_t/h(t)\right) = \infty \qquad a.s.;$$

(ii) $$\int_1^\infty \Pi[h(t),\infty)dt = \infty ;$$

(iii) $$\int_1^\infty \{\Phi(1/h(t)) - (1/h(t))\Phi'(1/h(t))\}dt = \infty.$$

Finally, if these assertions fail, then

$$\lim_{t\to\infty} \left(X_t/h(t)\right) = 0 \qquad a.s.$$

Finally, we state the law of the iterated logarithm for large times, referring to section 3 for a probabilistic interpretation of the condition that the Laplace exponent is regularly varying at 0+.

Theorem 14 *Suppose that the Laplace exponent Φ is regularly varying at 0+ with index $\alpha \in (0,1)$, let φ be the inverse function of Φ, and consider*

$$f(t) = \frac{\log\log t}{\varphi(t^{-1}\log\log t)} \qquad (e < t).$$

Then we have $\mathbb{P}$-a.s.

$$\liminf_{t\to\infty} \left(X_t/f(t)\right) = \alpha(1-\alpha)^{(1-\alpha)/\alpha}.$$

5. Dimension of the range

Loosely speaking, we are interested here in estimating the 'size' of the range of a subordinator on a given set of times. It is easy to derive from Fubini's theorem and Corollary II.20(ii) that the range $\{X_t, t \geq 0\}$ has zero Lebesgue measure a.s. if and only if the drift coefficient is zero. But this is quite a crude result, and in order to get more precise information, one is led to consider the notion of Hausdorff dimension.

For every subset E of a metric space and $\varepsilon > 0$, denote by $\mathscr{C}(\varepsilon)$ the set of all the coverings $C = \{B_i : i \in I\}$ of E with balls B_i of diameter $|B_i| < \varepsilon$. Then for every $\gamma > 0$, consider

$$m_\varepsilon^\gamma(E) = \inf_{C \in \mathscr{C}(\varepsilon)} \sum_{i \in I} |B_i|^\gamma.$$

Plainly $m_\varepsilon^\gamma(E)$ increases as ε decreases to $0+$, and the limit is denoted by

$$m^\gamma(E) = \sup_{\varepsilon > 0} \inf_{C \in \mathscr{C}(\varepsilon)} \sum_{i \in I} |B_i|^\gamma \in [0, \infty].$$

One can prove that for each fixed $\gamma > 0$, the mapping $E \to m^\gamma(E)$ defines a measure on Borel sets, called the γ-dimensional Hausdorff measure (see Rogers (1970)); but we will not need this in the sequel. When E is the d-dimensional Euclidean space, the d-dimensional Hausdorff measure is simply the Lebesgue measure (up to a constant multiplicative factor).

Clearly, for E fixed, the mapping $\gamma \to m^\gamma(E)$ decreases. Moreover, it is easy to see that if $m^\gamma(E) = 0$ then $m^{\gamma'}(E) = 0$ for every $\gamma' > \gamma$; and if $m^\gamma(E) > 0$ then $m^{\gamma'}(E) = \infty$ for every $\gamma' < \gamma$. This motivates the following definition.

Definition (Hausdorff dimension) *The critical value*

$$\dim(E) = \sup\{\gamma > 0 : m^\gamma(E) < \infty\} = \inf\{\gamma > 0 : m^\gamma(E) = 0\}$$

is called the Hausdorff dimension of E.

It can be immediately checked that the Hausdorff dimension is an increasing function on subsets, and that when the state space is Euclidean, the Hausdorff dimension of a set with positive Lebesgue measure coincides with the Euclidean dimension.

Theorem 15 below specifies the Hausdorff dimension of the range of a subordinator in terms of its Laplace exponent, and is due to Horowitz (1968). In particular, it shows that the Hausdorff dimension of the range of a stable subordinator with index $\alpha \in (0, 1)$ is α, a result which was mentioned first by Blumenthal and Getoor (1960, 1960-a, 1962).

Introduce

$$\sigma = \sup\{\alpha > 0 : \lim_{\lambda \to \infty} \lambda^{-\alpha}\Phi(\lambda) = \infty\},$$

which is known as the *lower index* of the subordinator.

Theorem 15 *For every $s > 0$, we have*

$$\dim(\{X_t : t \in [0, s]\}) = \sigma \qquad a.s.$$

Turning our attention to the ranges of a subordinator on arbitrary subsets $E \subseteq [0, \infty)$, we then obtain explicit bounds for the Hausdorff dimension of $\{X_t, t \in E\}$. The strength of the result lies in the fact that the same bounds hold for all subsets simultaneously. Introduce

$$\beta = \inf\{\alpha \geq 0 : \lim_{\lambda \to \infty} \lambda^{-\alpha}\Phi(\lambda) = 0\},$$

which is known as the *upper index* of the subordinator. Note that the lower index σ and the upper index β coincide whenever Φ is regularly varying at infinity.

Theorem 16 *The inequalities*

$$\sigma\dim(E) \leq \dim(\{X_t : t \in E\}) \leq \beta\dim(E)$$

hold for all subsets $E \subseteq [0, \infty)$, a.s.

The key step for the proof of the lower bound in Theorems 15 and 16 consists in establishing the regularity of the first-passage process, a result which is interesting in its own right.

Lemma 17 *Suppose that $\sigma > 0$. Then for every $\varepsilon > 0$, the first-passage process $T(\cdot) = \inf\{s \geq 0 : X_s > \cdot\}$, is a.s. Hölder-continuous on compact intervals with exponent $\sigma - \varepsilon$ for every $\varepsilon > 0$.*

Proof The argument relies on the classical criterion of Kolmogorov, see e.g. section XXIII.2 in Dellacherie, Maisonneuve and Meyer (1992). First, we note the following inequality that derives readily from the Markov property at the stopping time $T(t)$: For every $p > 0$ and $s, t \geq 0$

$$\mathbb{E}((T(t+s) - T(t))^p) \leq \mathbb{E}(T(s)^p).$$

We deduce

$$\begin{aligned}\mathbb{E}((T(t+s) - T(t))^p) &\leq p\int_0^\infty r^{p-1}\mathbb{P}(T(s) > r)dr \\ &= p\int_0^\infty r^{p-1}\mathbb{P}(X_r \leq s)dr.\end{aligned}$$

Using the obvious inequality

$$\mathbb{P}(X_r \leq s) \leq e\mathbb{E}(\exp\{-s^{-1}X_r\}) = \exp\{1 - r\Phi(s^{-1})\},$$

we get

$$\mathbb{E}((T(t+s) - T(t))^p) \leq e\Gamma(p+1)\Phi(s^{-1})^{-p}.$$

By the very definition of σ, we have $\Phi(s^{-1})^{-p} = o(s^{\sigma p - \varepsilon})$ as $s \to 0+$ for every $\varepsilon > 0$, so Kolmogorov's criterion applies and Lemma 17 is established. □

To prove the upper bound in Theorem 15, we establish first the following technical result.

Lemma 18 *We have*

$$\liminf_{x \to 0+} \mathbb{E}\left(1 - \exp\{-X_{T(x)}\}\right) \Phi(1/x) > 0.$$

Proof The statement is obvious when X is a pure drift, so we will assume with no loss of generality that $\Pi \not\equiv 0$. Using Proposition 2, we get

$$\mathbb{E}\left(1 - \exp\{-X_{T(x)}\}\right) \geq \int_{[0,x]} U(dy) \int_{(x-y,\infty)} \Pi(dz)(1 - e^{-z-y}).$$

It is easy to see that, provided that x is small enough, there is some constant $c > 0$ such that

$$\int_{(x-y,\infty)} \Pi(dz)(1 - e^{-z-y}) \geq c \qquad \text{for all } y \in [0,x].$$

Our assertion follows then from Proposition 1. □

We are now able to prove Theorem 15.

Proof of Theorem 15 We start with the lower bound

$$\dim(\{X_t : t \in [0,s]\}) \geq \sigma. \tag{13}$$

There is nothing to prove when $\sigma = 0$, so suppose that $\sigma > 0$, fix an arbitrary $x > 0$ and consider a covering of $\{X_t : t \in [0, T(x)]\}$ by finitely many intervals $[a_0, b_0], \cdots, [a_n, b_n]$, where $a_0 \leq b_0 \leq \cdots \leq a_n \leq b_n$. This forces $T(b_{i-1}) = T(a_i)$ for $i = 1, \cdots, n$. Then pick any $\gamma < \sigma$, recall from Lemma 17 that T is a.s. Hölder-continuous with exponent γ on $[0,x]$ and deduce that

$$\sum_{i=0}^{n} (b_i - a_i)^\gamma \geq K \sum_{i=0}^{n} (T(b_i) - T(a_i)) = K\,T(b_n) \geq K\,T(x) > 0 \quad \text{a.s.}$$

where $K > 0$ is a certain random variable. Because x is arbitrary, this entails (13).

We then establish the upper bound

$$\dim(\{X_t : t \geq 0\}) \leq \sigma, \tag{14}$$

by constructing an 'economic' cover of the range. Specifically, take any $\gamma > \sigma$, and pick a sequence of positive numbers $(x_n, n \in \mathbb{N})$ which converges to 0, such that

$$\lim_{n\to\infty} x_n^\gamma \Phi(x_n^{-1}) = 0. \tag{15}$$

Next, consider the following increasing family of stopping times defined by induction:

$$T_{n,0} = 0\,,\ T_{n,k+1} = \inf\{t > 0 : X_t > X_{T_{n,k}} + x_n\} \qquad (k = 0, 1, \cdots),$$

so that for each fixed n, the intervals $[X_{T_{n,k}}, X_{T_{n,k}} + x_n]$ have length x_n and form a cover of the range of X. Then introduce an independent exponential variable τ, say with parameter 1, and consider the variable

$$N_n = \mathrm{Card}\{k \in \mathbb{N} : T_{n,k} < \tau\}\,.$$

It is clear from the definition that

$$m^\gamma_{x_n}(\{X_t : 0 \leq t \leq \tau\}) \leq x_n^\gamma N_n,$$

in particular

$$\mathbb{P}(m^\gamma_{x_n}(\{X_t : 0 \leq t \leq \tau\}) > 1) \leq \mathbb{P}(N_n > x_n^{-\gamma}). \tag{16}$$

On the other hand, by the lack of memory of the exponential law and the Markov property at time $T_{n,k}$, we have $\mathbb{P}(N_n > k+1 \mid N_n > k) = \mathbb{P}(N_n > 1)$. Therefore

$$\mathbb{P}(N_n > k) = \mathbb{P}(N_n > 1)^k,$$

that is $N_n - 1$ has a geometric distribution with parameter $\mathbb{P}(N_n > 1)$. But according to Lemma 17,

$$\mathbb{P}(N_n > 1) = \mathbb{E}\left(\exp\{-X_{T(x_n)}\}\right) \leq 1 - 2c/\Phi(x_n^{-1})$$

for some constant $c > 0$, provided that x_n is small enough. Then (16) yields

$$-\log \mathbb{P}(m^\gamma_{x_n}(\{X_t : 0 \leq t \leq \tau\}) > 1) \geq c x_n^{-\gamma}/\Phi(x_n^{-1}).$$

Finally, we deduce from (15)

$$\lim_{n\to\infty} \mathbb{P}(m^\gamma_{x_n}(\{X_t : 0 \leq t \leq \tau\}) > 1) = 0,$$

and thus the γ-dimensional Hausdorff measure of $\{X_t : 0 \leq t \leq \tau\}$ is less than or equal to 1 a.s. Since γ may be taken arbitrarily close to σ, this establishes (14). □

It is interesting to point out that the result of Lemma 17 on the Hölder-continuity of the first-passage process is essentially optimal, in the sense that for every $\varepsilon > 0$, T is not Hölder-continuous with exponent $\sigma + \varepsilon$ on any non-trivial interval a.s. Indeed, if it were, then the argument that yields (13) would imply that the Hausdorff dimension of the range would be at least $\sigma + \varepsilon$ with positive probability.

The lower bound in Theorem 16 will again follow from Lemma 17; the upper bound requires the following lemma.

Lemma 19 *Fix any $\alpha > \beta$. Then there exists an integer $k(\alpha) > 0$ such that a.s., for all sufficiently large n and all $j = 1, \cdots, 2^n$, the range*

$$\{X_t : (j-1)2^{-n} \le t \le j2^{-n}\}$$

can be covered by $k(\alpha)$ intervals of length $2^{-n/\alpha}$.

Proof To start with, we note the easy inequality

$$\begin{aligned} \mathbb{P}(X_t \ge t^{1/\alpha}) &\le (1 - 1/e)^{-1}\mathbb{E}\left(1 - \exp\{-t^{-1/\alpha}X_t\}\right) \\ &= (1 - 1/e)^{-1}\left(1 - \exp\{-t\Phi(t^{-1/\alpha})\}\right). \end{aligned}$$

It follows from the definition of the upper index β that for $0 < \eta < 1 - \beta/\alpha$,

$$\mathbb{P}(X_t \ge t^{1/\alpha}) \le t^{-\eta} \qquad \text{for all } t > 0 \text{ small enough.} \tag{17}$$

Next, for any integer n and $j = 1, \cdots, 2^n$, consider the sequence of stopping times,

$$T_0 = (j-1)2^{-n}, \ T_{i+1} = \inf\{s > T_i : X_s - X_{T_i} > 2^{-n/\alpha}\}.$$

By the Markov property, the increments $T_{i+1} - T_i$ ($i = 0, 1, \cdots$) are independent and identically distributed. Moreover, the event $\{\{X_t : (j-1)2^{-n} \le t \le j2^{-n}\}$ cannot be covered by k intervals with length $2^{-n/\alpha}\}$ is contained in the event $\{T_k - T_0 \le 2^{-n}\}$, and by (17), its probability is therefore less than or equal to

$$\mathbb{P}(T_1 - T_0 \le 2^{-n}, \cdots, T_k - T_{k-1} \le 2^{-n}) = \mathbb{P}(T_1 - T_0 \le 2^{-n})^k \le 2^{-nk\eta}.$$

Thus, for a fixed n, the probability that for some integer $j \le 2^n$, $\{X_t : (j-1)2^{-n} \le t \le j2^{-n}\}$ cannot be covered by k intervals with length $2^{-n/\alpha}$ is less than $2^{-n(k\eta-1)}$. Now we choose $k = k(\alpha)$ so that $k\eta - 1 > 0$, and the lemma follows from the Borel-Cantelli lemma. □

Finally, we prove Theorem 16.

Proof of Theorem 16 We start with the lower bound. With no loss of generality, we may restrict our attention to subsets $E \subseteq [0, 1]$ with positive Hausdorff dimension. Pick any $k > 1/\sigma$ and let ω be any sample path for which the first-passage process $T(\cdot, \omega)$ has

$$T(t, \omega) - T(s, \omega) \leq K(\omega)(t - s)^{1/k} \qquad \text{for all } 0 \leq s \leq t \leq 1,$$

where $K(\omega)$ is some finite number. Then take any $\varepsilon > 0$ and $\theta < \dim(E)$, put $\eta = \varepsilon^{1/k}$ and cover $\{X_t(\omega) : t \in E\}$ by intervals I_i of length $a_i < \eta$. Next, introduce

$$b_i = \inf\{t \geq 0 : X_t(\omega) \in I_i\} \ , \ c_i = \sup\{t : X_t(\omega) \in I_i\} \ ,$$

and denote by J_i the interval centred at $\frac{1}{2}(b_i + c_i)$ with length $2(c_i - b_i)$. The intervals J_i cover E and $K(\omega)a_i^{1/k} \geq c_i - b_i$ for all i. Therefore,

$$(2K(\omega))^\theta \sum_i a_i^{\theta/k} \geq \sum_i (2(c_i - b_i))^\theta .$$

Recall the notation $m_\varepsilon^\gamma(\cdot)$ from the beginning of this section and deduce that

$$(2K(\omega))^\theta \, m_\eta^{\theta/k} \left(\{X_t : t \in E\}\right) \geq m_{2\varepsilon}^\theta(E).$$

When ε tends to $0+$, η tends to $0+$ and the right-hand side to ∞ because $\dim(E) > \theta$. Thus, $m^{\theta/k} \left(\{X_t : t \in E\}\right) = \infty$, that is

$$\dim \left(\{X_t : t \in E\}\right) \geq \theta/k,$$

and this gives the desired inequality by letting $k \downarrow 1/\sigma$ and $\theta \uparrow \dim(E)$ through a countable sequence.

We then focus on the upper bound. With no loss of generality, we may restrict our attention to subsets $E \subseteq [0, 1]$. Pick any $\alpha > \beta$ and let $k(\alpha)$ be as in Lemma 19. Suppose that $\dim(E) = \gamma$, choose $\varepsilon > 0$ and $\eta > 0$. Then cover E by intervals I_i of length $\ell_i < \varepsilon$ so that $\sum_i \ell_i^{\gamma+\eta} < \varepsilon$. Next, choose integers n_i such that $2^{-n_i-1} < \ell_i < 2^{-n_i}$, so that

$$I_i \subseteq [(j-1)2^{-n_i}, j2^{-n_i}] \cup [j2^{-n_i}, (j+1)2^{-n_i}] \qquad \text{for some } j = 1, \cdots, 2^{n_i}.$$

According to Lemma 19, if ε is small enough, $\{X_t : t \in I_i\}$ can be covered by $2k(\alpha)$ intervals of length $2^{-n_i/\alpha} \leq (2\ell_i)^{1/\alpha}$. Therefore

$$m_{(2\varepsilon)^{1/\alpha}}^{\alpha(\gamma+\eta)} \left(\{X_t : t \in E\}\right) \leq 2k(\alpha)\varepsilon,$$

hence $m^{\alpha(\gamma+\eta)} \left(\{X_t : t \in E\}\right) = 0$ and $\dim\left(\{X_t : t \in E\}\right) \leq \alpha(\gamma + \eta)$. This gives the desired inequality by letting $\eta \downarrow 0$ and $\alpha \downarrow \beta$ through a countable sequence. □

6. Exercises

1. *(Subordination of Lévy processes)* (a) Let $X = (X_t : t \geq 0)$ be a d-dimensional Lévy process, and $T = (T_t : t \geq 0)$ a subordinator independent of X. Prove that the time-changed process $X \circ T$ is a Lévy process and express its characteristic exponent in terms of those of X and T.
(b) Suppose henceforth that X is an isotropic stable process of index $\beta \in (0, 2]$, i.e. its characteristic exponent is proportional to $|\cdot|^\beta$, and T a stable subordinator of index $\alpha \in (0, 1)$. Check that then $X \circ T$ is an isotropic stable process of index $\alpha\beta$.
(c) Deduce that a polar set for an isotropic stable process of index $\beta \in (0, 2]$ is also polar for an isotropic stable process of index $\beta' \leq \beta$.
2. *(Potential measure)* Let $X^{(1)}$ and $X^{(2)}$ be two subordinators with respective drift coefficients $\mathrm{d}^{(1)}$ and $\mathrm{d}^{(2)}$, Lévy measures $\Pi^{(1)}$ and $\Pi^{(2)}$, and potential measure $U^{(1)}$ and $U^{(1)}$. Prove that if $\mathrm{d}^{(1)} = \mathrm{d}^{(2)}$ and $\overline{\Pi}^{(1)} = \overline{\Pi}^{(2)}$ on $(0, 1)$, then $U^{(1)} = U^{(2)}$ on $(0, 1)$ as well.
3. *(Renewal theorem for subordinators)* Suppose that $\mathbb{E}(X_1) = \mu < \infty$ and that hypothesis (H) of chapter I holds. Check that the renewal theorem (Theorem I.21) can be rewritten as
$$\lim_{x \to \infty} U([x, x + h)) = h/\mu, \qquad \forall h \geq 0.$$
As an application, prove that $X_{T(x)} - x$ and $\Delta X_{T(x)} = X_{T(x)} - X_{T(x)-}$ both converge in law as x goes to ∞ and characterize the limit distributions in terms of the mean μ, the drift coefficient and the Lévy measure.
4. *(Complement to the arcsine laws)* Assume that the conditions of Theorem 6 are fulfilled. Prove that $x^{-1} X_{T(x)}$ converges in law as x goes to ∞ and identify the limit distribution.
5. *(Law of the iterated logarithm for passage times)* Consider the first-passage process $T(x) = \inf\{t \geq 0 : X_t > x\}$, $x \geq 0$, and assume that the hypothesis of Theorem 11 holds. Prove that
$$\limsup_{x \to 0+} \frac{T(x)\Phi(x^{-1} \log|\log x|)}{\log|\log x|} = \alpha^{-\alpha}(1-\alpha)^{-(1-\alpha)}.$$
This extends Khintchine's celebrated law of the iterated logarithm, as will be stressed in Exercise VII.8.
6. *(Comparison of the size of a subordinator and its jumps)* Let $f : [0, \infty) \to [1, \infty)$ be an increasing function and $\overline{\Pi}$ the tail of the Lévy measure.

(a) Prove that the process

$$\sum_{s\leq t} \mathbf{1}_{\{\Delta X_s \geq f(X_{s-})\}} - \int_0^t \overline{\Pi} \circ f(X_s) ds \qquad (t \geq 0)$$

is a martingale, and deduce from the convergence theorem of martingales that the events $\{\Delta X_t \geq f(X_{t-})$ infinitely often as t goes to $\infty\}$ and $\{\int_0^\infty \overline{\Pi} \circ f(X_s)ds = \infty\}$ coincide up to a set of zero probability.

(b) Let $g : [0, \infty) \to [0, \infty)$ be an increasing function. Check that $\int^\infty g(X_s)ds = \infty$ a.s. if and only if $\int^\infty g(x)U(dx) = \infty$, where U is the potential measure. [Hint: use the Markov property to show that if $\mathbb{P}(\int_0^\infty g(X_s)ds < \infty) > 0$, then $\int_0^\infty g(X_s)ds$ is bouded from above by a geometric random variable.]

(c) Conclude that $\Delta X_t \geq f(X_{t-})$ infinitely often as t goes to ∞ a.s. if and only if the integral $\int^\infty \overline{\Pi} \circ f(x)U(dx)$ diverges.

7. *(Positive increase for the integrated tail)* The motivation for this exercise stems from Proposition 10. Prove that if the drift coefficient is zero, then the following assertions are equivalent:

(a) $$\liminf_{x\to 0+} \left(x\overline{\Pi}(x)/I(x)\right) > 0\,;$$

(b) $$\liminf_{x\to 0+} \left(I(2x)/I(x)\right) > 1\,;$$

(c) $$\limsup_{\lambda\to\infty} \left(\lambda\Phi'(\lambda)/\Phi(\lambda)\right) < 1\,;$$

(d) $$\limsup_{\lambda\to\infty} \left(\Phi(2\lambda)/\Phi(\lambda)\right) < 2\,.$$

Deduce that if they hold, then condition (iii) of Theorem 9 and condition (iii) of Proposition 10 are the same.

7. Comments

Meyer (1969) gives a self-contained introduction to subordinators; we also refer to Bertoin (1998) for a comprehensive account including applications to some other random processes.

The proof of Proposition 1 is inspired by a theorem of de Haan and Stadtmüller, see Bingham, Goldie and Teugels (1987) on page 118. This result combined with that for the integrated tail of the Lévy measure can be viewed as an analogue of an estimate for increasing random walks due to Erickson (1973); see also Horowitz (1968).

Section 2 is from Kesten (1969) who established similar results for general Lévy processes. Here, the monotonicity of the paths eases Kesten's arguments. More generally, passage times in sets have been investigated by Horowitz (1972) and Hawkes (1975). The latter gives a detailed account of potential theory for subordinators.

Our presentation of the arcsine laws is incomplete, the convergence in law of Theorem 6 can be reinforced, involving simultaneously the values of X immediately before and after it crosses a large (or small) level. See Bingham, Goldie and Teugels (1987) on pages 360-7 for further results in discrete times. Recent developments connected with random partitions of the time have been made by Pitman and Yor (1992), see also Perman, Pitman and Yor (1992) and Perman (1993). We refer to Bertoin (1995-d) for sample-path results connected with the arcsine laws.

There exists a huge variety of results on rates of growth of subordinators, we refer to Fristedt (1974) for a survey. Proposition 8 is from Shtatland (1965), Theorem 9 from Fristedt (1967) (see also Gihman and Skorohod (1975)), and Proposition 10 from Bertoin (1995-d). The law of the iterated logarithm is due to Fristedt (1964) in the stable case. Breiman (1968-a) obtains a so-called delicate law by a clever adaptation of an argument of Mootoo. Fristedt and Pruitt (1971) establish a more general but less precise law of the iterated logarithm which is valid for a wide class of subordinators. It seems that Theorem 11 first appeared in a slightly different form in Barlow, Perkins and Taylor (1986), Pruitt (1991) contains an even sharper result. The present proof via the estimate of Lemma 12 is implicit in Jain and Pruitt (1987). See also Bertoin and Caballero (1995) for the case when the Laplace exponent is slowly varying. One can complete Theorem 11 with a *uniform* result on the lower envelope of subordinators, see Hawkes (1971) in the stable case, Fristedt and Pruitt (1972) and Bertoin (1995) for the general case. Some of the results presented in this section can be extended to stable processes which are not subordinators (cf. chapter VIII and the references therein) and general Lévy processes, see Fristedt (1971), Pruitt (1981) and Wee (1988, 1990). Finally, we mention that Kallenberg (1974) proved similar theorems for processes with exchangeable increments, and we refer to Hu and Taylor (1997) and to Jaffard (1996) for interesting developments involving the so-called multifractal structure. Sonia Fourati raised an interesting problem on the rate of growth of subordinators which seems still open even in the stable case: Given an increasing function $f : (0,\infty) \to (0,\infty)$, is it possible, for almost every sample path, to pick a (random) instant $t > 0$ such that $X_{t+\varepsilon} - X_t \le f(\varepsilon)$ for all $\varepsilon > 0$ small enough?

The upper and lower indices β and σ were introduced by Blumenthal and Getoor (1961), they often enable one to compare the sample path behaviour of a given Lévy process to that of a stable process with index β or σ. There is a close connection between the law of the iterated logarithm and the Hausdorff measure of the range of a Lévy process. Taylor and Wendel (1966) determine the exact Hausdorff function of the range of a stable subordinator, this was extended by Fristedt and Pruitt (1971) to arbitrary subordinators. The sharpest result in this field is in Barlow, Perkins and Taylor (1986). Pruitt (1969) generalized Theorem 15, specifying the Hausdorff dimension of the range of a general Lévy process. Results on the exact Hausdorff measure of the range Lévy processes appear in Taylor (1967) and Pruitt and Taylor (1969) in the stable case, and Dupuis (1974) for symmetric Lévy processes. See also Jain and Pruitt (1968) for the graph of stable processes. The lower bound in Theorem 16 is from Hawkes (1974) and the upper bound from Hawkes and Pruitt (1974). Finally, we refer to Fristedt and Taylor (1992), Pruitt and Taylor (1996) and Rezakhanlou and Taylor (1988) for results on the packing measure of subordinators and Lévy processes.

IV

Local Time and Excursions of a Markov Process

The purpose of this chapter is to investigate the structure of the successive lengths of the intervals of excursion of a Markov process M away from a point. To this end, we construct an increasing process L, called the local time, which stays constant on the excursion intervals. The right-continuous inverse of L is a (possibly killed) subordinator whose jumps correspond to the lengths of the excursion intervals. The study culminates with the description of the process of the excursions of M in terms of a Poisson point process.

1. Framework

Our main goal in this chapter is to investigate the distribution of the intervals of time during which a Markov process makes an excursion away from a given point. At first sight, this topic does not seem directly related to Lévy processes, but the connection is much closer than it may appear. Specifically, we will see in section 3 that subordinators actually play a most important rôle in the study. On the other hand, excursion theory will be one of the key tools of chapter VI, when we will develop fluctuation theory for real-valued Lévy processes.

Throughout this chapter, M denotes a 'nice' Markov process taking values in $\mathbb{R}^n$ (more generally, one can replace the Euclidean space for instance by any Polish space). Below is a formal definition, note that most processes which are usually referred to as Markov processes in the literature (such as for instance Feller processes) are 'nice' in our sense.

Let $(\Omega', \mathscr{G}, \mathbf{P})$ be some probability space endowed with a right-continuous complete filtration $(\mathscr{G}_t)_{t \geq 0}$. We consider a stochastic process $M = (M_t, t \geq 0)$ taking values in $\mathbb{R}^d$ having right-continuous sample paths and with $\mathbf{P}(M_0 = 0) = 1$. We assume that M is adapted, that is for every $t \geq 0$, M_t is $\mathscr{G}_t$-measurable. This implies the apparently stronger property that M is progressively measurable, i.e. $(M_s, 0 \leq s \leq t)$ is $\mathscr{G}_t$-measurable for each $t \geq 0$, see e.g. Theorem IV.15 in Dellacherie and Meyer (1975). We suppose that there exists a family of probability measures $(\mathbf{P}_x, x \in \mathbb{R}^d)$, which can be thought of as the laws of M started from an arbitrary point, such that

(MP) *For every stopping time* $T < \infty$, *under the conditional law* $\mathbf{P}(\cdot \mid M_T = x)$, *the shifted process* $(M_{T+t}, t \geq 0)$ *is independent of* $\mathscr{G}_T$ *and has the law* $\mathbf{P}_x$.

As usual, we will refer in the sequel to (MP) as the *Markov property*. Recall that it entails the Blumenthal zero-one law, i.e. the initial sigma field $\mathscr{G}_0$ is $\mathbf{P}$-trivial. We will also use the property that for every stopping time T, the first return time to the origin after T, $R_T = \inf\{t > T : M_t = 0\}$, is a stopping time. See Theorem IV.50 in Dellacherie and Meyer (1975).

To start with, note that according to the Blumenthal zero-one law, the probability that M returns to the origin at arbitrarily small times, $\mathbf{P}(R_0 = 0)$, necessarily equals 0 or 1. We say that 0 is *irregular* in the first case, and *regular* in the second case. When 0 is regular, we introduce the first exit time from 0, $S = \inf\{t \geq 0 : M_t \neq 0\}$. Then S is a stopping time, and again $\mathbf{P}(S = 0)$ necessarily equals 0 or 1. We say that 0 is a *holding point* in the first case, and an *instantaneous point* in the second case.

The analysis of the excursions of M away from 0 is only delicate when 0 is regular and instantaneous, because then the closed zero set of M is perfect and nowhere dense (whereas it is discrete in the irregular case, and a union of disjoint intervals in the case of the holding point). The study relies on the existence of a remarkable increasing process L called the *local time* , which increases precisely on the closed zero set of M. The notion of local times of Markov processes was introduced by Blumenthal and Getoor (1968) using arguments of potential theory. There exists another approach based on the Doob-Meyer decomposition of supermartingales, see e.g. section XV.3 in Dellacherie and Meyer (1987). In the next section, we will construct this local time by a bare-hands method which is closer to the general spirit of this text. The definition of the local time is straightforward when 0 is either irregular or a holding point; however, it will be more convenient to postpone the discussion of these cases to section 5.

Without further mention, we will assume throughout sections 2-4 that 0 *is regular and instantaneous.*

2. Construction of the local time

We will define the local time by approximations involving the numbers of excursion intervals of certain types. We start with some definitions. We denote the zero set of M by $\mathscr{Z} = \{t : M_t = 0\}$ and its closure by $\overline{\mathscr{Z}}$. An open interval (g, d) is called an *excursion interval* if it is maximal among the open intervals on which $M \neq 0$. In other words, (g, d) is an excursion interval if $M_t \neq 0$ for all $t \in (g, d)$, $g \in \overline{\mathscr{Z}}$ and $d \in \overline{\mathscr{Z}} \cup \{\infty\}$. In this case, g, d and $\ell = d - g$ are known as the *left-end point* , *right-end point* and the *length* of the excursion interval. The notation g and d refers to the French 'gauche' (left) and 'droite' (right). The set of excursion intervals is endowed with a natural total order, namely $(g, d) < (g', d')$ if $g < d \leq g' < d'$. Note also that the excursion intervals are precisely the intervals that appear in the canonical representation of the open set $[0, \infty) - \overline{\mathscr{Z}}$ as the countable union of disjoint open intervals.

Next, we fix a real number $c > 0$ such that with positive probability, there is at least one excursion interval with length $\ell > c$. Because the sample paths of M are right-continuous, there is always such a constant provided that the Markov process is not identically zero, a case ruled out by the assumption that 0 is instantaneous. The choice of the constant c is arbitrary and changing it affects most of the results of this chapter in a simple way, usually by the introduction of some deterministic multiplicative factor. This is not relevant for the problems that we are concerned with, and we will briefly indicate at each major step the effect of changing c into some other constant.

It is then easy to see that

$$\mathbf{P}(\text{there is at least one excursion interval with length } > c) = 1.$$

Specifically, consider for every $t > 0$ the event $\Lambda_t = \{$all the excursion intervals with right-end point $d < t$ have length $\ell \leq c\}$, and pick t large enough such that $\mathbf{P}(\Lambda_t) < 1$. The first return to 0 after time t is a stopping time, possibly infinite. If it is infinite, then the process has an excursion interval of infinite length, *a fortiori* it has an excursion interval with length $\ell > c$. If it is finite, we can apply the Markov property. The inequality $\mathbf{P}(\Lambda_{3t}) \leq \mathbf{P}(\Lambda_t)^2$ follows readily, and by iteration

$\mathbf{P}(\Lambda_{3^n t}) \leq \mathbf{P}(\Lambda_t)^{2^n}$ for every integer n. Hence $\lim_{s\to\infty} \mathbf{P}(\Lambda_s) = 0$, which entails our assertion.

For every $a > 0$ and $n = 1, 2, \cdots$, we denote by $\ell_n(a)$, $g_n(a)$ and $d_n(a)$, respectively the length, the left-end point and the right-end point of the n-th excursion interval with length $\ell > a$, where 'n-th' refers to the natural order on excursion intervals. When the total number of excursion intervals with length $\ell > a$ is strictly less than n, we decide that $\ell_n(a) = 0, g_n(a) = d_n(a) = \infty$. One sees by induction that every $d_n(a)$ is a stopping time.

We observe that $\mathbf{P}(\ell_1(a) > c) > 0$ for every $a \in (0, c)$. Indeed, if this failed for some $a > 0$, then $d_1(a) < g_1(c)$ a.s. In particular, $d_1(a)$ is a finite stopping time, and we deduce, iterating the Markov property at $d_n(a)$, that $d_n(a) < g_1(c)$ a.s. for every integer $n > 0$. Thus $\mathbf{P}(g_1(c) < \infty) = 0$, which is absurd. This enables us to introduce a decreasing and right-continuous function $\overline{\Pi} : (0, \infty] \to (0, \infty)$ which is meant to describe the distribution of the lengths of the excursion intervals. Specifically, for every $a \in (0, \infty]$, we put

$$\overline{\Pi}(a) = \begin{cases} 1/\mathbf{P}(\ell_1(a) > c) & \text{if } a \leq c, \\ \mathbf{P}(\ell_1(c) > a) & \text{if } a > c. \end{cases} \tag{1}$$

Clearly $\overline{\Pi}(c) = 1$, and it is easy to check (successively applying the Markov property at the right-end points of excursion intervals with length $\ell > c$) that the zero set $\mathscr{Z}$ of M is a.s. bounded or unbounded according as $\overline{\Pi}(\infty)$ is positive or null. One says that 0 is *transient* in the first case, and *recurrent* in the second. We also observe that

$$\lim_{a\to 0+} \overline{\Pi}(a) = \infty .$$

Indeed, if this failed, then with positive probability, there would be no excursion intervals in $[0, g_1(c))$. Because 0 is regular and the paths are right-continuous, M would then stay at 0 on some neighbourhood of the origin of time with positive probability as well. This would contradict the assumption that 0 is instantaneous.

Lemma 1 *For every $a \in (0, \infty)$ and $b \leq a$ such that $\overline{\Pi}(b) > 0$, we have*

$$\mathbf{P}(\ell_1(b) > a) = \overline{\Pi}(a)/\overline{\Pi}(b).$$

Proof For $b = c$ or $a = c$, the statement merely rephrases the definition. First, suppose that $b < c < a$. Then

$$\begin{aligned} \mathbf{P}(\ell_1(b) > a) &= \mathbf{P}(\ell_1(b) > c, \ell_1(c) > a) \\ &= \mathbf{P}(\ell_1(c) > a) - \mathbf{P}(\ell_1(b) \leq c, \ell_1(c) > a) . \end{aligned}$$

Applying the Markov property at the right-end point of the first excursion with length $\ell > b$, we see that the events $\{\ell_1(b) \leq c\}$ and $\{\ell_1(c) > a\}$

are independent,

$$\mathbf{P}(\ell_1(b) \le c, \ell_1(c) > a) = \mathbf{P}(\ell_1(b) \le c)\mathbf{P}(\ell_1(c) > a)$$

and then, from the definition of $\overline{\Pi}$, we get

$$\mathbf{P}(\ell_1(b) > a) = \overline{\Pi}(a) - (1 - 1/\overline{\Pi}(b))\overline{\Pi}(a) = \overline{\Pi}(a)/\overline{\Pi}(b).$$

Next, suppose that $b < a < c$. Then we have

$$\begin{aligned}\mathbf{P}(\ell_1(b) > c) &= \mathbf{P}(\ell_1(b) > a, \ell_1(a) > c)\\ &= \mathbf{P}(\ell_1(a) > c) - \mathbf{P}(\ell_1(b) \le a, \ell_1(a) > c)\,.\end{aligned}$$

The same argument as before based on the Markov property yields

$$\begin{aligned}1/\overline{\Pi}(b) = \mathbf{P}(\ell_1(b) > c) &= 1/\overline{\Pi}(a) - (1 - \mathbf{P}(\ell_1(b) > a))/\overline{\Pi}(a)\\ &= \mathbf{P}(\ell_1(b) > a)/\overline{\Pi}(a)\,.\end{aligned}$$

The case $c < b < a$ is similar. □

As a consequence of Lemma 1, we observe that changing the arbitrary constant c only alters the function $\overline{\Pi}$ by a constant multiplicative factor.

Next, we introduce for every $a > 0$ and $t > 0$ the total number of excursion intervals with length $\ell > a$ which started strictly before time t,

$$N_a(t) = \sup\{n : g_n(a) < t\}. \tag{2}$$

Note that if $M_t \neq 0$, then the excursion interval straddling t is counted provided that its length is larger than a. On the other hand, if t is the left-end point of an excursion interval, then this interval is always discarded. The following lemma gives the distribution of this number evaluated at the left-end point of certain excursion intervals.

Lemma 2 *Let $a \in (0, \infty]$ and $b < a$ be such that $\overline{\Pi}(b) > 0$. Then $N_b(g_1(a))$ is independent of the shifted process $(M_{g_1(a)+t}, t \ge 0)$ (by convention, the shifted process is taken identical to ∂ when $g_1(a) = \infty$) and*

$$\mathbf{P}(N_b(g_1(a)) \ge k) = (1 - \overline{\Pi}(a)/\overline{\Pi}(b))^k \qquad \textit{for } k = 0, 1, \cdots,$$

i.e. $N_b(g_1(a))$ has a geometric distribution with parameter $1 - \overline{\Pi}(a)/\overline{\Pi}(b)$.

Proof Applying the Markov property at the right-end point of the n-th excursion interval with length $\ell > b$, we see that conditionally on $\{d_n(b) < \infty\}$, the shifted process $M^{(n)} = (M_{d_n(b)+t}, t \ge 0)$ has the same law as M and is independent of $\mathcal{G}_{d_n(b)}$. In particular, for every measurable functional $F \ge 0$,

$$\begin{aligned}\mathbf{E}(F(M^{(n)}), N_b(g_1(a)) \ge n) &= \mathbf{E}(F(M^{(n)}), d_n(b) < g_1(a))\\ &= \mathbf{P}(d_n(b) < g_1(a))\ \mathbf{E}(F(M)).\end{aligned}$$

On the event $\{d_n(b) < g_1(a)\}$, the shifted process $(M_{g_1(a)+t}, t \geq 0)$ can be obtained from $M^{(n)}$ by translation of the origin of time to $g_1^{(n)}(a)$, the left-end point of the first excursion interval of $M^{(n)}$ with length $\ell > a$. As a consequence, $N_b(g_1(a))$ and $(M_{g_1(a)+t}, t \geq 0)$ are independent.

On the other hand, applying the Markov property again at $d_n(b)$, we get

$$\mathbf{P}(d_{n+1}(b) < g_1(a) \mid d_n(b) < g_1(a)) = \mathbf{P}(d_1(b) < g_1(a)) = \mathbf{P}(\ell_1(b) \leq a).$$

We now deduce from Lemma 1

$$\mathbf{P}(N_b(g_1(a)) \geq n+1 \mid N_b(g_1(a)) \geq n) = 1 - \overline{\Pi}(a)/\overline{\Pi}(b),$$

which establishes the lemma. □

We are now able to state the main technical lemma of this section.

Lemma 3 *For every $u \in (0, \infty]$ such that $\overline{\Pi}(u) > 0$, the process*

$$N_a(d_1(u))/\overline{\Pi}(a), \qquad a \in (0, u),$$

is a left-continuous uniformly integrable backwards martingale. It converges a.s. and in $L^1(\mathbf{P})$ as a goes to $0+$. Its limit has an exponential distribution with parameter $\overline{\Pi}(u)$ and is independent of $\ell_1(u)$.

Proof For simplicity, we will write $d = d_1(u)$. For $a < c$, introduce the sigma-field generated by the lengths of the intervals of excursion with length $\ell > a$ which were completed before time d,

$$\mathscr{L}_a = \sigma(\ell_k(a), k = 1, \cdots, N_a(d)).$$

Plainly, $N_a(d)$ is measurable with respect to $\mathscr{L}_a$, and $\mathscr{L}_a \subseteq \mathscr{L}_b$ if $b < a$, that is $(\mathscr{L}_a)_{a>0}$ is a reversed filtration.

Then, we decompose the path of M at the right-end points $d_k(a)$ of the excursions intervals with length $\ell > a$, and use the convention $d_0(a) = 0$. On the one hand, the Markov property at $d_k(a)$ shows that conditionally on $N_a(d) = n$ and $\ell_k(a) = \lambda_k$ $(k = 1, \cdots, n)$, the shifted processes

$$Y^{(k)} = (M_{d_{k-1}(a)+t}, 0 \leq t \leq d_k(a) - d_{k-1}(a)), \qquad k = 1, \cdots, n,$$

are independent, and each $Y^{(k)}$ has the same law as $(M_t, 0 \leq t \leq d_1(a))$ conditioned on $\ell_1(a) = \lambda_k$. On the other hand, we know from Lemma 2 that for $b < a$, $N_b(d_1(a)) = \xi_1 + 1$, where $\xi_1 = N_b(g_1(a))$ is independent of $\ell_1(a)$ and has a geometric law with parameter $1 - \overline{\Pi}(a)/\overline{\Pi}(b)$ (one has to add 1 to take account of the excursion interval $(g_1(a), d_1(a))$ which has length $\ell_1(a) > a > b$). Therefore, the distribution of $N_b(d)$ conditionally on $\mathscr{L}_a$ is that of $(\xi_1 + 1) + \cdots + (\xi_n + 1)$, with $n = N_a(d)$, where the ξ_k's are independent geometric variables with parameter $1 - \overline{\Pi}(a)/\overline{\Pi}(b)$. Because $\xi_k + 1$ has the expectation $\overline{\Pi}(b)/\overline{\Pi}(a)$, this entails

$$\mathbf{E}(N_b(d) \mid \mathscr{L}_a) = N_a(d)\overline{\Pi}(b)/\overline{\Pi}(a),$$

and the martingale property is established.

The left-continuity is obvious from the construction, and almost sure convergence follows, see section O.3. Because according to Lemma 2, $N_a(d) = N_a(g) + 1$ is independent of $\ell_{N_a(d)}(a) = \ell_1(u)$, the limit is independent of $\ell_1(u)$ as well. Finally, since $N_a(d) - 1$ has a geometric law with parameter $1 - \overline{\Pi}(u)/\overline{\Pi}(a)$, the limit has an exponential distribution with parameter $\overline{\Pi}(u)$. In particular, it has expectation $1/\overline{\Pi}(u) = \lim_{a\to 0+} \mathbf{E}(N_a(d)/\overline{\Pi}(a))$, which implies convergence in $L^1(\mathbf{P})$ and that the martingale is uniformly integrable (see e.g. Theorem VI.7 in Dellacherie and Meyer (1980)). □

We are now able to construct the local time. Recall that the quantities $\overline{\Pi}(a)$ and $N_a(t)$ have been defined in (1) and (2) respectively.

Theorem 4 *The following assertions hold a.s.:*

(i) *For all $t \geq 0$, $N_a(t)/\overline{\Pi}(a)$ converges as a tends to $0+$, the limit is denoted by $L(t)$.*

(ii) *The mapping $t \to L(t)$ is increasing and continuous, it is called the local time of M at 0.*

(iii) *The support of the Stieltjes measure dL is $\overline{\mathscr{Z}}$.*

Proof (i)–(ii) We know from Lemma 3 that $N_a(t)/\overline{\Pi}(a)$ converges to L_t for $t = d_1(u)$. Applying the Markov property successively at each $d_k(u)$, $k = 1, 2, \cdots$, and then letting $u \to 0+$, we see that the convergence holds whenever $t \in \mathscr{D}$, where

$$\mathscr{D} = \{d_k(u), u > 0 \text{ and } k = 1, 2, \cdots\}$$

stands for the set of right-end points of excursion intervals.

The mapping $L : \mathscr{D} \to [0, \infty)$ is clearly increasing, we now prove that it has a unique increasing extension to $[0, \infty)$. To this end, we fix $\varepsilon > 0$ and consider the event that the increment of L between the instants $d_{k-1}(a)$ and $d_k(a)$ does not exceed ε for any $k \leq N_a(d_1(c))$ (the number of excursion intervals with length $\ell > a$ completed before the end of the first excursion with length $\ell > c$):

$$\Lambda = \{L(d_k(a)) - L(d_{k-1}(a)) \leq \varepsilon, \forall k \leq N_a(d_1(c))\}.$$

On the one hand, we know from Lemma 2 that $N_a(d_1(c)) - 1$ has a geometric distribution with parameter $1 - \overline{\Pi}(c)/\overline{\Pi}(a)$. On the other hand, it follows from Lemma 3 and the Markov property that, conditionally on $N_a(d_1(c)) = n$, the variables $L(d_k(a)) - L(d_{k-1}(a))$ $(k = 1, \cdots, n)$ are independent and have an exponential distribution with parameter $\overline{\Pi}(a)$.

The probability of Λ is therefore

$$\sum_{n=0}^{\infty} \frac{\overline{\Pi}(c)}{\overline{\Pi}(a)} \left(1 - \frac{\overline{\Pi}(c)}{\overline{\Pi}(a)}\right)^n (1 - \exp\{-\varepsilon\overline{\Pi}(a)\})^{n+1}$$

$$= (1 - \exp\{-\varepsilon\overline{\Pi}(a)\}) \frac{\overline{\Pi}(c)}{\overline{\Pi}(a)} \left[1 - \left(1 - \frac{\overline{\Pi}(c)}{\overline{\Pi}(a)}\right)(1 - \exp\{-\varepsilon\overline{\Pi}(a)\})\right]^{-1}.$$

Recall that $\lim_{a\to 0+} \overline{\Pi}(a) = \infty$, so that the foregoing quantity converges to 1 as $a \to 0+$. This shows that there is no gap larger than ε in the range of L on the set $\mathscr{D} \cap [0, d_1(c)]$, a.s. Letting ε tend to 0+ and applying the Markov property successively at the $d_k(c)$'s, this implies that the range of L on $\mathscr{D}$ is dense a.s. Therefore, we have a.s.

$$\inf\{L(t) : t \in \mathscr{D}, t > s\} = \sup\{L(t) : t \in \mathscr{D}, t < s\} \qquad \text{for all } s \in [0, \infty). \tag{3}$$

Denoting the quantity in (3) by $L(s)$, we now see that $L : [0, \infty) \to [0, \infty)$ is continuous and is the unique increasing extension of $L : \mathscr{D} \to [0, \infty)$.

An immediate monotonicity argument shows that

$$\sup\{N_a(t) : t \in \mathscr{D}, t < s\} \le N_a(s) \le \inf\{N_a(t) : t \in \mathscr{D}, s < t\}$$

and we deduce from (3), Lemma 3 and the continuity of L that $N_a(s)/\overline{\Pi}(a)$ converges to $L(s)$ for all $s \in [0, \infty)$, a.s.

(iii) It is plain from (i) that the support of dL is contained in the closure of the zero set of M. We still have to check that a.s., for all $0 < s < t$, $L(s) < L(t)$ whenever M visits 0 in the open interval (s, t). With no loss of generality, we may restrict our attention to the case when s and t vary in some countable dense set (e.g. rational numbers), and thus it is sufficient to show that for any fixed $0 < s < t$,

$$\mathbf{P}(L(s) = L(t) \text{ and } M_u = 0 \text{ for some } u \in (s, t)) = 0.$$

Applying the Markov property at the first return to 0 after time s, all that is needed is to verify that $\mathbf{P}(L(v) = 0) = 0$ for every $v > 0$. This is an immediate consequence of Lemma 3. Specifically, for any $\varepsilon > 0$, there is $a > 0$ such that $\mathbf{P}(d_1(a) < v) > 1 - \varepsilon$ (because 0 is an instantaneous point) and we know from Lemma 3 that $\mathbf{P}(L(d_1(a)) = 0) = 0$. Since L increases, it follows that $\mathbf{P}(L(v) = 0) < \varepsilon$. □

We point out that the construction of L given in Theorem 4 depends on the arbitrary constant $c > 0$. But changing the constant merely affects L by a deterministic multiplicative factor (because the same holds for $\overline{\Pi}$), and this is not relevant to our purposes. So to be completely rigorous, we should say that L is *a* local time and indicate the dependence upon c; but we will not do so, for the sake of conciseness.

The construction of the local time when M starts from 0 immediately extends to the case when M starts from an arbitrary point, by application of the Markov property at the first return to the origin. This enables us to note the following important property. For every stopping time $T < \infty$, denote by $L' = (L'(t), t \geq 0)$ the local time at 0 of the shifted process $M' = (M_{T+t}, t \geq 0)$. Then it is plain from Theorem 4(i) that

$$L(T+t) = L(T) + L'(t) \qquad \text{for all } t \geq 0 \text{ a.s.} \tag{4}$$

One says that L is an *additive functional* . In particular, the Markov property implies that for every stopping time T with $M_T = 0$ a.s. on $\{T < \infty\}$, under $\mathbf{P}(\cdot \mid T < \infty)$, the shifted process $((M_{T+t}, L(T+t) - L(T)), t \geq 0)$ is independent of $\mathcal{G}_T$ and has the same law as (M, L).

It is clear that L is an adapted process, i.e. for every $t \geq 0$, $L(t)$ is $\mathcal{G}_t$-measurable. We conclude this section with the observation that the local time process which has been constructed in Theorem 4 is essentially the unique continuous adapted process that increases only on the closure of the zero set of M and which has the additive property.

Proposition 5 *Let $A = (A(t), t \geq 0)$ be a continuous increasing $(\mathcal{G}_t)$-adapted process such that the following conditions hold:*

(i) *The support of the Stieltjes measure dA is included in $\overline{\mathcal{Z}}$, a.s.*
(ii) *For every stopping time with $M_T = 0$ a.s. on $\{T < \infty\}$, the shifted process $((M_{T+t}, A(T+t) - A(T)), t \geq 0)$ is independent of $\mathcal{G}_T$ under $\mathbf{P}(\cdot \mid T < \infty)$, and has the same law as (M, A) under $\mathbf{P}$.*

Then there exists a constant $k \geq 0$ such that $A \equiv kL$ a.s.

Proof We start by observing that for every $b > 0$, $A(d_1(b))$ has an exponential distribution. Specifically, fix $s, t > 0$ and consider the random time $T = \inf\{v : A(v) > t\}$. For every $v \geq 0$, the events $\{T < v\}$ and $\{A(v) > t\}$ coincide because A is continuous, and since A is adapted, we see that T is a stopping time (recall that the filtration is right-continuous). We now work conditionally on $\{T < \infty\}$. Since A is continuous and increases only on the zero set of M, we have $A(T) = t$ and $M_T = 0$. Put $M' = (M_{T+v}, v \geq 0)$ and $A' = (A(T+v) - t, v \geq 0)$. By condition (ii), the pair (M', A') has the same distribution as (M, A) under $\mathbf{P}$. Then note that the events $\{A(d_1(b)) > t + s\}$ and $\{A(d_1(b)) > t, A'(d'_1(b)) > s\}$ are identical, where $d'_1(b)$ stands for the right-end point of the first excursion interval of M' with length $\ell > b$. Therefore, by condition (ii), we have

$$\begin{aligned}\mathbf{P}(A(d_1(b)) > t+s) &= \mathbf{P}(A(d_1(b)) > t)\mathbf{P}(A'(d'_1(b)) > s \mid T < d_1(b)) \\ &= \mathbf{P}(A(d_1(b)) > t)\mathbf{P}(A(d_1(b)) > s)\,,\end{aligned}$$

which shows that $A(d_1(b))$ has an exponential distribution. Denote its parameter by $F(b)$. With no loss of generality, we may assume that $F(b) < \infty$ for all $b \in (0, c]$, because otherwise A would be identically zero and there would be nothing to prove.

Next, we verify that F and $\overline{\Pi}$ are proportional. To this end, we first deduce from (ii) (applied to the stopping time $T = \inf\{v : A(v) > t\}$) that $A(d_1(b))$ is independent of $\ell_1(b)$, the length of the first excursion interval with length $\ell > b$. Then we fix $a > b$. Applying (ii) again to the right-end points of the $N_b(g_1(a))$ excursion intervals with length $\ell > b$ which were completed before $g_1(a)$, we deduce that

$$A(d_1(a)) = \xi_1 + \cdots + \xi_n \quad , \quad n = N_b(g_1(a)) + 1, \tag{5}$$

where the ξ_i's are independent exponential variables with parameter $F(b)$ which are also independent of $N_b(g_1(a))$. Recall from Lemma 2 that $N_b(g_1(a))$ has a geometric distribution with parameter $1 - \overline{\Pi}(a)/\overline{\Pi}(b)$. Taking expectations of both terms in the first equality of (5) yields

$$F(a) = \overline{\Pi}(a)F(b)/\overline{\Pi}(b),$$

that is $kF = \overline{\Pi}$ for some constant $k > 0$.

Finally, we deduce from (5) that

$$\mathbf{E}(|A(d_1(a)) - N_b(d_1(a))/F(b)|^2) = F(b)^{-2}\left(\overline{\Pi}(b)/\overline{\Pi}(a)\right).$$

The right-hand side converges to 0 as $b \to 0+$, and we deduce from Theorem 4(i) that $A(d_1(a)) = kL(d_1(a))$. An argument based on the Markov property similar to that of the beginning of the proof of Theorem 4 shows that $A(t) = kL(t)$ for every $t \in \mathscr{D}$, the set of right-end points of excursion intervals. Since both A and L only increase on the zero set of M, this identity extends to $\mathscr{Z}^c$, the complement of the zero set of M. But this set is dense (because 0 is an instantaneous point), and since A and L are both continuous, we have $A(t) = kL(t)$ everywhere. □

Corollary 6 *There exists a constant* $\mathrm{d} \geq 0$ *such that a.s.*

$$\int_0^t \mathbf{1}_{\{M_s=0\}} ds = \int_0^t \mathbf{1}_{\{s \in \overline{\mathscr{Z}}\}} ds = \mathrm{d}L(t) \qquad \text{for all } t \geq 0.$$

Proof The closed set $\overline{\mathscr{Z}}$ differs from the zero set of M by at most countably many points (because M has at most countably many jumps), so the two integrals in the statement coincide. The condition (i) of Proposition 5 is clearly fulfilled, and (ii) follows readily from the Markov property for M and the additivity property of the integral. □

3. Inverse local time

The main purpose of this section is to describe the distribution of the local time process L introduced in the preceding section, characterizing the law of its inverse,

$$L^{-1}(t) = \inf\{s \geq 0 : L(s) > t\} \qquad (t \geq 0).$$

In the sequel, we will also use the notation

$$L^{-1}(t-) = \inf\{s \geq 0 : L(s) \geq t\} = \lim_{s \to t-} L^{-1}(s), \qquad t > 0.$$

We start the study of the inverse local time with the following elementary properties.

Proposition 7 (i) *For every $t \geq 0$, both $L^{-1}(t)$ and $L^{-1}(t-)$ are stopping times.*

(ii) *The process L^{-1} is increasing, right-continuous and adapted to the filtration $(\mathcal{G}_{L^{-1}(t)})$.*

(iii) *We have a.s. for all $t > 0$*

$$L^{-1}(L(t)) = \inf\{L^{-1}(u) : L^{-1}(u) > t\} = \inf\{s > t : M_s = 0\}$$

and

$$L^{-1}(L(t)-) = \sup\{L^{-1}(u) : L^{-1}(u) < t\} = \sup\{s < t : M_s = 0\}.$$

In particular $L^{-1}(t) \in \mathcal{Z}$ on $\{L^{-1}(t) < \infty\}$.

Proof (i) For every $s, t > 0$, the events $\{L^{-1}(t) < s\}$ and $\{L(s) > t\}$ are identical (because L is continuous) and hence $L^{-1}(t)$ is a stopping time. This shows that $L(t-)$ is a limit of stopping times, and thus is a stopping time as well.

(ii) is immediate.

(iii) The identity $L^{-1}(L(t)) = \inf\{L^{-1}(u) : L^{-1}(u) > t\}$ derives readily from the definition of L^{-1}. Then denote by $D_t = \inf\{s > t : M_s = 0\}$, the first return to 0 strictly after time t.

First, suppose that $D_t > t$. By Theorem 4(iii), L stays constant on the time interval $[t, D_t)$ and therefore we have $D_t \leq L^{-1}(L(t))$. We may now assume that $D_t < \infty$, since otherwise there is nothing to prove. Then D_t belongs to the support of the measure dL and is isolated on the left (again by Theorem 4(iii)). Since L is continuous, D_t cannot be isolated on the right, that is to say that $L(s) > L(D_t) = L(t)$ for all $s > D_t$; and hence $D_t \geq L^{-1}(L(t))$.

Then suppose that $D_t = t$, so that t belongs to the support of dL and is not isolated on the right. Just as in the foregoing, this implies that $L(s) > L(t)$ for all $s > t$, and hence $t \geq L^{-1}(L(t))$. The converse inequality is obvious.

The second identity in (iii) follows from similar arguments. Finally, on the event $\{L^{-1}(t) < \infty\}$, there exists s such that $L(s) = t$ and the first identity shows that $L^{-1}(t)$ is a zero of M. □

Proposition 7 immediately yields the following characterization of the excursion intervals. Since L is continuous and the support of dL coincides with the closure of the zero set of M, $\overline{\mathscr{Z}}$ has no isolated point, and then (iii) shows that the excursion intervals are precisely the open intervals of the type $(L^{-1}(t-), L^{-1}(t))$ whenever $L^{-1}(t-) < L^{-1}(t)$.

Our next result identifies the law of the process L^{-1} as that of a subordinator killed at rate $\overline{\Pi}(\infty)$. The intuitive explanation for this killing is that the origin is recurrent when $\overline{\Pi}(\infty) = 0$, and transient when $\overline{\Pi}(\infty) > 0$. Recall that the real number $\mathrm{d} \geq 0$ has been introduced in Corollary 6, and consider the measure Π on $(0, \infty)$ given by

$$\Pi(s,t] = \overline{\Pi}(s) - \overline{\Pi}(t) \qquad (0 < s < t < \infty).$$

Theorem 8 *The inverse local time L^{-1} is a subordinator with Lévy measure Π, drift coefficient d, and killed at rate $\overline{\Pi}(\infty)$. One has for all $t, \lambda > 0$*

$$\mathbf{E}(\exp\{-\lambda L^{-1}(t)\}) = \exp\{-t\Phi(\lambda)\}$$

(with the convention $\mathrm{e}^{-\infty} = 0$), where the Laplace exponent Φ is given by

$$\Phi(\lambda) = \lambda \left(\mathrm{d} + \int_0^\infty \mathrm{e}^{-\lambda r}\, \overline{\Pi}(r) dr \right).$$

As a consequence, note that the integral $\int (1 \wedge a)\Pi(da)$ must be finite.

Proof For the sake of clarity, we will treat the cases $\overline{\Pi}(\infty) = 0$ and $\overline{\Pi}(\infty) > 0$ separately, though the arguments are similar.

(i) Assume first that $\overline{\Pi}(\infty) = 0$, so there is no infinite excursion interval a.s. In particular, $d_1(c) < \infty$ a.s., and we deduce by iteration of the Markov property that $d_n(c) < \infty$ a.s. for every integer n. It follows now from Lemma 3, the Markov property and the additive property of local time that $L(d_n(c))$ can be expressed as the sum of n independent exponential variables with parameter $\overline{\Pi}(c) = 1$. In particular, $L(\infty) = \lim_{n\to\infty} L(d_n(c)) = \infty$ a.s.

By Proposition 7(i), we may apply the Markov property at $L^{-1}(t)$. The shifted process $\widetilde{M} = (\widetilde{M}_s = M_{L^{-1}(t)+s}, s \geq 0)$ has the same law as M and is independent of $\mathscr{G}_{L^{-1}(t)}$. It follows from the additivity of L that the local time $\widetilde{L}$ of $\widetilde{M}$ is given by $\widetilde{L}(s) = L(L^{-1}(t) + s) - t$ and hence the inverse local time of $\widetilde{M}$ is specified by

$$\widetilde{L}^{-1}(s) = L^{-1}(t+s) - L^{-1}(t).$$

This shows that L^{-1} has homogeneous independent increments, and since its sample paths are increasing and right-continous, L^{-1} is a subordinator.

The Lévy measure Π of L^{-1} is the characteristic measure of the Poisson point process of its jumps, ΔL^{-1}. For each $a > 0$, denote by $T_a = \inf\{t \geq 0 : \Delta L^{-1}(t) > a\}$, the instant of the first jump with length $\ell > a$. We see from Proposition 7(iii) that T_a coincides with the local time L evaluated on the first excursion interval with length $\ell > a$ (since L is continuous, the zero set of M has no isolated points). Thus, according to Lemma 3, T_a has an exponential distribution with parameter $\overline{\Pi}(a)$. The comparison with Proposition O.2(i) implies that $\overline{\Pi}(a) = \Pi(a, \infty)$.

Finally, recall that the excursion intervals are the open intervals which appear in the canonical decomposition of the open set $[0, \infty) - \overline{\mathscr{Z}}$. Using again the correspondence between the jumps of L^{-1} and the length of the excursion intervals of M specified in Proposition 7(iii), we have

$$L^{-1}(t) = \int_0^{L^{-1}(t)} \mathbf{1}_{\{s \in \overline{\mathscr{Z}}\}} ds + \sum_{s \leq t} \Delta L^{-1}(s).$$

According to Corollary 6, the foregoing integral coincides with $\mathrm{d}L(L^{-1}(t)) = \mathrm{d}t$. That is

$$L^{-1}(t) = \mathrm{d}t + \sum_{s \leq t} \Delta L^{-1}(s),$$

which shows that the drift coefficient of L^{-1} is d.

(ii) Next suppose that $\overline{\Pi}(\infty) > 0$. The Markov property and the additivity of L imply by the same arguments as in the case $\overline{\Pi}(\infty) = 0$ that for every $0 < t < t'$, the law of $(L^{-1}(s), 0 \leq s \leq t)$ is the same conditionally on $L^{-1}(t) < \infty$ as conditionally on $L^{-1}(t') < \infty$, and coincides with the law of a subordinator S restricted to the time interval $[0, t]$. Since the events $\{L^{-1}(t) < \infty\}$ and $\{L(\infty) > t\}$ are the same, we may rephrase the preceding assertion by claiming that $(L^{-1}(s), s < L(\infty))$ is distributed as $(S_s, s < \tau)$, where τ is a random variable independent of S and distributed as $L(\infty)$. Finally, we know from Lemma 3 that $L(\infty) = L(d(\infty))$ has an exponential distribution with parameter $\overline{\Pi}(\infty)$.

We denote the Lévy measure of S by Π. As in (i), introduce for each $a > 0$, $T_a = \inf\{t \geq 0 : \Delta L^{-1}(t) > a\}$, so that $T_a = L(d_1(a))$. Therefore we have for every $a > 0$

$$\begin{aligned} 1 - \exp\{-t\Pi(a, \infty)\} &= \mathbf{P}(\exists s < t : \Delta S_s > a) \\ &= \mathbf{P}(T_a < t \mid L(\infty) > t) \\ &= \exp\{t\overline{\Pi}(\infty)\}\mathbf{P}(L(d_1(a)) < t, L(\infty) > t). \end{aligned}$$

On the one hand, according to Lemma 3, the law of $L(d_1(a))$ conditionally on $d_1(a) < \infty$ is the exponential distribution with parameter $\overline{\Pi}(a)$. On the other hand, according to Lemma 1,

$$\mathbf{P}(d_1(a) < \infty) = \mathbf{P}(d_1(a) < d(\infty)) = 1 - \overline{\Pi}(\infty)/\overline{\Pi}(a).$$

We now deduce from the Markov property applied at $d_1(a)$ that

$$\begin{aligned}
&\mathbf{P}(L(d_1(a)) < t, L(\infty) > t)\\
&= (1 - \overline{\Pi}(\infty)/\overline{\Pi}(a)) \int_0^t ds\, \overline{\Pi}(a) \exp\{-s\overline{\Pi}(a)\} \exp\{-(t-s)\overline{\Pi}(\infty)\}\\
&= (1 - \exp\{-t(\overline{\Pi}(a) - \overline{\Pi}(\infty))\}) \exp\{-t\overline{\Pi}(\infty)\}.
\end{aligned}$$

Hence $\Pi(a, \infty) = \overline{\Pi}(a) - \overline{\Pi}(\infty)$, and finally we check as in (i) that the drift coefficient of S is d.

(iii) The identity for the Laplace exponent follows from the Lévy-Khintchine formula for a (possibly killed) subordinator by an integration by parts. □

Theorem 8 emphasizes the important connection between local times of Markov processes and subordinators. In particular, the results of Chapter III have interesting applications to Markov processes and their local times; here are two typical examples. First, the local time corresponds to the first-passage process of the inverse local time, and information on its asymptotic behaviour at $0+$ and at ∞ follows from the results in section III.4. Second, the closure of the range of the inverse local time coincides with the closure of the zero set of M (by Proposition 7). Hence Theorem III.15 provides a formula for the Hausdorff dimension of the zero set of a Markov process. We refer to Exercise 5 for another example of such an application. It is also worth mentioning that any subordinator can be thought of as the inverse local time of some Markov process, see Exercise 2.

4. Excursion measure and excursion process

We are now concerned with the excursions of M away from 0, that is the pieces of path of the type $(M_{g+t}, 0 \le t < d - g)$ corresponding to each excursion interval (g, d). In particular, lifetimes of excursions coincide with lengths of excursion intervals. To start with, we introduce a sigma-finite measure which is meant to describe the distribution of the excursions of M in the same way as $\overline{\Pi}$ is related to the length of the excursion intervals.

For every $a > 0$, denote by $\mathscr{E}^{(a)}$ the set of excursions with lifetime $\zeta > a$, that is

$$\mathscr{E}^{(a)} = \{\omega \in \Omega : \zeta > a \text{ and } \omega(t) \ne 0 \text{ for all } 0 < t < \zeta\},$$

and by $\mathscr{E} = \bigcup_{a>0} \mathscr{E}^{(a)}$ the set of excursions. These sets are endowed with the topology induced by Skorohod's topology. For each $a > 0$ with $\overline{\Pi}(a) > 0$, denote by $n(\cdot \mid \zeta > a)$ the probability measure on $\mathscr{E}^{(a)}$ corresponding to the law of M on the first excursion interval with length $\ell > a$, that is the law of the process $(M_{g_1(a)+t}, 0 \le t < \ell_1(a))$ under $\mathbf{P}$

(recall that the assumption $\overline{\Pi}(a) > 0$ ensures that $g_1(a) < \infty$ a.s.). This probability is called the law of the excursions of M with lifetime $\zeta > a$.

Lemma 9 *Pick any $a > 0$ with $\overline{\Pi}(a) > 0$. For any $b \in (0, a)$ and measurable event $\Lambda \subseteq \mathscr{E}^{(a)}$, we have*

$$\overline{\Pi}(a)n(\Lambda \mid \zeta > a) = \overline{\Pi}(b)n(\Lambda \mid \zeta > b).$$

Proof Recall from Lemma 2 that

$$\overline{\Pi}(a)/\overline{\Pi}(b) = \mathbf{P}(N_b(g_1(a)) = 0)$$

is the probability that the first excursion interval with length $\ell > b$ actually has length $\ell > a$. We thus see that the probability measure

$$(\overline{\Pi}(b)/\overline{\Pi}(a))\, n(\cdot, \zeta > a \mid \zeta > b)$$

is simply the law of the first excursion with lifetime $\zeta > a$ conditioned on $N_b(g_1(a)) = 0$. But again according to Lemma 2, the first excursion with lifetime $\zeta > a$ is independent of $N_b(g_1(a))$, in particular the preceding conditional law is the same as the unconditioned, viz. $n(\cdot \mid \zeta > a)$. □

Lemma 9 shows that there is a unique measure n on $\mathscr{E} = \bigcup_{a>0} \mathscr{E}^{(a)}$, called the *excursion measure* of M, such that

$$n(\Lambda) = \overline{\Pi}(a)n(\Lambda \mid \zeta > a) \quad \text{for every } \Lambda \subseteq \mathscr{E}^{(a)}.$$

In particular, $n(\zeta > a) = \overline{\Pi}(a)$ and the law of the excursions of M with lifetime $\zeta > a$ can be thought of as n conditioned on $\zeta > a$ (this justifies the notation *a posteriori*). An important consequence is that the excursion measure has the simple Markov property. More precisely, it is easy to check that for every $a > 0$, the random time

$$g_1(a) + a = \inf\{t \geq a : M_s \neq 0 \text{ for all } s \in [t - a, t]\}$$

is a $(\mathscr{G}_t)$-stopping time. The Markov property for M and the very definition of the excursion measure then imply that under n, conditionally on $\{\epsilon(a) = x, a < \zeta\}$ (where ϵ denotes the generic excursion and ζ its lifetime), the shifted process $(\epsilon(t + a), 0 \leq t < \zeta - a)$ is independent of $(\epsilon(t), 0 \leq t \leq a)$ and is distributed as $(M_t, 0 \leq t < R)$ under $\mathbf{P}_x$. In other words, it has the law of the Markov process started at b and killed at its first return to the origin. We also point out that the latter is a 'nice' Markov process (the argument is similar to that in Proposition II.4). Finally we mention that our definition of the excursion measure actually depends on the arbitrary constant c. But changing the constant would only affect the excursion measure by a constant multiplicative factor, and again this is not relevant to our purposes.

We pointed out in section 3 that the excursion intervals are precisely the open intervals of the type $(L^{-1}(t-), L^{-1}(t))$ for $L^{-1}(t-) < L^{-1}(t)$. This suggests the use of the inverse local time as a new time scale to analyse the excursions of M. To this end, we introduce the so-called *excursion process* of M, $e = (e_t, t \geq 0)$ which takes values in the space of excursions with an additional isolated point Υ, $\mathscr{E} \cup \{\Upsilon\}$, and is given by

$$e_t = \left(M_{s+L^{-1}(t-)}, 0 \leq s < L^{-1}(t) - L^{-1}(t-)\right) \text{ if } L^{-1}(t-) < L^{-1}(t) \quad (6)$$

and $e_t = \Upsilon$ otherwise. The key result is the following description of the excursion process, essentially due to Itô (1970). Recall from section O.5 the definition of a stopped Poisson point process.

Theorem 10 (i) *If* 0 *is recurrent, then* $(e_t, t \geq 0)$ *is a Poisson point process with characteristic measure* n.
(ii) *If* 0 *is transient, then* $(e_t, t \leq L(\infty))$ *is a Poisson point process with characteristic measure* n*, stopped at the first point in* $\mathscr{E}^{(\infty)}$*, the space of excursions with infinite lifetime.*

Proof The argument is essentially the same as for Theorem 8, except that we now deal with excursions themselves rather than merely the lengths of excursions intervals. We present a detailed proof of (i) and leave (ii) to the reader.

So assume that 0 is recurrent. Since for each $t \geq 0$, $L^{-1}(t)$ is a stopping time, we may introduce the sigma-field $\mathscr{H}_t = \mathscr{G}_{L^{-1}(t)}$, and we see that $(\mathscr{H}_t)$ is a filtration. All that is needed is to check that for every $\varepsilon > 0$ and measurable set $B \subset \mathscr{E}^{(\varepsilon)}$, the counting processes $N_t^B = \text{card}\{0 < s \leq t : e_s \in B\}$ $(t \geq 0)$ is an $(\mathscr{H}_t)$-Poisson process with intensity $n(B)$. Indeed, if $B_1, \cdots, B_k$ are disjoint measurable sets, then their respective counting processes never jump simultaneously and therefore will be independent. For every $s, t \geq 0$, $N_{t+s}^B - N_t^B$ is the number of excursions of M in B which were completed during the time interval $(L^{-1}(t), L^{-1}(t+s)]$. Introduce the shifted process $\widetilde{M} = (\widetilde{M}_u = M_{L^{-1}(t)+u}, u \geq 0)$, recall that according to the Markov property and the fact that $L^{-1}(t)$ is a zero of M (by Proposition 7(iii)), $\widetilde{M}$ is independent of $\mathscr{H}_t$ and has the same law as M. Denote its local time by $\widetilde{L}$ and inverse local time by $\widetilde{L}^{-1}$. The additivity of L implies that for every $u \geq 0$, $\widetilde{L}^{-1}(u) = L^{-1}(t+u) - L^{-1}(t)$, and therefore $N_{t+s}^B - N_t^B = \widetilde{N}_s^B$ is the number of excursions of $\widetilde{M}$ in B which were completed during the time interval $(0, \widetilde{L}^{-1}(s)]$. As a consequence $N_{t+s}^B - N_t^B$ has the same law as N_s^B and is independent of $\mathscr{H}_t$. This shows that N^B is an $(\mathscr{H}_t)$-Poisson process and hence e is a Poisson point process.

Let ν be the characteristic measure of e. We see from Proposition O.2 that for every $u > 0$, the conditional law $\nu(\cdot \mid \mathscr{E}^{(u)})$ is the law of the excursions with lifetime $\zeta > u$, that is

$$\nu(\cdot, \zeta > u)/\nu(\zeta > u) = n(\cdot \mid \zeta > u) = n(\cdot, \zeta > u)/n(\zeta > u)$$

On the other hand, the local time evaluated on the first excursion interval with length $\ell > u$, $L(d_1(u))$, is the instant of the first point of e in $\mathscr{E}^{(u)}$, and we know from Lemma 3 that $L(d_1(u))$ has an exponential distribution with parameter $\overline{\Pi}(u)$. Hence $\nu(\zeta > u) = \overline{\Pi}(u)$ and the measures ν and n coincide on $\mathscr{E}^{(u)}$. Since $\bigcup_{u>0} \mathscr{E}^{(u)} = \mathscr{E}$, the proof is now complete. □

Theorem 10 is interesting not only because it gives a nice description of the excursions of M, but more significantly, because many explicit calculations can now be carried out using formulas for Poisson point processes (see section O.5). To this end, it is most convenient to rephrase the compensation formula in the framework of excursion theory. For every left-end point $g < \infty$ of an excursion interval, denote by $\epsilon_g = (M_{g+t}, 0 \le t < d - g)$ the excursion starting at time g. Consider a measurable function $F : \mathbb{R}_+ \times \Omega' \times \mathscr{E} \to [0, \infty)$ such that for every $\epsilon \in \mathscr{E}$, the process $t \to F_t(\epsilon) = F(t, \omega', \epsilon)$ is left-continuous and adapted. We now claim:

Corollary 11 (Compensation formula in excursion theory) *We have*

$$\mathbf{E}\left(\sum_g F_g(\epsilon_g)\right) = \mathbf{E}\left(\int_0^\infty dL(s)\left(\int_{\mathscr{E}} F_s(\epsilon) n(d\epsilon)\right)\right),$$

where the sum in the left-hand side is taken over all the left-end points of excursion intervals.

Proof Because the left-end points of excursion intervals are precisely of the type $L^{-1}(t-)$ with t such that $L^{-1}(t) - L^{-1}(t-) > 0$, we have

$$\sum_g F_g(\epsilon_g) = \sum_t \mathbf{1}_{\{t \le L(\infty)\}} F_{L^{-1}(t-)}(e_t),$$

with the convention $F_s(\Upsilon) = 0$ for every s. The process $t \to F_{L^{-1}(t-)}$ is left-continuous and adapted to the filtration $(\mathscr{H}_t)$ of the excursion process, and thus is predictable. On the other hand, $L(\infty)$ is an $(\mathscr{H}_t)$-stopping time, because $L(\infty)$ is the instant of the first point taking values in the set of excursions with infinite lifetime, see Proposition O.2. Thus $s \to \mathbf{1}_{\{t \le L(\infty)\}}$ is a predictable process as well. The application of the compensation formula for Poisson point processes is legitimate and we get the desired result. □

Of course, by a monotone class theorem, the compensation formula extends to the case when F is only predictable.

Calculations in excursion theory are often a bit puzzling for beginners, mostly because one has to work with two different time scales: the natural time scale for the Markov process M and the scale of the inverse local time for the excursion process. Let us give a detailed presentation of two typical examples of applications.

In the first example, we intend to specify the so-called *entrance law* of the excursion measure, that is to calculate $n(\epsilon(t) \in \cdot, t < \zeta)$. For this purpose, we consider a Borel function $f : \mathbb{R}^d \cup \{\partial\} \to [0,\infty)$ with $f(0) = f(\partial) = 0$. Our basic data are the q-resolvent of f,

$$\mathbf{E}\left(\int_0^\infty \exp\{-qt\} f(M_t)dt\right),$$

and the Laplace exponent of the inverse local time process, Φ. Making use of the assumption that $f(0) = 0$, we can express the q-resolvent of f as

$$\mathbf{E}\left(\sum \int_g^d \exp\{-qt\} f(M_t)dt\right)$$
$$= \mathbf{E}\left(\sum_g \exp\{-qg\} \int_0^\ell \exp\{-qt\} f(\epsilon_g(t))dt\right),$$

where the sum is taken over all the left-end points g of excursion intervals, $\ell = d - g$ and ϵ_g denotes the excursion starting at time g. An application of the compensation formula (Corollary 11) enables us to rewrite the foregoing quantity as

$$\mathbf{E}\left(\int_0^\infty e^{-qs} dL(s)\right) n\left(\int_0^\zeta \exp\{-qt\} f(\epsilon(t))dt\right),$$

where ϵ stands for the generic excursion and ζ for its lifetime. On the other hand, a straightforward calculation based on Theorem 8 shows that the first factor in the product equals

$$\mathbf{E}\left(\int_0^{L(\infty)} ds \exp\{-qL^{-1}(s-)\}\right) = 1/\Phi(q).$$

In conclusion, we have obtained the following formula that characterizes the entrance law:

$$n\left(\int_0^\zeta \exp\{-qt\} f(\epsilon(t))dt\right) = \Phi(q)\mathbf{E}\left(\int_0^\infty \exp\{-qt\} f(M_t)dt\right). \quad (7)$$

The second example relies on the exponential formula of section O.5. For simplicity, we shall assume that 0 is recurrent. We consider a locally bounded Borel function $f : \mathbb{R}^d \to [0,\infty)$ and the increasing process

$$A(t) = \int_0^t f(M_s)ds \qquad (t \geq 0).$$

This is an additive functional of M in the sense that an identity similar to (4) holds for A, and the very same argument as in the proof of Theorem 8 shows that the process $A \circ L^{-1} = S$ obtained after time-changing A by the inverse local time L^{-1} is a subordinator. We aim at specifying its drift coefficient and Lévy measure. To this end, we first re-express S, splitting the time interval $[0, L^{-1}(t)]$ into excursion intervals and the closure of the zero set of M, and using the additivity of A. We get

$$S(t) = A(L^{-1}(t)) = f(0)\int_0^{L^{-1}(t)} \mathbf{1}_{\{s\in\overline{\mathcal{Z}}\}}ds + \sum_{0\le r\le t}\int_{L^{-1}(r-)}^{L^{-1}(r)} f(M_s)ds.$$

Next, we use Corollary 6 and rewrite this quantity in terms of the excursion process and the inverse local time, as

$$t\mathrm{d}f(0) + \sum_{0\le r\le t}\int_0^{\zeta} f(e_r(s))ds,$$

where we denoted by $\zeta = \zeta(e_r)$ and $e_r(s)$ the lifetime and the value at time s of the excursion e_r, respectively. Then we deduce from Theorem 10 and the exponential formula of section O.5 that for every $\lambda > 0$,

$$\mathbf{E}\left(\exp\{-\lambda S(1)\}\right) = \exp\left\{-\lambda \mathrm{d}f(0) - n\left(\int_0^{\zeta}(1-\exp\{-\lambda f(\epsilon(s))\})ds\right)\right\}$$

This shows that the drift coefficient of S is $\mathrm{d}f(0)$, and its Lévy measure can be expressed in terms of the occupation measure under n as

$$n\left(\int_0^{\zeta} \mathbf{1}_{\{f(\epsilon(s))\in\cdot\}}ds\right).$$

5. The cases of holding points and of irregular points

The most important features that we obtained in the last three sections are the connections between the excursions of a nice Markov process away from a regular instantaneous point on the one hand, and subordinators and Poisson point processes on the other hand. Our purpose here is to show that similar connections also exist when the origin is either irregular or a holding point. This case is much easier to analyse than the preceding, so we will merely depict the situation and leave the elementary arguments to the reader.

First, assume that 0 is a holding point and consider the sequence of successive exits from/returns to 0, $R_0 < S_1 < R_1 \cdots$, where $R_0 = 0$, $R_n = \inf\{t > S_n : M_t = 0\}$, $S_{n+1} = \inf\{t > R_n : M_t \neq 0\}$. The Markov property implies that $M_{S_1} \neq 0$, and that S_1 has an exponential distribution and is independent of the first excursion $(M_{S_1+t}, 0 \le t < R_1 - S_1)$. On the event $\{R_1 < \infty\}$, we have $M_{R_1} = 0$ and the Markov property enables us to iterate the argument. We can thus express the zero set of M as $\mathcal{Z} = [R_0, S_1) \cup [R_1, S_2) \cup \cdots$, and there exists an obvious continuous additive functional (in the sense of (4)) which increases exactly on $\overline{\mathcal{Z}}$, namely

$$\int_0^t \mathbf{1}_{\{M_s=0\}}ds \qquad (t \ge 0).$$

We call a local time of M at 0 any process $L = (L(t), t \geq 0)$ such that

$$\mathrm{d}L(t) = \int_0^t \mathbf{1}_{\{M_s=0\}} ds \qquad (t \geq 0)$$

for some constant $\mathrm{d} > 0$. The right-continuous inverse L^{-1} of L is then continuous except at $L(S_i)$ $(i = 1, \cdots)$. It can be immediately checked that the analogue of Proposition 7 holds and that L^{-1} is a subordinator with drift coefficient d (killed at a certain rate in the transient case). Moreover, the excursion process defined by (6) is a Poisson point process (stopped at the instant of the first point in the set of paths with infinite length in the transient case). Finally, the characteristic measure of this Poisson point process is simply proportional to the law of the first excursion of M, $(M_{S_1+t}, 0 \leq t < R_1 - S_1)$.

Then suppose that 0 is irregular, and consider the sequence $(R_n, n \in \mathbb{N})$ of successive return times to 0, defined by $R_0 = 0$, $R_{n+1} = \inf\{t > R_n : M_t = 0\}$. We know by hypothesis that $R_1 \in (0, \infty]$ a.s. On the event $\{R_1 < \infty\}$, we can apply the Markov property at R_1. We see that conditionally on $\{R_1 < \infty\}$, the shifted process $(M_{R_1+t}, t \geq 0)$ is independent of the first excursion $(M_t, 0 \leq t \leq R_1)$ and has the law $\mathbf{P}$. This argument can be repeated verbatim for $n = 1, 2, \cdots$, and we see in particular that $(R_n, n \in \mathbb{N})$ is an increasing random walk, killed at some independent geometric variable in the transient case. The most natural thing to do is then to define the local time at some instant $t \geq 0$ as the number of returns to the origin before t. The problem with this definition is that one gets a process taking integer values and its inverse $(R_n, n \in \mathbb{N})$ is a process in discrete time. It would be more convenient to have the same relation between excursions and subordinators even in the irregular case, and this incites us to use the following artifice. Since time substitution by an independent Poisson process transforms a random walk into a Lévy process (actually, a compound Poisson process), we introduce a sequence of independent exponential variables with the same parameter, $\tau_0, \tau_1, \cdots$, which is also independent of the Markov process M. We call the local time of M at 0 the process $L = (L(t), t \geq 0)$ given by

$$L(t) = \sum_{i=0}^{n(t)} \tau_i \,, \; n(t) = \max\{i : R_i < t\}.$$

Clearly L increases exactly on $\{t \geq 0 : M_t = 0\}$, but it is not adapted to the filtration $(\mathscr{G}_t)$ and is only right-continuous. Discontinuity is not really annoying, and to circumvent the first problem, we simply replace $(\mathscr{G}_t)$ by $(\mathscr{G}'_t)$, with $\mathscr{G}'_t = \mathscr{G}_t \vee \sigma(L(s), 0 \leq s \leq t)$. Then the analogue of Proposition 7 holds, the right-continuous inverse L^{-1} is a subordinator (killed at a certain rate in the transient case), and the excursion process defined by (6) is a Poisson point process (stopped at the instant of the first point in the set of paths with infinite length in the transient case). Finally, the characteristic measure of this Poisson point process is again simply proportional to the law of the first excursion of M, $(M_t, 0 \leq t < R_1)$.

6. Exercises

1. *(Local time and first entrance times)* Let B be a Borel set in $\mathbb{R}^d$, and suppose that the first passage time in B, $T_B = \inf\{t \geq 0 : M_t \in B\}$, is a positive $(\mathcal{G}_t)$-stopping time. Prove that $L(T_B)$ has an exponential distribution and specify its parameter in terms of the excursion measure.
2. *(A Markov process with given local time)* The purpose of this exercise is to show that any subordinator with infinite Lévy measure can be thought of as the inverse local time of a nice recurrent Markov process. Assume that the canonical process X is a subordinator with infinite Lévy measure under the probability measure $\mathbb{P}$, and denote the first passage time above the level $t \geq 0$ by $T(t) = \inf\{s \geq 0 : X_s > t\}$. Recall that $T(t)$ is an $(\mathcal{F}_t)$-stopping time, and consider the sigma-field $\mathcal{G}_t = \mathcal{F}_{T(t)}$.
(a) Check that the filtration $(\mathcal{G}_t)$ is right-continuous.
(b) Prove that the process $M = (M_t, t \geq 0)$ given by $M_t = \inf\{X_s - t : X_s > t\}$ is a nice Markov process under $(\mathbb{P}_x, (\mathcal{G}_t))$ and that 0 is a regular and recurrent point.
(c) Check that the zero set of M coincides with the range of X a.s., and deduce that the inverse local time of M is proportional to X.
3. *(Reconstruction from the excursions)* Give an explicit formula for M_t in terms of the drift coefficient of the inverse local time d and the excursion process $e = (e_s, s \geq 0)$. [Hint: first express the inverse local time as a function of d and e.]
4. *(Decomposition at the last passage at the origin)* Assume that 0 is transient and denote the last passage time at the origin by $g = g(\infty) = \sup\{t \geq 0 : M_t = 0\}$. Prove that the shifted process $(M_{g+t}, t \geq 0)$ is independent of $(M_t, t < g)$ and has the law $n(\cdot \mid \zeta = \infty)$. [Hint: use either the compensation formula or Exercise 3.]
5. *(An arcsine law for Markov processes)* For every $t \geq 0$, let $\gamma(t) = \sup\{s < t : M_s = 0\}$ denote the last zero of M before time t. Making use of the arcsine laws for subordinators of section III.3, prove the following limit theorem. The random variables $t^{-1}\gamma(t)$ converge in distribution as $t \to \infty$ if and only if $\mathbf{E}(\gamma(t))/t$ converges to some $\alpha \in [0, 1]$. In that case, the limit distribution is the generalized arcsine law with parameter α. [Hint: Use Proposition 7(iii).]

7. Comments

The concept and construction of the local time were introduced by Lévy (1965) in the Brownian case. The present approach is essentially an adaptation of that of Lévy, the idea of using the convergence of martingales is due to Itô and McKean (1965). One should note that the Markov property only appears through the regenerative property of the zero set $\mathscr{Z}$, and the same arguments apply to the construction of a local time on a general regenerative closed set. See Hoffmann-Jørgensen (1969) and Meyer (1970). Nonetheless, the gain of generality is illusory, see Maisonneuve (1971) and Horowitz (1972). We refer to Hawkes (1977) and Fitzsimmons, Fristedt and Maisonneuve (1985) for further developments, in particular in connection with potential theory for subordinators. Different pathwise constructions on the local time are proposed by Fristedt and Pruitt (1971), who identify the Stieltjes measure dL with some deterministic Hausdorff measure restricted to $\mathscr{Z}$, Greenwood and Pitman (1980-a) and Fristedt and Taylor (1983).

The connection between local times and subordinators was noted by Blumenthal and Getoor (1968), and hints at excursion theory via Theorem I.1. The formalization of excursion theory is due to Itô (1970), but the main ideas may have been already known to Lévy; see Chung (1988). The theory has been further developed by Maisonneuve (1975), see Blumenthal (1992) and Dellacherie, Maisonneuve and Meyer (1992) for a recent exposition.

V

Local Times of a Lévy Process

The occupation measure of a Lévy process on any finite time interval is a.s. absolutely continuous with respect to the Lebesgue measure whenever single points have positive capacity. When moreover the origin is a regular point, the density defines a family of increasing continuous processes, which are local times in the sense of chapter IV. A wide class of continuous additive functionals can then be expressed as integrals of local times. We present a method for investigating the distribution of the Hilbert transform of local times evaluated at certain random times, and point out a remarkable connection with the standard Cauchy process. Finally, we study the regularity in the time and space variables of the local times; in particular, a sufficient condition for the joint continuity a.s. is given.

1. Occupation measure and local times

Applying the results of section IV.2 and the spatial homogeneity, one can define the local time of a real-valued Lévy process at any given point of the line, provided that the origin is regular for itself. The first purpose of this section is to show that one can construct these local times simultaneously at all points. The construction is based on the analysis of a natural random measure supported by the closed range of the Lévy process, called the occupation measure.

Definition (Occupation measure) *For every $t > 0$, the occupation measure on the time interval $[0, t]$ is the measure μ_t given for every measurable function $f : \mathbb{R} \to [0, \infty)$ by*

$$\int_{\mathbb{R}} f(x)\mu_t(dx) = \int_0^t f(X_s)\, ds.$$

Our first concern is to investigate the Lebesgue decomposition of the occupation measure as the sum of its absolutely continuous and its singular components. The following result is a remarkable complement to Theorem II.16; it claims that the occupation measure is singular a.s. if single points are essentially polar, and absolutely continuous a.s. otherwise. (The reason why we concentrate on the real-valued case is then clear: single points are always essentially polar in dimension $d \geq 2$; see Corollary II.17.)

Theorem 1 *The following assertions are equivalent:*

(i) $$\int_{-\infty}^{\infty} \Re\left(\frac{1}{1+\Psi(\xi)}\right) d\xi < \infty ;$$

(ii) *For every $t \geq 0$, μ_t is absolutely continuous with respect to the Lebesgue measure, with density in $L^2(dx \otimes d\mathbb{P})$.*

Moreover, if (i) fails, then μ_t is singular for every $t > 0$, a.s.

Proof We first check that (i)$\Rightarrow$(ii). The argument is a variation of that used by Berman (1969) to investigate the occupation measure of Gaussian processes, it relies on Plancherel's theorem.

Introduce the measure μ by

$$\int_{\mathbb{R}} f(x)\mu(dx) = \int_0^{\infty} e^{-s} f(X_s)ds = \int_0^{\infty} dt e^{-t} \int_{\mathbb{R}} f(x)\mu_t(dx).$$

In particular, μ_t is absolutely continuous with respect to μ with densities bounded from above by e^t. We thus simply have to check

$$\int_{-\infty}^{\infty} \mathbb{E}\left(|\mathcal{F}\mu(\xi)|^2\right) d\xi < \infty, \tag{1}$$

where $\mathcal{F}\mu(\xi) = \int e^{ix\xi}\mu(dx)$ denotes the Fourier transform of μ. Indeed in that case, a combination of Fubini's theorem and Plancherel's theorem shows that almost surely, μ has a density with respect to the Lebesgue measure which is square-integrable.

From the very definition of μ, one has

$$\begin{aligned}\mathbb{E}\left(|\mathscr{F}\mu(\xi)|^2\right) &= \mathbb{E}\left[\mathscr{F}\mu(\xi)\mathscr{F}\mu(-\xi)\right] \\ &= \mathbb{E}\left[\left(\int_0^\infty e^{-s}\exp\{i\xi X_s\}ds\right)\left(\int_0^\infty e^{-t}\exp\{-i\xi X_t\}dt\right)\right] \\ &= \mathbb{E}\left(\int_0^\infty ds\int_0^\infty dt\, e^{-(s+t)}\exp\{i\xi(X_s - X_t)\}\right).\end{aligned}$$

The integrand is plainly integrable, Fubini's theorem applies and we can re-express the last displayed quantity as

$$2\Re\left(\int_0^\infty ds\, e^{-s}\int_s^\infty dt\, e^{-t}\exp\{-(t-s)\Psi(\xi)\}\right) = \Re\left(\frac{1}{1+\Psi(\xi)}\right).$$

In conclusion,

$$\int_{-\infty}^{\infty}\mathbb{E}\left(|\mathscr{F}\mu(\xi)|^2\right)d\xi = \int_{-\infty}^{\infty}\Re\left(\frac{1}{1+\Psi(\xi)}\right)d\xi < \infty$$

and (1) is proven.

Finally, suppose that (i) fails. Then by Theorem II.16, single points are essentially polar, that is

$$0 = \int_{-\infty}^{\infty}\mathbb{P}_x(\exists t > 0 : X_t = 0)\,dx = \int_{-\infty}^{\infty}\mathbb{P}(\exists t > 0 : X_t = x)\,dx,$$

and an application of Fubini's theorem shows that the Lebesgue measure of the range $\{X_t : t \geq 0\}$ is zero a.s. Because the occupation measure is supported by the closure of the range (which differs from the range by at most countably many points since the paths are right-continuous with left limits), this implies that μ_t is singular a.s. for every t. □

Suppose now that the assertions of Theorem 1 hold. Lebesgue's differentiation theorem enables us to define a particular version of the density of the occupation measure, called the *local times*.

Definition (Local times) *For every $t \geq 0$ and $x \in \mathbb{R}$, the quantity*

$$\limsup_{\varepsilon\to 0+}\frac{1}{2\varepsilon}\int_0^t \mathbf{1}_{\{|X_s - x| < \varepsilon\}}ds$$

is denoted by $L(x,t)$ and called the local time at level x and time t.

There is nothing special about the 'limsup', replacing it e.g. by 'liminf' would have no effect on the sequel. Noting that ε can be restricted to rational numbers in the definition, we see that $(L(x,t), x \in \mathbb{R})$ serves as an $(\mathscr{F}_t)$-measurable version of the density of μ_t. Observe that for every $x \in \mathbb{R}$, $L(x,\cdot)$ is an increasing process, which increases only when $X = x$.

In particular, the mapping $x \to L(x,t)$ has compact support for every $t > 0$, a.s.

The identity

$$\int_0^t f(X_s)ds = \int_{-\infty}^{\infty} f(x)L(x,t)dx \tag{2}$$

holds for all measurable bounded functions $f \geq 0$, a.s., and is known as the *occupation density formula* . We point out that (2) holds a.s. simultaneously for all $t \geq 0$. Specifically, it holds a.s., simultaneously for all rational numbers $q \geq 0$. Then consider for every $x \in \mathbb{R}$ and $t > 0$, $\underline{L}(x,t) = \sup\{L(x,q) : q \in \mathbb{Q} \text{ and } q < t\}$ and $\overline{L}(x,t) = \inf\{L(x,q) : q \in \mathbb{Q} \text{ and } q > t\}$. Plainly, $\underline{L}(x,t) \leq L(x,t) \leq \overline{L}(x,t)$, and, considering a monotone sequence of rational numbers converging to t and applying the monotone convergence theorem,

$$\int_0^t f(X_s)ds = \int_{-\infty}^{\infty} f(x)\underline{L}(x,t)dx = \int_{-\infty}^{\infty} f(x)\overline{L}(x,t)dx.$$

Our assertion follows.

The analysis of local times when 0 is irregular for itself is fairly easy. Typically, X then has bounded variation (cf. Bretagnolle (1971), see also Corollary II.20), and if $\mathrm{d} \neq 0$ denotes the drift coefficient, then the following identity between random measures holds for every $x \neq 0$:

$$dL(x,\cdot) = \mathrm{d}^{-1} \sum_{t:X_t=x} \delta_t,$$

where δ_t stands for the Dirac point mass at t. We refer to Fitzsimmons and Port (1991) for an extensive account, and henceforth concentrate on the regular case.

We will assume throughout the rest of this chapter that assertion (i) of Theorem 1 holds and that 0 *is regular for itself. We will implicitly work with the probability measure* $\mathbb{P} = \mathbb{P}_0$.

The main part of this section is devoted to the study of the smoothness of local times in the time variable, that is when the level has been fixed. We first observe that one can replace 'limsup' by 'lim' in the definition of local times.

Proposition 2 *For every* $x \in \mathbb{R}$

$$\lim_{\varepsilon \to 0+} \frac{1}{2\varepsilon} \int_0^t \mathbf{1}_{\{|X_s - x| < \varepsilon\}} ds = L(x,t)$$

uniformly on compact intervals of time, in $L^2(\mathbb{P})$. *As a consequence, the process* $L(x,\cdot)$ *is continuous a.s.*

Proof Let τ be an independent random time, with an exponential distribution of parameter 1. It follows from (1) that local times evaluated at time τ form a random variable in $L^2(dy \otimes d\mathbb{P})$. By the maximal inequality of Hardy and Littlewood (cf. Theorem 1 on page 5 in Stein (1970)), we have

$$\int_{-\infty}^{\infty} \sup_{\varepsilon>0} \left|(2\varepsilon)^{-1} \int_{y-\varepsilon}^{y+\varepsilon} L(v,\tau)dv\right|^2 dy \leq 4 \int_{-\infty}^{\infty} L(y,\tau)^2 dy.$$

The theorem of dominated convergence thus applies and we deduce from the occupation density formula (2) and Fubini's theorem that for a.e. $y \in \mathbb{R}$, the following convergence holds in $L^2(\mathbb{P})$:

$$\lim_{\varepsilon\to0+} \frac{1}{2\varepsilon} \int_0^\tau \mathbf{1}_{\{|X_s-y|<\varepsilon\}} ds = \lim_{\varepsilon\to0+} \frac{1}{2\varepsilon} \int_{y-\varepsilon}^{y+\varepsilon} L(v,\tau)dv = L(y,\tau). \tag{3}$$

Now pick any y for which (3) is fulfilled and consider for every $\varepsilon > 0$ the martingale

$$M_t^\varepsilon = \mathbb{E}\left(\frac{1}{2\varepsilon} \int_0^\tau \mathbf{1}_{\{|X_s-y|<\varepsilon\}} ds \mid \mathscr{F}'_t\right), \qquad t \geq 0,$$

where $\mathscr{F}'_t = \mathscr{F}_t \vee \sigma(t \wedge \tau)$. Plainly, for $t < \tau$,

$$\int_0^\tau \mathbf{1}_{\{|X_s-y|<\varepsilon\}} ds = \int_0^t \mathbf{1}_{\{|X_s-y|<\varepsilon\}} ds + \int_0^{\tau-t} \mathbf{1}_{\{|X_{s+t}-y|<\varepsilon\}} ds,$$

so by the Markov property and the lack of memory of the exponential law, we have a.s.

$$M_t^\varepsilon = \frac{1}{2\varepsilon} \int_0^{t\wedge\tau} \mathbf{1}_{\{|X_s-y|<\varepsilon\}} ds + \mathbf{1}_{\{t<\tau\}} f_\varepsilon(X_t), \tag{4}$$

where

$$f_\varepsilon(x) = \mathbb{E}_x\left(\frac{1}{2\varepsilon} \int_0^\tau \mathbf{1}_{\{|X_s-y|<\varepsilon\}} ds\right), \qquad x \in \mathbb{R}.$$

Applying Fubini's theorem, we can re-express $f_\varepsilon(x)$ as

$$\frac{1}{2\varepsilon} \int_0^\infty e^{-t} \mathbb{P}_x(|X_t - y| < \varepsilon)dt = \frac{1}{2\varepsilon} \int_{y-\varepsilon}^{y+\varepsilon} u^1(v-x)dv,$$

where u^1 is the continuous version of the 1-resolvent density (recall Theorem II.19). We stress that $u^1(x)$ tends to 0 as x goes to infinity (by Corollary II.18), so u^1 is uniformly continuous.

Note that f_ε is continuous, so the right-hand side of (4) is the right-continuous version of M^ε. By (3) and Doob's maximal inequality, M_t^ε converges as $\varepsilon \to 0+$, uniformly on $t \in [0,\infty)$, in $L^2(\mathbb{P})$. On the other hand, because u^1 is continuous, $f_\varepsilon(x)$ converges uniformly on $x \in \mathbb{R}$, and then $f_\varepsilon(X_t)$ converges uniformly on $t \in [0,\infty)$, in $L^2(\mathbb{P})$.

In conclusion,

$$\frac{1}{2\varepsilon}\int_0^t \mathbf{1}_{\{|X_s-y|<\varepsilon\}}ds$$

converges uniformly on $t \in [0, \tau)$, in $L^2(\mathbb{P})$.

We thus have proved the proposition for every $x = y$ for which (3) holds. An application of the Markov property at the first passage time at such a y entails the proposition for $x = 0$. The case of a general x follows similarly, applying now the Markov property at the first passage time at $-x$ (recall from Corollary II.18 that these passage times are finite with positive probability). □

Proposition 2 yields a simple connection between the local times and the resolvent densities. Recall from Theorem II.19 that for every $q > 0$, one always considers the canonical resolvent density u^q that is continuous.

Lemma 3 *For every $x \in \mathbb{R}$, let $dL(x,t)$ denote the Stieltjes measure of the increasing function $L(x,\cdot)$. We have*

$$\mathbb{E}\left(\int_0^\infty e^{-qt}dL(x,t)\right) = u^q(x).$$

Proof One deduces from the proof of Proposition 2 that for

$$\mathbb{E}\left(\int_0^\infty e^{-qt}dL(x,t)\right) = \lim_{\varepsilon\to 0+}\frac{1}{2\varepsilon}\int_0^\infty e^{-qt}\mathbb{P}(|X_t - x| < \varepsilon)dt.$$

This entails our claim. □

More generally, one can of course replace the left-hand side of the displayed formula in Lemma 3 by

$$\mathbb{E}_y\left(\int_0^\infty e^{-qt}dL(x+y,t)\right)$$

for any $y \in \mathbb{R}$.

We next present an important application of Proposition 2, which justifies *a posteriori* the denomination 'local time' for the occupation densities.

Proposition 4 (i) *The process $L(0,\cdot) = (L(0,t), t \geq 0)$ fulfils the conditions of Proposition IV.5. As a consequence, it coincides up to a constant factor with the local time L which has been defined in section IV.2.*

(ii) *The inverse local time*

$$\sigma(\cdot) = L^{-1}(0,\cdot) = \inf\{t \geq 0 : L(0,t) > \cdot\}$$

is a subordinator, killed at an independent exponential time if X is transient. Its Laplace exponent is given for $\lambda > 0$ by

$$\mathbb{E}\left(\exp\{-\lambda\sigma(t)\}\right) = \exp\{-t/u^{\lambda}(0)\}, \qquad t > 0,$$

where we used the convention $e^{-\infty} = 0$.

Proof (i) We know from Proposition 2 that $L(0,\cdot)$ is a continuous increasing adapted process that increases only when $X = 0$. Moreover $L(0,\cdot)$ appears as the limit of processes which satisfy equation (IV.4). It follows that $L(0,\cdot)$ satisfies equation (IV.4) as well, and Proposition IV.5 applies.

(ii) We know from (i) and Theorem IV.8 that the inverse local time σ satisfies

$$\mathbb{E}\left(\exp\{-\lambda\sigma(t)\}\right) = \exp -t\Phi(\lambda),$$

where Φ is its Laplace exponent. Taking $q = \lambda$ in Lemma 3 and then changing variables yields $\Phi(\lambda) = 1/u^{\lambda}(0)$. □

In the sequel, it will be convenient to denote the Laplace exponent by $1/\kappa$, that is, according to Theorem II.19(iii),

$$\kappa(q) = u^q(0) = \frac{1}{2\pi}\int_{-\infty}^{\infty} \Re\left(\frac{1}{q + \Psi(\xi)}\right) d\xi, \qquad q > 0.$$

An immediate and useful consequence of Proposition 4 is that if $\tau(q)$ stands for an independent exponential time with parameter $q > 0$, then $L(0, \tau(q))$ follows an exponential distribution with parameter $1/\kappa(q)$. Because $\kappa(q)$ tends to 0 as $q \to \infty$, we deduce that for every fixed $t > 0$, all the exponential moments of $L(0, t)$ are finite.

We conclude this section with a correspondence between finite measures and certain additive functionals which is a companion to the correspondence between finite measures and integrable excessive functions developed in section I.3. A *positive continuous additive functional* is a mapping on Ω, $A : \omega \to A(\omega)$, taking values in the space of continuous increasing functions $a : [0, \infty) \to [0, \infty)$, such that $\omega \to A_t(\omega)$ is $\mathcal{F}_t$-adapted for every $t \geq 0$, and which satisfies the *additive property* : For every $x \in \mathbb{R}$ and $s \geq 0$,

$$A(\omega)_{T+s} = A(\omega)_T + A(\theta_T \circ \omega)_s \qquad \mathbb{P}_x\text{-a.s.}$$

where $T < \infty$ stands for a generic stopping time and θ. for the shift operator. For instance, if $f : \mathbb{R} \to [0, \infty)$ is a locally bounded measurable function, then

$$A^f(\omega)_t = \int_0^t f(X_s(\omega))ds, \qquad t \geq 0, \tag{5}$$

is a positive continuous additive functional. In the sequel, the dependence on the path ω will be omitted.

For every $q > 0$, the q-resolvent of A is the function $U^q A$ given by

$$U^q A(x) = \mathbb{E}_x\left(\int_0^\infty e^{-qt} dA_t\right), \qquad x \in \mathbb{R}.$$

For instance, when $A = A^f$ is given by (5), one has simply $U^q A = U^q f$. We say that A is *integrable* if $\int U^q A(x)dx < \infty$; it is easy to check that this property does not depend on $q > 0$. The following representation of an integrable positive continuous additive functional in terms of local times extends the occupation density formula.

Theorem 5 *Let A be an integrable positive continuous additive functional.*

(i) *For every $q > 0$, $U^q A$ is a q-excessive function.*

(ii) *There exists a unique finite measure ν such that for every $q > 0$ and $x \in \mathbb{R}$,*

$$U^q A(x) = U^q \nu(x) = \int_{\mathbb{R}} u^q(y - x)\nu(dy).$$

(iii) *For every $x \in \mathbb{R}$, $\mathbb{P}_x$-a.s.,*

$$A_t = \int_{-\infty}^{\infty} L(a,t)\nu(da), \qquad \textit{for all } t \geq 0.$$

Conversely, for every finite measure ν, $\int_{-\infty}^{\infty} L(a,\cdot)\nu(da)$ is an integrable positive continuous additive functional.

Proof (i) The additive property and the Markov property readily imply the *resolvent equation*, viz.

$$U^q A - U^r A + (q - r)U^r(U^q A) = 0,$$

for every $q, r > 0$. As a consequence, $(r - q)U^r(U^q A) \leq U^q A$ for every $r > 0$. On the other hand, $\mathbb{P}_x$-a.s. for every $x \in \mathbb{R}$, $\int_0^\infty e^{-rt} dA_t$ decreases to 0 as $r \to \infty$, and it follows that $\lim_{r\to\infty} U^r A(x) = 0$. Thanks to the resolvent equation, this shows that $(r - q)U^r(U^q A)$ converges pointwise to $U^q A$ as $r \to \infty$.

(ii) This follows from (i) and Proposition I.14. That ν does not depend on q will be clear from (iii).

(iii) Introduce

$$A_t^\nu = \int_{\mathbb{R}} L(a,t)\nu(da), \qquad t \geq 0.$$

A straightforward calculation shows that if τ is an independent exponential time with parameter 1, then

$$\mathbb{E}_x(A_\tau^\nu) = \int_{\mathbb{R}} u^1(y-x)\nu(dy) < \infty,$$

so that A_t^ν is finite for every $t \geq 0$, $\mathbb{P}_x$-a.s. for every x. It is then immediately checkable that A^ν has continuous paths, using the monotone convergence theorem and the property that for every $a \in \mathbb{R}$, $L(a,\cdot)$ is continuous and increasing a.s. On the other hand, it is plain that A^ν is a positive additive functional.

To prove that $A = A^\nu$, we first use the additivity to write

$$\int_0^\infty e^{-s}d(A_s - A_s^\nu) = \int_0^t e^{-s}d(A_s - A_s^\nu) + e^{-t}\int_0^\infty e^{-s}d(A_s(\theta_t) - A_s^\nu(\theta_t)).$$

Applying the Markov property, we see that the martingale

$$M_t = \mathbb{E}\left(\int_0^\infty e^{-t}d(A_t - A_t^\nu) \mid \mathcal{F}_t\right)$$

can be expressed as

$$M_t = \int_0^t e^{-s}d(A_s - A_s^\nu) + e^{-t}\mathbb{E}_{X_t}\left(\int_0^\infty e^{-s}d(A_s - A_s^\nu)\right).$$

But for every $x \in \mathbb{R}$,

$$\mathbb{E}_x\left(\int_0^\infty e^{-s}d(A_s - A_s^\nu)\right) = U^q A(x) - U^q A^\nu(x) = 0,$$

and in conclusion $M_t = \int_0^t e^{-s}d(A_s - A_s^\nu)$. Thus M is a continuous martingale of bounded variation, so it must be constant (see e.g. Proposition IV.1.2 in Revuz and Yor (1994)). This entails that $A = A^\nu$. The last assertion of Theorem 5 is obvious from the foregoing. □

It is now an easy matter to extend by localization the representation of integrable positive continuous additive functionals as integrals of local times, to certain non-integrable additive functionals. Specifically, suppose that A is a positive continuous additive functional with

$$U^q A(y) = \mathbb{E}_y\left(\int_0^\infty e^{-qt}dA_t\right) < \infty$$

for some $y \in \mathbb{R}$ and $q > 0$. Applying the Markov property at the first passage time at $x \in \mathbb{R}$ and the additivity, we get the following inequality:

For every $x \in \mathbb{R}$

$$U^q A(y) \geq \mathbb{E}_y \left(\int_{T_x}^{\infty} e^{-qt} dA_t \right) = \mathbb{E}_y(\exp\{-qT_x\}) U^q A(x).$$

Recall from Corollary II.18 that $\mathbb{E}_y(\exp\{-qT_x\}) = u^q(y-x)/u^q(0) > 0$, this implies that $U^q A$ is finite everywhere and bounded on every compact set. Next, introduce for every $k > 0$

$$A_t^{(k)} = \int_0^t \mathbf{1}_{\{|X_s|<k\}} dA_s, \qquad t \geq 0,$$

so $A^{(k)}$ is again a positive continuous additive functional. Moreover, the Markov property entails that, in the obvious notation, for every $x \in \mathbb{R}$,

$$U^q A^{(k)}(x) \leq \mathbb{E}_x(\exp\{-qT_{[-k,k]}\}) \sup\{U^q A(y) : -k \leq y \leq k\}.$$

On the other hand, we know from section II.2 that

$$\int_{-\infty}^{\infty} \mathbb{E}_x(\exp\{-qT_{[-k,k]}\}) dx = C^q([-k,k])/q < \infty$$

where C^q denotes the q-capacity. Hence $A^{(k)}$ is integrable and we deduce from Theorem 5 that there exists a finite measure $\nu^{(k)}$ such that

$$A_t^{(k)} = \int_{-\infty}^{\infty} L(a,t) \nu^{(k)}(da), \qquad \text{for all } t \geq 0.$$

It is immediately checkable that for every $k < k'$, $\nu^{(k)} = \mathbf{1}_{[-k,k]} \nu^{(k')}$, and hence there is a Radon measure ν such that $\nu^{(k)} = \mathbf{1}_{[-k,k]} \nu$ for every k. We conclude that

$$A_t = \int_{-\infty}^{\infty} L(a,t) \nu(da), \qquad \text{for all } t \geq 0,$$

and for every $x \in \mathbb{R}$,

$$U^q A(\cdot) = U^q \nu(\cdot) = \int_{\mathbb{R}} u^q(y - \cdot) \nu(dy).$$

2. Hilbert transform of local times

The Hilbert transform $\mathscr{H}$ is an operator obtained as the limit as $\varepsilon \to 0+$ of the convolution by the bounded function

$$h^{\varepsilon}(y) = \frac{1}{\pi y} \mathbf{1}_{\{|y|>\varepsilon\}}, \qquad y \in \mathbb{R}.$$

When $f \in L^2(\mathbb{R})$, the limit $\mathscr{H} f = \lim_{\varepsilon \to 0+} h^{\varepsilon} \star f$ exists almost everywhere and in the L^2-sense as well. Because the Fourier transform of h^{ε},

$$\mathscr{F} h^{\varepsilon}(\xi) = \frac{2i}{\pi} \int_{\varepsilon}^{\infty} \frac{\cos \varepsilon\xi - \cos v\xi}{v^2} dv, \qquad \xi \in \mathbb{R},$$

converges pointwise as ε tends to $0+$ to $i \operatorname{sgn}(\xi) = i \left(\mathbf{1}_{\{\xi>0\}} - \mathbf{1}_{\{\xi<0\}} \right)$ and has $|\mathscr{F}h^{\varepsilon}(\xi)| \leq 3$, one obtains the following simple formula:

$$\mathscr{F}(\mathscr{H}f)(\xi) = \mathrm{i}\ \mathrm{sgn}(\xi)\mathscr{F}f(\xi).$$

By Plancherel's theorem, the Hilbert transform thus induces an involutive isomorphism of $L^2(\mathbb{R}, dx)$, which lies at the heart of the theory of singular integrals and has many important applications to harmonic analysis. We refer to chapters II-III in Stein (1970) for a detailed account.

Our purpose here is to investigate the Hilbert transform of the local times,

$$\mathscr{H}L(\cdot,t)(x) = \lim_{\varepsilon\to 0+} \frac{1}{\pi}\int_{-\infty}^{\infty} \mathbf{1}_{\{|a-x|>\varepsilon\}} \frac{L(a,t)}{x-a}\,da$$

whenever the limit exists. Observe that the occupation density formula (2) then allows us to re-express this quantity as

$$\mathscr{H}L(\cdot,t)(x) = \lim_{\varepsilon\to 0+} \frac{1}{\pi}\int_0^t \mathbf{1}_{\{|X_s-x|>\varepsilon\}} \frac{ds}{x-X_s}.$$

We shall impose throughout this section the following condition on the characteristic exponent of the Lévy process: For some (and then all) $q > 0$

$$\int_{-\infty}^{\infty} \frac{d\xi}{|q+\Psi(\xi)|} < \infty. \tag{6}$$

By Fourier inversion, this implies that the q-resolvent kernel has a continuous density u^q, in particular 0 is regular for itself (see Corollary II.20).

To start with, we observe that for every fixed x, $\mathscr{H}L(\cdot,t)(x)$ exists for all $t \geq 0$, a.s. More precisely, we have the following.

Proposition 6 *For every $x \in \mathbb{R}$,*

$$H^{\varepsilon}(x,t) = \frac{1}{\pi}\int_0^t \mathbf{1}_{\{\varepsilon<|X_s-x|\}} \frac{ds}{X_s - x}$$

converges as $\varepsilon \to 0+$ uniformly on compact intervals of time, in $L^2(\mathbb{P})$. The limit is denoted by $H(x,t)$ and $H(x,t) = -\mathscr{H}L(\cdot,t)(x)$ a.s. As a consequence, $H(x,\cdot)$ has continuous paths a.s.

Proof The argument is essentially a variation on that in Proposition 2. Let τ be an independent exponential time with parameter 1. Because a.s. the function $L(\cdot,\tau)$ is square-integrable and has compact support, we deduce from Theorem 4 of chapter II in Stein (1970) that a.s.,

$$\lim_{\varepsilon\to 0+} \frac{1}{\pi}\int_{\varepsilon<|v-y|} \frac{L(v,\tau)}{v-y}\,dv = -\mathscr{H}L(\cdot,\tau)(y) \qquad \text{for a.e. } y \in \mathbb{R}$$

and for some numerical constant $c > 0$

$$\int_{-\infty}^{\infty} \sup_{\varepsilon>0} \left| \int_{\varepsilon<|v-y|} \frac{L(v,\tau)}{v-y} dv \right|^2 dy \leq c \int_{-\infty}^{\infty} L(v,\tau)^2 dv.$$

Recall from (1) that the expectation of the right-hand side is finite, so by Fubini's theorem, for a.e. y,

$$\mathbb{E}\left(\sup_{\varepsilon>0} \left| \int_{\varepsilon<|v-y|} \frac{L(v,\tau)}{v-y} dv \right|^2 \right) < \infty.$$

We now deduce by dominated convergence and the occupation density formula (2) that for a.e. y, the following convergence holds in $L^2(\mathbb{P})$:

$$\lim_{\varepsilon \to 0+} H^\varepsilon(y,\tau) = -\mathscr{H}L(\cdot,\tau)(y). \tag{7}$$

Next, pick any y for which (7) is fulfilled and consider for every $\varepsilon > 0$ the martingale $M_t^\varepsilon = \mathbb{E}\left(H^\varepsilon(y,\tau) \mid \mathscr{F}'_t\right)$, $t \geq 0$, where $\mathscr{F}'_t = \mathscr{F}_t \vee \sigma(t \wedge \tau)$. Note that for $t < \tau$,

$$H^\varepsilon(y,\tau) = H^\varepsilon(y,t) + \frac{1}{\pi} \int_0^{\tau-t} \mathbf{1}_{\{\varepsilon<|X_{t+s}-y|\}} \frac{ds}{X_{t+s}-y}.$$

So by the Markov property and the lack of memory of the exponential law,

$$M_t^\varepsilon = H^\varepsilon(y, t\wedge\tau) + \mathbf{1}_{\{t<\tau\}} f_\varepsilon(X_t), \tag{8}$$

where

$$f_\varepsilon(x) = \mathbb{E}_x\left(H^\varepsilon(y,\tau)\right) = \frac{1}{\pi} \int_{\varepsilon<|v-y|} u^1(v-x) \frac{dv}{v-y}. \tag{9}$$

Note that f_ε is continuous, so the right-hand side of (8) is the right-continuous version of M^ε. By (7) and Doob's maximal inequality, M_t^ε converges as $\varepsilon \to 0+$, uniformly on $t \in [0,\infty)$, in $L^2(\mathbb{P})$.

On the other hand, we deduce from (9) that the Fourier transform of f_ε is

$$\mathscr{F} f_\varepsilon(\xi) = \frac{\exp\{i\xi(x+y)\}}{1+\Psi(\xi)} \mathscr{F} h^\varepsilon(\xi).$$

Recall that $|\mathscr{F} h^\varepsilon| \leq 3$, so $|\mathscr{F} f_\varepsilon(\xi)| \leq 3|1+\Psi(\xi)|^{-1}$, and by (6), the theorem of dominated convergence applies. By Fourier inversion, this implies that f_ε converges uniformly as $\varepsilon \to 0+$ and thus $f_\varepsilon(X_t)$ converges uniformly on $t \in [0,\infty)$, in $L^2(\mathbb{P})$.

This entails the proposition for every $x = y$ for which (7) holds. An application of the Markov property at the first passage time at such a

y establishes the proposition for $x = 0$. The case of a general x follows similarly, applying the Markov property at the first passage time at $-x$. $\square$

Plainly, the spatial homogeneity enables us to reduce the study of $H(x, \cdot)$ to that of $H(0, \cdot)$ and we thus concentrate on the latter. In the Brownian case, Biane and Yor (1987) have obtained remarkable formulas for the distribution of $H(0, \cdot)$ evaluated at certain random times such as an independent exponential time or the first instant when the local time at level 0 reaches a given value. Their arguments rely on excursion theory and properties of Bessel processes. Fitzsimmons and Getoor (1992) extended these formulas to a wide class of symmetric Lévy processes, using the method of moments and a combinatorial lemma on Euler numbers. The main purpose of this section is to present a direct approach to the determination of the distribution of $H(0, \cdot)$, which applies for general Lévy processes subject to (6).

To facilitate the reading, we first outline an informal argument for the computation of the characteristic function of $H(0, \tau)$, where $\tau = \tau(q)$ is an independent exponential time with parameter $q > 0$. For a fixed $\lambda \in \mathbb{R}$, we should like to evaluate

$$V(x) = \mathbb{E}_x\left(\int_0^\infty e^{-qt} \exp\{i\lambda H(0, t)\} dt\right)$$

at $x = 0$. For this, we use the celebrated Feynman-Kac formula (see e.g. section III.39 in Williams (1979)) and get the identity

$$-i\lambda U^q(hV) = q^{-1} - V,$$

where $h(x) = (\pi x)^{-1}$. We then apply the Fourier transform and deduce from Proposition I.9 the equation

$$\frac{-i\lambda}{q + \Psi(-\xi)} \mathcal{F}(hV)(\xi) = 2\pi q^{-1} \delta_0(\xi) - \mathcal{F}V(\xi) \tag{10}$$

where δ_0 denotes the Dirac point mass at 0. Because $\mathcal{F}h(\xi) = i\,\mathrm{sgn}(\xi)$, we can re-express $\mathcal{F}(hV)(\xi)$ as

$$\begin{aligned}\frac{1}{2\pi}\mathcal{F}h \star \mathcal{F}V(\xi) &= \frac{i}{2\pi}\left[2\int_{-\infty}^{\xi} \mathcal{F}V(t)dt - \int_{-\infty}^{\infty} \mathcal{F}V(t)dt\right] \\ &= \frac{i}{\pi}\int_{-\infty}^{\xi} \mathcal{F}V(t)dt - iV(0).\end{aligned}$$

Thus (10) yields

$$\frac{\lambda}{\pi(q + \Psi(-\xi))}(F(\xi) - \pi V(0)) = \frac{2\pi}{q}\delta_0(\xi) - F'(\xi) \tag{11}$$

where $F(\xi) = \int_{-\infty}^{\xi} \mathcal{F}V(t)dt$ and $F'(\xi) = \mathcal{F}V(\xi)$. We can think of (11) as a linear differential equation and its generic solution is

$$F(\xi) = c\exp\{-2\lambda\kappa_\xi(q)\} + \pi V(0) + \frac{2\pi}{q}\exp\{-2\lambda\kappa_0(q)\}\,\mathbf{1}_{\{\xi>0\}}$$

where c is a complex number and

$$\kappa_\xi(q) = \frac{1}{2\pi}\int_{-\infty}^{\xi}\frac{dt}{q+\Psi(-t)}, \qquad \xi \in \mathbb{R}.$$

We have on the one hand $F(-\infty) = 0$, which forces $c = -\pi V(0)$. On the other hand, we have by Fourier inversion $F(\infty) = 2\pi V(0)$, and the latter gives

$$V(0) = 2q^{-1}\frac{\exp\{-2\lambda\kappa_0(q)\}}{1+\exp\{-2\lambda\kappa(q)\}},$$

because $\kappa_\infty(q) = \kappa(q)$. In conclusion, we obtain

$$\mathbb{E}(\exp\{i\lambda H(0,\tau(q))\}) = 2\frac{\exp\{-2\lambda\kappa_0(q)\}}{1+\exp\{-2\lambda\kappa(q)\}}.$$

We emphasize that this reasoning is only heuristic, in particular we have been very liberal in deriving and solving (11). The main problem stems from the fact that the application of the Feynman-Kac formula to the Hilbert transform is illegal, because $\mathscr{H}$ is not a Radon measure. Nonetheless, it is possible to make the argument rigorous, working with approximations. We will not give any detail and will rather focus on distributions related to the inverse of the local time, $\sigma = L^{-1}(0,\cdot)$, which require similar reasoning but are perhaps more delicate to tackle. Recall the convention $e^{-\infty} = 0$ and that the Laplace exponent of σ is $1/\kappa$.

Theorem 7 *For every $q > 0$ and $\lambda \in \mathbb{R}$, one has*

$$\mathbb{E}(\exp\{-q\sigma(t) + i\lambda H(0,\sigma(t))\}) = \exp\{-t\lambda\coth\lambda\kappa(q)\}, \qquad t > 0$$

(by convention, the indicator function of $\sigma(t) < \infty$ is implicit in the left-hand side).

A standard application of the Markov property and the additivity property of $H(0,\cdot)$ show that the pair $(\sigma, H(0,\sigma))$ is a Lévy process taking values in $[0,\infty)\times\mathbb{R}$ and killed at an independent exponential time if X is transient (see the end of section IV.4 for a detailed argument), and Theorem 7 thus specifies its characteristic exponent. Now recall that X is recurrent if and only if $\lim_{q\to 0+}\kappa(q) = \infty$. We see that in that case, the right-hand side in the formula of Theorem 7 converges to $\exp\{-t|\lambda|\}$ as q goes to $0+$. The latter is the characteristic function of the symmetric Cauchy variable with parameter t. In other words, one observes the following striking identity.

Corollary 8 *When X is recurrent, $H(0, \sigma(\cdot))$ is a standard Cauchy process.*

The rest of this section is devoted to the proof of Theorem 7, which we now outline.

For every fixed $q > 0$ and $\lambda \in \mathbb{R}$, we will establish the existence of a complex number c and a bounded continuous function $g : \mathbb{R} \to \mathbb{R}$ such that $g(X_\cdot)\exp\{cL(0,\cdot) - qt + i\lambda H(0,\cdot)\}$ is a martingale. An application of the optional sampling theorem and the fact that $X_{\sigma(t)} = 0$ whenever $\sigma(t) < \infty$ then show that

$$\mathbb{E}(\exp\{-q\sigma(t) + i\lambda H(0, \sigma(t))\}) = \exp\{-tc\}.$$

To specify c, we will use the Feynman-Kac formula to get an equation involving c, the unknown function g, the resolvent operators of the Lévy process and the Hilbert transform $\mathscr{H}$. The key point is that when one applies the Fourier transform $\mathscr{F}$ to this equation, one obtains a *linear differential equation* which we are then able to solve explicitly. This is due to the remarkable identity $\mathscr{F}\mathscr{H} = \mathrm{i}\,\mathrm{sgn}$. Again, this sketch of the approach is only heuristic because the Hilbert transform is not a Radon measure, and this forces us to work with the approximations $H^{\varepsilon}(0,\cdot)$ rather than with $H(0,\cdot)$ directly.

To start with, we establish the Feynman-Kac formula. Consider a complex Radon measure ν on the real line and denote its absolute variation by $|\nu|$. Suppose the r-resolvent of $|\nu|$ is finite for some fixed $r > 0$, that is

$$\int_{\mathbb{R}} u^r(x-y)|\nu(dy)| < \infty \qquad \text{for every } x \in \mathbb{R}, \tag{12}$$

and introduce the continuous additive functional with bounded variation,

$$A_t^\nu = \int_{\mathbb{R}} L(x,t)\nu(dx), \qquad t \geq 0.$$

Then for every complex-valued bounded measurable function g, we can define the r-resolvent of the complex Radon measure $g\nu$ by

$$U^r(g\nu)(x) = \int_{-\infty}^{\infty} u^r(y-x)g(y)\nu(dy) = \mathbb{E}_x\left(\int_0^\infty e^{-rt}g(X_t)dA_t^\nu\right).$$

We then make the additional assumption that

$$\sup_{x\in\mathbb{R}} \mathbb{E}_x\left(\int_0^\infty e^{-rt}|\exp\{-A_t^\nu\}|\,dt\right) < \infty \tag{13}$$

and put for every bounded measurable function f

$$V_\nu^r f(x) = \mathbb{E}_x\left(\int_0^\infty e^{-rt}f(X_t)\exp\{-A_t^\nu\}\,dt\right), \qquad x \in \mathbb{R}.$$

Note that $V^r_\nu f$ is a bounded measurable function, so the r-resolvent of the complex Radon measure $\left(V^r_\nu f\right)\nu$ is well defined.

Feynman-Kac formula $\qquad U^r((V^r_\nu f)\nu) = U^r f - V^r_\nu f.$

Proof Applying the Markov property and the additivity of A^ν, we have

$$\begin{aligned}
&U^r((V^r_\nu f)\nu)(x)\\
&= \mathbb{E}_x\left(\int_0^\infty e^{-rt}\mathbb{E}_{X_t}\left(\int_0^\infty e^{-rs} f(X_s)\exp\{-A^\nu_s\}ds\right) dA^\nu_t\right)\\
&= \mathbb{E}_x\left(\int_0^\infty e^{-rt}\int_t^\infty e^{-r(s-t)} f(X_s)\exp\{A^\nu_t - A^\nu_s\}ds\, dA^\nu_t\right)\\
&= \mathbb{E}_x\left(\int_0^\infty e^{-rs} f(X_s)\int_0^s \exp\{A^\nu_t - A^\nu_s\}\, dA^\nu_t\, ds\right)\\
&= \mathbb{E}_x\left(\int_0^\infty e^{-rs} f(X_s)\left(1-\exp\{-A^\nu_s\}\right) ds\right).
\end{aligned}$$

We can rewrite the last displayed quantity as $U^r f(x) - V^r_\nu f(x)$, and the Feynman-Kac formula is proven. □

We then turn our attention to distributions related to the first passage time at 0, T_0. Fix $q > 0$ and $\lambda \in \mathbb{R}$, and introduce for every $x \in \mathbb{R}$

$$g^\varepsilon(x) = \mathbb{E}_x\left(\exp\{-qT_0 + i\lambda H^\varepsilon(0, T_0)\}\right).$$

Recall that we are using the convention $e^{-\infty} = 0$, so the characteristic function of $T_0 < \infty$ is implicit in this definition. Note that $g^\varepsilon(0) = 1$ and that

$$|g^\varepsilon(x)| \le \mathbb{E}_x(\exp -qT_0) = u^q(-x)/u^q(0) \qquad \text{[by Corollary II.18]},$$

in particular g^ε is integrable.

We aim at applying the Feynman-Kac formula to the function $f = g^\varepsilon$ and the Radon measure

$$\nu^\varepsilon(dx) = -c^\varepsilon\delta_0(dx) + q\,dx - i\lambda h^\varepsilon(x)dx,$$

where c^ε is some complex number that will be chosen in the forthcoming Lemma 9, $\delta_0(dx)$ the Dirac point mass at 0, and $h^\varepsilon(x) = \mathbf{1}_{\{\varepsilon<|x|\}}(\pi x)^{-1}$. Plainly, (12) holds. The additive functional corresponding to ν^ε is

$$A^\varepsilon_t = -c^\varepsilon L(0,t) + qt - i\lambda H^\varepsilon(0,t).$$

In particular, $|\exp\{-A^\varepsilon_t\}| \le \exp\{|c^\varepsilon|L(0,t)\}$ and we deduce from Proposition 4 that (13) holds provided that r is large enough.

The Feynman-Kac formula would be much simpler if we had $V^r_{\nu^\varepsilon} f = kf$

for some constant k, and the purpose of the next lemma is to show that c^{ε} may be chosen in such a way.

Lemma 9 *For every $\varepsilon > 0$, there exists a complex number c^{ε} such that the process*

$$g^{\varepsilon}(X_t)\exp\{c^{\varepsilon}L(0,t) - qt + \mathrm{i}\lambda H^{\varepsilon}(0,t)\}, \qquad t \geq 0,$$

is a $\mathbb{P}_x$-martingale for every $x \in \mathbb{R}$.

Proof A standard application of the Markov property and the additivity show that the time-changed process $(A^{\varepsilon} \circ \sigma(t), t < L(0,\infty))$, is a Lévy process, killed at the independent exponential time $L(0,\infty)$ whenever X is transient. This guarantees the existence of a (unique) complex number c^{ε} such that for every $t > 0$

$$\mathbb{E}\left(\exp\{-A^{\varepsilon}_{\sigma(t)}\}\right) = \exp\{-c^{\varepsilon}t\}, \tag{14}$$

with the convention $e^{-\infty} = 0$. As a consequence, $\exp\{c^{\varepsilon}t - A^{\varepsilon}_{\sigma(t)}\}$, $t \geq 0$, is a $\left(\mathbb{P}_0, \mathscr{F}_{\sigma(t)}\right)$-martingale. We claim that this implies that for every $x \in \mathbb{R}$,

$$\mathbb{E}_x(g^{\varepsilon}(X_t)\exp\{-A^{\varepsilon}_t\}) = g^{\varepsilon}(x). \tag{15}$$

Specifically, suppose first that $x = 0$ and recall from Proposition IV.7 that $\sigma(L(0,t)) = t + T_0 \circ \theta_t$ is the first return to 0 after time t. We then deduce from the Markov property, the additivity and the very definition of g^{ε} that

$$\begin{aligned}\mathbb{E}(g^{\varepsilon}(X_t)\exp\{-A^{\varepsilon}_t\}) &= \mathbb{E}\left(\exp\{-A^{\varepsilon}_t\}\mathbb{E}_{X_t}(\exp\{-qT_0 + \mathrm{i}\lambda H^{\varepsilon}(0,T_0)\})\right)\\ &= \mathbb{E}(\exp\{-A^{\varepsilon}_t - A^{\varepsilon}_{T_0}(\theta_t)\})\\ &= \mathbb{E}(\exp\{-A^{\varepsilon}_{\sigma(L(0,t))}\}).\end{aligned}$$

On the other hand, the events $\{L(0,t) > s\}$ and $\{\sigma(s) \leq t\}$ coincide, so $L(0,t)$ is an $\mathscr{F}_{\sigma(\cdot)}$-stopping time, and recall that the exponential moments of $L(0,t)$ are finite, see the remark after Proposition 4. We can thus apply the optional sampling theorem to the martingale $\exp\{c^{\varepsilon}\cdot - A^{\varepsilon}_{\sigma(\cdot)}\}$, which entails (15) when $x = 0$. The general case follows from the Markov property applied at the first hitting time of 0.

Finally, we deduce from the Markov property and the additivity that for $t < t'$

$$\begin{aligned}\mathbb{E}(g^{\varepsilon}(X_{t'})\exp\{-A^{\varepsilon}_{t'}\} \mid \mathscr{F}_t) &= \exp\{-A^{\varepsilon}_t\}\mathbb{E}_{X_t}(g^{\varepsilon}(X_{t'-t})\exp\{-A^{\varepsilon}_{t'-t}\})\\ &= \exp\{-A^{\varepsilon}_t\}g^{\varepsilon}(X_t) \qquad \text{[by (15)]}.\end{aligned}$$

This establishes the lemma. □

We next combine the Feynman-Kac formula and the Fourier transform to investigate g^{ε}.

Lemma 10 *Provided that q is large enough, there exists a constant $k > 0$ such that for every $\varepsilon > 0$, the Fourier transform of g^ε satisfies the inequality*

$$|\mathscr{F}g^\varepsilon(\xi)| \leq \frac{k}{|q+\Psi(\xi)|}, \qquad \xi \in \mathbb{R}.$$

Proof We apply the Feynman-Kac formula for $f = g^\varepsilon$, and observe that $V^r_{\nu\varepsilon} g^\varepsilon = g^\varepsilon/r$, thanks to Lemma 9. We get $U^r(g^\varepsilon \nu^\varepsilon) = rU^r g^\varepsilon - g^\varepsilon$. Taking the Fourier transform (in the sense of tempered distributions, i.e. when these bounded functions are viewed as elements of the Schwartz space $\mathscr{S}'$) and using Proposition I.9, we obtain

$$\frac{\mathscr{F}(g^\varepsilon \nu^\varepsilon)(\xi)}{r+\Psi(-\xi)} = r\frac{\mathscr{F}g^\varepsilon(\xi)}{r+\Psi(-\xi))} - \mathscr{F}g^\varepsilon(\xi),$$

and after simplification $\mathscr{F}(g^\varepsilon \nu^\varepsilon)(\xi) = -\Psi(-\xi)\mathscr{F}g^\varepsilon(\xi)$. As $g^\varepsilon \nu$ is a complex measure with finite total variation, we note from (7) that $\mathscr{F}g^\varepsilon \in L^1(dx)$. Recall the definition of ν^ε, that $g^\varepsilon(0) = 1$ and the identity $\mathscr{F}(ab) = \frac{1}{2\pi}\mathscr{F}a \star \mathscr{F}b$ whenever a and b are square-integrable functions. We finally get

$$(\Psi(-\xi)+q)\mathscr{F}g^\varepsilon(\xi) = c^\varepsilon + \mathrm{i}\frac{\lambda}{2\pi}\mathscr{F}g^\varepsilon \star \mathscr{F}h^\varepsilon(\xi). \tag{16}$$

Because $|\mathscr{F}h^\varepsilon(\xi)| \leq 3$ for every ξ and ε,

$$|\mathscr{F}g^\varepsilon \star \mathscr{F}h^\varepsilon(\xi)| \leq 3\int_{-\infty}^{\infty} |\mathscr{F}g^\varepsilon(x)|\,dx, \tag{17}$$

and we deduce from (16) that

$$\int_{-\infty}^{\infty} |\mathscr{F}g^\varepsilon(\xi)|(\xi)\,d\xi \leq \left(|c^\varepsilon| + |\lambda|\int_{-\infty}^{\infty} |\mathscr{F}g^\varepsilon(\xi)|d\xi\right)\int_{-\infty}^{\infty}\frac{d\xi}{|q+\Psi(\xi)|}.$$

Hence, if q is such that

$$\int_{-\infty}^{\infty} |q+\Psi(\xi)|^{-1}d\xi < \frac{1}{2|\lambda|} \wedge 1$$

(which holds provided that q is large enough), then

$$\int_{-\infty}^{\infty} |\mathscr{F}g^\varepsilon(\xi)|\,d\xi \leq 2|c^\varepsilon|.$$

Note from Proposition 6 that c^ε (which is given by (14)) is bounded in ε, so feeding the foregoing inequality into (16) and (17) establishes our second assertion. □

We are finally able to prove Theorem 7.

Proof of Theorem 7 Suppose first that q is large enough. It follows from (14) and Proposition 6 that c^{ε} converges as ε tends to 0+ to a certain complex number, say c. Similarly, g^{ε} converges pointwise to some function g. Because $g^{\varepsilon}(x) \leq u^q(-x)/u^q(0)$, the theorem of dominated convergence applies and g^{ε} converges to g also in $L^1(\mathbb{R})$. This implies that $\mathscr{F}g^{\varepsilon}$ converges uniformly to $\mathscr{F}g$, and by Lemma 10, (6), and dominated convergence, in $L^1(\mathbb{R})$ as well. Recall that $\mathscr{F}h^{\varepsilon}(\xi)$ converges pointwise to i sgn(ξ) as ε tends to 0+, so taking the limit in (16) yields

$$(\Psi(-\xi)+q)\mathscr{F}g(\xi) = c - \frac{\lambda}{2\pi}\text{sgn} \star \mathscr{F}g(\xi).$$

Consider the indefinite integral of $\mathscr{F}g$, $G(\cdot) = \int_{-\infty}^{\cdot} \mathscr{F}g(\xi)d\xi$, so that (by Fourier inversion) $G(\infty) = 2\pi g(0) = 2\pi$ and $\text{sgn} \star \mathscr{F}g(\xi) = 2G(\xi) - 2\pi$. We thus observe that G satisfies the following linear differential equation:

$$(\Psi(-\xi)+q)G'(\xi) = (c+\lambda) - \frac{\lambda}{\pi}G(\xi).$$

The general solution is $G(\xi) = \pi(c+\lambda)\lambda^{-1} - c'\exp\{-2\lambda\kappa_{\xi}(q)\}$, with

$$\kappa_{\xi}(q) = \frac{1}{2\pi}\int_{-\infty}^{\xi}\frac{dx}{\Psi(-x)+q}.$$

Since $G(-\infty) = 0$, we must have $c' = \pi(c+\lambda)/\lambda$. Moreover, we know that $G(\infty) = 2\pi$, so $(c+\lambda)(1-\exp\{-2\lambda\kappa(q)\}) = 2\lambda$. Because $\kappa_{\infty}(q) = \kappa(q)$, this gives

$$c = \lambda \coth \lambda\kappa(q).$$

On the other hand, we deduce from Lemma 9 that $g(X_\cdot)\exp\{cL(0,\cdot) - q\cdot + \mathrm{i}\lambda H(0,\cdot)\}$ is a martingale. Applying Doob's optional sampling theorem at $\sigma(t) \wedge s$ gives

$$\begin{aligned}&\mathbb{E}(\exp\{ct - q\sigma(t) + \mathrm{i}\lambda H(0,\sigma(t))\}, \sigma(t) < s)\\ &= 1 - \mathbb{E}(g(X_s)\exp\{cL(0,s) - qs + \mathrm{i}\lambda H(0,s)\}, \sigma(t) \geq s),\end{aligned}$$

because $X_{\sigma(t)} = 0$ when $\sigma(t) < \infty$. Then let s tend to ∞ to get

$$\mathbb{E}(\exp\{-q\sigma(t) + \mathrm{i}\lambda H(0,\sigma(t))\}) = \exp\{-t\lambda \coth \lambda\kappa(q)\}.$$

This proves Theorem 7 provided that q is large enough. Both terms in the preceding identity are analytic in the variable $q \in (0,\infty)$, and the formula extends for arbitrary $q > 0$. □

3. Jointly continuous local times

We next turn our attention to the study of the smoothness of local times viewed as random functions on $\mathbb{R} \times [0,\infty)$. The approach relies crucially on metric entropy and majorizing measures, and is due to Barlow (1985) and Barlow and Hawkes (1985). We will assume throughout this section that X *is recurrent*. We emphasize that the main result (Theorem 15) holds in the transient case as well (it should be intuitively clear that the recurrence assumption is irrelevant, because the long-time behaviour of the Lévy process does not affect its local properties). Actually, it is easy to reduce the study of the transient case to the recurrent. For instance, a slight modification of the Lévy measure of X enables us to construct a centred Lévy process X' whose paths coincide with those of X up to an exponential time with arbitrarily small parameter, and X' is recurrent according to the test of Chung and Fuchs (see Exercise I.10). Because the arguments in the recurrent case are easier to follow in the framework of this text (at least at first reading), we shall only develop this part.

For every real number a, observe that the first passage time T_a is finite a.s. (since X is recurrent and single points have positive capacity) and that $L(0, T_a)$ has an exponential distribution. Indeed, for every $s, t \geq 0$, a standard application of the Markov property at time $\sigma(t)$ and the additivity property of the local time give

$$\begin{aligned}\mathbb{P}(L(0,T_a) > t+s) = \mathbb{P}(\sigma(t+s) < T_a) &= \mathbb{P}(\sigma(t) < T_a)\mathbb{P}(\sigma(s) < T_a) \\ &= \mathbb{P}(L(0,T_a) > t)\mathbb{P}(L(0,T_a) > s)\end{aligned}$$

which establishes our claim (see also Exercise IV.1). We denote its parameter by $1/h(a)$, that is

$$h(a) = \mathbb{E}(L(0, T_a)) .$$

We start with a pair of elementary lemmas.

Lemma 11 *The function h is symmetric, and more precisely, for every $a \in \mathbb{R}$,*

$$h(a) = \frac{1}{\pi}\int_{-\infty}^{\infty}(1-\cos\xi a)\Re\left(\frac{1}{\Psi(\xi)}\right)d\xi .$$

Proof Applying the Markov property at time T_a and using the addivity of the local time, we have for every $q > 0$,

$$\begin{aligned}&\mathbb{E}\left(\int_0^{\infty} e^{-qt}dL(0,t)\right)\\ &= \mathbb{E}\left(\int_0^{T_a} e^{-qt}dL(0,t)\right) + \mathbb{E}(e^{-qT_a})\mathbb{E}_a\left(\int_0^{\infty} e^{-qt}dL(0,t)\right).\end{aligned}$$

Then we deduce from Lemma 3 and Theorem II.19 that

$$\mathbb{E}\left(\int_0^{T_a} e^{-qt}dL(0,t)\right) = \left(u^q(0)^2 - u^q(a)u^q(-a)\right)/u^q(0). \quad (18)$$

On the one hand, the left term in (18) increases to $h(a)$ as q decreases to $0+$. On the other hand, because $T_a < \infty$ a.s., one deduces from Corollary II.18 that $u^q(a) \sim u^q(-a) \sim u^q(0)$ as $q \to 0+$ and it follows that

$$h(a) = \lim_{q \to 0+} (2u^q(0) - u^q(a) - u^q(-a)).$$

Recall now the formula (iii) in Theorem II.19:

$$2u^q(0) - u^q(a) - u^q(-a) = \frac{1}{\pi} \int_{-\infty}^{\infty} (1 - \cos \xi a) \Re \left(\frac{1}{q + \Psi(\xi)} \right) d\xi .$$

As $1/(1+\Psi)$ is the Fourier transform of an integrable function, namely the 1-resolvent density u^1, we know from the Riemann-Lebesgue theorem that $\lim_{|\xi| \to \infty} |\Psi(\xi)| = \infty$. Moreover, it is immediately deducible from the Lévy-Khintchine formula that $|\xi|^2 = O(\Re\Psi(\xi))$ as $|\xi| \to 0$. As a consequence, there is a constant $c > 0$ (which depends only on a and Ψ), such that the integrand in the last displayed formula is bounded by $c\Re\left(1/(1 + \Psi(\xi))\right)$, for all $\xi \in \mathbb{R}$ and $q \in (0, 1)$. The lemma now follows by dominated convergence. □

Recall that $\sigma_\cdot = \inf\{s \geq 0 : L(0, s) > \cdot\}$ denotes the inverse local time at level 0. The time-changed process $L(a, \sigma_\cdot)$ will play an important rôle in the sequel, thanks to its simple probabilistic structure.

Lemma 12 *For every $a \in \mathbb{R}$, $(L(a, \sigma_t), t \geq 0))$ is a subordinator. Its characteristic exponent is given by*

$$-\log \mathbb{E}(\exp\{-\lambda L(a, \sigma_1)\}) = \frac{\lambda}{\lambda h(a) + 1}, \qquad \lambda > 0.$$

Proof The additivity of the local time and the argument at the end of section IV.4 show that $L(a, \sigma_\cdot)$ is a subordinator. Because $L(a, \cdot)$ increases exactly when $X = a$, $L(a, \sigma_\cdot)$ stays at 0 until time $L(0, T_a)$ at which it performs its first jump. In particular, $L(a, \sigma_\cdot)$ has zero drift. Recall that $L(0, T_a)$ has an exponential distribution with parameter $1/h(a)$. Next, according to Proposition IV.7, $\sigma(L(0, T_a)) = \inf\{t > T_a : X_t = 0\}$ is the first return to 0 after time T_a. Denote this quantity by R. By the Markov property and the fact that $L(a, T_a) = 0$, we deduce that $L(a, R)$ has the same distribution as $L(0, T_{-a})$, that is an exponential distribution with parameter $1/h(-a) = 1/h(a)$. In conclusion, the Lévy measure of $L(a, \sigma_\cdot)$ is $h(a)^{-2} \exp\{-x/h(a)\} dx$, $x > 0$, and the formula for the Laplace exponent follows.

Lemma 12 provides a simple argument for the following crucial upper bound.

Lemma 13 *For every $a, b \in \mathbb{R}$ and $x, y > 0$*

$$\mathbb{P}\left(\exists s \le \sigma_y^b : L(b,s) - L(a,s) > x\right) \le \exp\left\{-\frac{x^2}{4yh(a-b)}\right\},$$

where $\sigma^b = L^{-1}(b,\cdot)$ is the inverse of the local time at level b.

Proof Using the Markov property at the first hitting time of b and the spatial homogeneity, we need only treat the case $b = 0$.

Consider the stopping time $T = \inf\{s \ge 0 : s - L(a, \sigma_s) > x\}$, so $\mathbb{P}(T \le y)$ is an alternative expression for the probability which appears in the statement. Observe that the subordinator $L(a, \sigma_\cdot)$ cannot jump at time T, so that $L(a, \sigma_T) = T - x$ provided that $T < \infty$. In particular, $\mathbb{P}(T \le y) = 0$ for $y < x$, and we now focus on the case $x \ge y$.

According to Lemma 12, for every $\lambda > 0$, the process

$$\exp\left\{-\lambda L(a, \sigma_s) + \frac{s\lambda}{\lambda h(a) + 1}\right\}, \qquad s \ge 0,$$

is a martingale. We deduce from Doob's optional sampling theorem applied at $T \wedge y$ that

$$\mathbb{E}\left(\exp\left\{\lambda(x - T) + \frac{T\lambda}{\lambda h(a) + 1}\right\}, T \le y\right) \le 1,$$

and then

$$\mathbb{P}(T \le y) \le \exp\left\{-\lambda x + \lambda y\left(1 - \frac{1}{\lambda h(a) + 1}\right)\right\}.$$

Evaluating the right-hand side for $\lambda = x/(2yh(a))$ establishes the desired upper bound. □

Our interest in Lemma 13 stems from the fact that it yields the Lipschitz property for the increments of the local times in a certain Orlicz space. This Lipschitz property is a most powerful tool for investigating the regularity of processes taking values in Banach spaces, see Ledoux and Talagrand (1991). Specifically, consider the space C_b of continuous bounded real-valued functions on $[0, \infty)$, equipped with the uniform norm $|\cdot|_u$. For every random variable Y taking values in C_b, we denote by

$$\|Y\|_\psi = \inf\{c > 0 : \mathbb{E}(\exp\{|Y|_u^2/c\}) - 1 \le 1\}$$

its Orlicz norm associated to the Young function $\psi(x) = \exp(x^2) - 1$.

Then for every $y > 0$ and $a \in \mathbb{R}$, we introduce the C_b-valued random variable $y_a = (y_a(t), t \geq 0)$, where

$$y_a(t) = y \wedge L(a,t), \qquad t \geq 0.$$

Corollary 14 *Put $d(a,b) = h(a-b)$ for every $a, b \in \mathbb{R}$. Then*

(i) *d is a metric on $\mathbb{R}$ equivalent to the Euclidean, i.e. the induced topology coincides with the Euclidean topology;*

(ii) *for every $y > 0$ and $a, b \in \mathbb{R}$, $\|y_a - y_b\|_{\psi} \leq 12yd(a,b)$.*

Proof (i) The symmetry is plain from Lemma 11, and we first turn our attention to the triangular inequality. We deduce from Lemma 12 that $\mathbb{E}(L(a,\sigma_1)) = 1$, hence the compensated subordinator $t - L(a,\sigma_t)$, $t \geq 0$, is a martingale in the time-changed filtration $(\mathscr{F}_{\sigma_t})$. Note also that the events $\{L(0,T_b) > t\}$ and $\{\sigma_t < T_b\}$ coincide, so $L(0,T_b)$ is an $(\mathscr{F}_{\sigma_t})$-stopping time. By monotone convergence and Doob's optional sampling theorem, we thus have

$$\begin{aligned} h(b) = \mathbb{E}(L(0,T_b)) &= \lim_{t\to\infty} \mathbb{E}(L(0,T_b) \wedge t) \\ &= \lim_{t\to\infty} \mathbb{E}\left(L(a,\sigma_{L(0,T_b)\wedge t})\right) \\ &= \mathbb{E}\left(L(a,\sigma_{L(0,T_b)})\right). \end{aligned} \tag{19}$$

Recall that $\sigma(L(0,T_b))$ is the first return to 0 after time T_b (by Proposition IV.7), and for short, denote this quantity by R.

Next, consider the Lévy process started at a, its first passage at 0, T_0, its first passage at b after T_0, T'_b, and finally its first return to 0 after T'_b, R'. Plainly, $T_b \leq R'$, and applying the Markov property at T_0 gives

$$\begin{aligned} h(b-a) = \mathbb{E}_a(L(a,T_b)) &\leq \mathbb{E}_a(L(a,R')) \\ &= \mathbb{E}_a(L(a,T_0)) + \mathbb{E}(L(a,R)) \\ &= h(-a) + h(b) \qquad \text{[by (19)]}. \end{aligned}$$

This entails the triangular inequality.

Recall that $L(0,T_a)$ has an exponential distribution, so $d(0,a) = \mathbb{E}(L(0,T_a))$ converges to 0 if and only if $L(0,T_a)$ converges to 0 in probability. Because $L(0,\cdot)$ is continuous and positive immediately after the origin of time a.s., the foregoing is equivalent to the convergence of T_a to 0 in probability. By Theorem II.19, this squares with the convergence of a to 0 in the Euclidean sense.

(ii) Fix a, b and y and observe the following. If $y \wedge L(b,t) - y \wedge L(a,t) \geq x$ for some $t \geq 0$, then the first instant for which the (in)equality occurs is necessarily bounded from above by $\sigma_y^b = \inf\{s \geq 0 : L(b,s) > y\}$. We can thus re-express the event $\{|y_b - y_a|_u > x\}$ as

$$\left\{\exists s \leq \sigma_y^b : L(b,s) - L(a,s) > x\right\} \cup \left\{\exists s \leq \sigma_y^a : L(a,s) - L(b,s) > x\right\}.$$

Applying Lemma 13, we deduce

$$\mathbb{P}\left(|y_b - y_a|_u > x\right) \leq 2\exp\left\{-\frac{x^2}{4yd(a,b)}\right\}.$$

This entails that for an arbitrary $c > 0$

$$\begin{aligned}
\mathbb{E}(\exp\{|y_b - y_a|_u^2/c\} - 1) &= \frac{1}{c}\int_0^\infty e^{x/c}\mathbb{P}(|y_b - y_a|_u^2 > x)dx \\
&\leq \frac{2}{c}\int_0^\infty e^{x/c}\exp\left\{-\frac{x}{4yd(a,b)}\right\}dx \\
&= 2\left(\frac{c}{4yd(a,b)} - 1\right)^{-1}.
\end{aligned}$$

The last quantity is less than 1 whenever $c > 12yd(a,b)$, which proves our assertion. □

Corollary 14 enables us to use the method of the majorizing measure to study the smoothness of the process $(y_a, a \in \mathbb{R})$. Specifically, denote for every $\varepsilon > 0$ by $m(\varepsilon)$ the Lebesgue measure of the ball in the metric d with radius ε, so that

$$m(\varepsilon) = m\left(\left\{a \in \mathbb{R} : \frac{1}{\pi}\int_{-\infty}^{\infty}(1 - \cos a\xi)\Re\left(\frac{1}{\Psi(\xi)}\right)d\xi < \varepsilon\right\}\right)$$

(because d is homogeneous and the Lebesgue measure translation invariant, $m(\varepsilon)$ does not depend on the centre of the ball). We now particularize to our setting Theorem 11.14 in Ledoux and Talagrand (1991), using also the remark at the bottom of page 300 there. If the majorizing measure condition

$$\int_{0+}\sqrt{\log 1/m(\varepsilon)}d\varepsilon < \infty \tag{20}$$

is satisfied, then for every $y > 0$, there exists a d-continuous version of $(y_a, a \in \mathbb{R})$. That is to say that there exists a process $(\tilde{y}_a, a \in \mathbb{R})$ with continuous sample paths for the metric d, and such that for every a, $y_a = \tilde{y}_a$ a.s. Because d is equivalent to the Euclidean metric, the sample paths of $\tilde{y}$ are also continuous in the usual sense. It is now an easy matter to verify that $\tilde{y}_a(t)$ coincides with $L(a,t)$ provided that t is small enough. This yields the main result of this section, which provides a sufficient condition for the almost sure joint continuity of the local times. Barlow (1988) has proven that this condition is also necessary; however, we will not develop this part here.

Theorem 15 *If (20) holds, then the mapping $(a,t) \to L(a,t)$ is continuous a.s.*

For instance, when X is a stable process with index $\alpha \in (1,2]$, one gets that $h(a) = c|a|^{\alpha-1}$ and $m(\varepsilon) = c'\varepsilon^{1/(\alpha-1)}$. Plainly, (20) holds and X possesses almost surely jointly continuous local times. This result is originally due to Trotter (1958) in the Brownian case $\alpha = 2$ and to Boylan (1964) for $\alpha \in (1,2]$.

Proof Fix $y > 0$ and introduce

$$\tilde{y}_*(t) = \sup\{\tilde{y}_a(t) : a \in \mathbb{R}\} \quad \text{and} \quad \vartheta(y) = \inf\{t \geq 0 : \tilde{y}_*(t) = y\}.$$

We first check that $\vartheta(y)$ is an $(\mathcal{F}_t)$-stopping time with $\vartheta(y) > 0$ a.s.

Because $\tilde{y}_a$ depends continuously on a, a.s., we have for every fixed $t \geq 0$

$$\tilde{y}_*(t) = \sup\{\tilde{y}_q(t) : q \in \mathbb{Q}\} \qquad \text{a.s.},$$

so $\tilde{y}_*$ is an increasing adapted process and this entails that $\vartheta(y)$ is a stopping time. Suppose that $\vartheta(y) = 0$ a.s. Then there exists a sequence of random variables $a(n)$, $n = 1, \cdots$, with

$$\tilde{y}_{a(n)}(1/n) > y/2 \quad \text{for all } n, \text{ a.s.}$$

On the other hand, we know from Proposition 2 that a.s., simultaneously for all rational numbers q and all $t \geq 0$

$$\tilde{y}_q(t) = y \wedge L(q,t) = \lim_{\varepsilon \to 0+} y \wedge \left(\frac{1}{2\varepsilon} \int_0^t \mathbf{1}_{\{|X_s - q| < \varepsilon\}} ds \right).$$

We deduce by the continuity of $a \to \tilde{y}_a$ that a.s., for all $a \in \mathbb{R}$ and $t > 0$,

$$\tilde{y}_a(t) = 0 \quad \text{whenever} \quad |a| > \sup\{|X_s| : s \leq t\}.$$

This forces the sequence $a(n)$ to converge to 0 a.s. Then, again by continuity and the fact that $\tilde{y}_0(\cdot) = y \wedge L(0,\cdot)$ a.s., we must have $L(0,t) \geq y/2$ a.s. for every $t > 0$. This disagrees with Proposition 4, so according to the Blumenthal zero-one law, $\vartheta(y) > 0$ a.s.

Next we prove that $\tilde{y}_a(t) = L(a,t)$ simultaneously for all $a \in \mathbb{R}$ and $t < \vartheta(y)$, a.s. Specifically, a.s. for each $a \in \mathbb{R}$, $L(a,t) = \tilde{y}_a(t)$ for all $t < \vartheta(y)$ and we deduce from Fubini's theorem and the occupation density formula that for each $f \in \mathcal{C}_c$ (i.e. f continuous with compact support),

$$\int_0^t f(X_s)ds = \int_{-\infty}^{\infty} f(a)\tilde{y}_a(t)da, \qquad \text{for all } t < \vartheta(y), \text{ a.s.}$$

Letting f range over a countable dense subset of $\mathcal{C}_c$, this entails that the foregoing identity holds simultaneously for all bounded measurable functions with compact support. In particular, a.s., for all $x \in \mathbb{R}$, $\varepsilon > 0$

and $t < \vartheta(y)$,

$$\frac{1}{2\varepsilon}\int_0^t \mathbf{1}_{\{|X_s - x|<\varepsilon\}} ds = \frac{1}{2\varepsilon}\int_{x-\varepsilon}^{x+\varepsilon} \widetilde{y}_a(t) da,$$

and because $\widetilde{y}_a(t)$ depends continuously on a, the right-hand side converges to $\widetilde{y}_x(t)$ as ε tends to $0+$. By the very definition of local times, we thus have $L(x,t) = \widetilde{y}_x(t)$ simultaneously for all $x \in \mathbb{R}$ and $t < \vartheta(y)$, a.s.

In particular, the mapping $(a,t) \to L(a,t)$ is continuous on $\mathbb{R} \times [0, \vartheta(y))$, a.s. Applying the Markov property at $\vartheta(y)$, one readily sees that $\vartheta(2y) \geq \vartheta(y) + \vartheta'(y)$, where $\vartheta'(y)$ is an independent copy of $\vartheta(y)$. This entails that $\vartheta(y)$ tends to ∞ a.s. as y goes to ∞, and completes the proof. □

One can also express the majorizing measure condition (20) in terms of the so-called increasing rearrangement of h, that is the inverse function of $\varepsilon \to m(\varepsilon)$; see Barlow and Hawkes (1985). Alternatively, (20) can also be viewed as a metric entropy condition, see chapter 11 in Ledoux and Talagrand (1991), and Bass and Khoshnevisan (1992). We finally mention that sharp properties on the modulus of continuity of the local times process (in the space variable) are also available via metric entropy arguments, see in particular Dudley (1973). A more elementary result is proposed as Exercise 3.

4. Exercises

1. *(Occupation measure of a perturbed Lévy process)* Adapt the argument of the proof of Theorem 1 to show that the following three assertions are equivalent (recall that m denotes the Lebesgue measure on $\mathbb{R}$):

   (i) For every measurable function $f : [0,\infty) \to \mathbb{R}$,
   $$m\left(\{X_t + f(t) : t \geq 0\}\right) > 0 \quad \text{a.s.}$$
   (ii) For every measurable function $f : [0,\infty) \to \mathbb{R}$,
   $$\mathbb{P}\left(m\left(\{X_t + f(t) : t \geq 0\}\right) > 0\right) > 0.$$
   (iii)
   $$\int_{-\infty}^{\infty} \frac{d\xi}{1 + \Re\Psi(\xi)} < \infty.$$
   [Hint: to prove the implication (ii)⇒(iii), take f random distributed as an independent copy of $-X$.]
2. *(Hilbert transform of local times at a last zero)* Let $\tau = \tau(q)$ be an

exponential time with parameter $q > 0$, independent of X. Denote by $g_\tau = \sup\{t < \tau : X_t = 0\}$ the last passage time at 0 before τ. Deduce from Theorem 7 and the compensation formula of excursion theory that for every $\lambda \in \mathbb{R}$,

$$\mathbb{E}\left(\exp\{i\lambda H(0, g_\tau)\}\right) = \frac{\tanh \lambda\kappa(q)}{\lambda\kappa(q)}.$$

3. *(Hölder-continuity of local times)* Let $\alpha > 1$. Prove that if

$$\Re\left(1/\Psi(\lambda)\right) = O(|\lambda|^{-\alpha}) \qquad (|\lambda| \to \infty),$$

then, in the notation of section 3, $h(a) = O(a^{\alpha-1})$ as $a \to 0+$. Deduce from Kolmogorov's criterion that for every $t > 0$ and $\varepsilon > 0$, $a \to L(a, t)$ is Hölder-continuous with exponent $(\alpha - 1)/2 - \varepsilon$, a.s. [Hint: use Lemma 13.]

4. *(A law of the iterated logarithm for the local time at* 0*)* Let $\alpha \in (1, 2]$, $\beta \in [-1, 1]$ and $r : (0, \infty) \to (0, \infty)$ be an increasing function that is regularly varying at ∞ with index α. Denote the inverse function of r by r^{-1} and suppose that

$$\lim_{t\to\infty} \frac{\Psi(at)}{r(t)} = |a|^\alpha \left(1 - i\beta \tan(\pi\alpha/2)\mathrm{sgn}(a)\right), \qquad a \in \mathbb{R}.$$

Note that this condition is equivalent to the convergence as t tends to 0+ of $r^{-1}(1/t)X_t$ to a standard stable law with index α and asymmetry parameter β; see section VIII.1.

(a) Prove that the Laplace exponent $1/\kappa$ of the inverse local time at 0, σ, has

$$\kappa(q) \sim c(\alpha, \beta) q / r^{-1}(q) \qquad (q \to \infty),$$

where $c(\alpha, \beta) > 0$ is some numerical constant. Deduce that κ is regularly varying at ∞ with index $1 - 1/\alpha$.

(b) Derive from Exercise III.5 the following law of the iterated logarithm for $L(0, \cdot)$:

$$\limsup_{t\to 0+} \frac{L(0, t)}{\kappa(t^{-1} \log|\log t|) \log|\log t|} = \alpha^{1/\alpha}(1 - 1/\alpha)^{-1+1/\alpha} \qquad \text{a.s.}$$

5. *(Common points in the ranges of independent Lévy processes)* In this exercise, $X^1, \cdots, X^k$ are independent d-dimensional Lévy processes with respective characteristic exponents $\Psi_1, \cdots, \Psi_k$. Our purpose is to exhibit a simple condition for the existence of common points in their ranges.

Consider the random measure ν on $(\mathbb{R}^d)^{k-1}$ given by

$$\int_{(\mathbb{R}^d)^{k-1}} f\,d\nu = \int_0^\infty dt_1 e^{-t_1}\cdots\int_0^\infty dt_k e^{-t_k} f\left(X^2_{t_2}-X^1_{t_1},\cdots,X^k_{t_k}-X^{k-1}_{t_{k-1}}\right)$$

where $f:(\mathbb{R}^d)^{k-1}\to[0,\infty)$ is a measurable function.
(a) For every $\lambda=(\lambda_1,\cdots,\lambda_{k-1})\in(\mathbb{R}^d)^{k-1}$ denote the Fourier transform of ν evaluated at λ by

$$\mathscr{F}\nu(\lambda)=\int_{(\mathbb{R}^d)^{k-1}}\exp i\{\langle\lambda_1,x_1\rangle+\cdots+\langle\lambda_{k-1},x_{k-1}\rangle\}\nu(dx_1,\cdots,dx_k).$$

Check that

$$\mathbb{E}\left(|\mathscr{F}\nu(\lambda)|^2\right)=\prod_{j=1}^k\Re\left(\frac{1}{1+\Psi_j(\lambda_j-\lambda_{j-1})}\right),$$

with $\lambda_0=\lambda_k=0$.
(b) Deduce that ν is a.s. absolutely continuous whenever

$$\int_{(\mathbb{R}^d)^{k-1}}\prod_{j=1}^k\Re\left(\frac{1}{1+\Psi_j(\lambda_j-\lambda_{j-1})}\right)d\lambda_1\cdots d\lambda_{k-1}<\infty.$$

Prove that in that case, there exists a measurable set $B\subseteq(\mathbb{R}^d)^k$ with positive Lebesgue measure, such that for every $b=(b_1,\cdots,b_k)\in B$,

$$\mathbb{P}\left((b_1+\mathscr{R}_1)\cap\cdots\cap(b_k+\mathscr{R}_k)\neq\emptyset\right)>0,$$

where $\mathscr{R}_j=\overline{\{X^j_t:t\geq 0\}}$ denotes the closure of the range of X^j.
(c) Suppose now that the X^j's are identically distributed, viz. $\Psi_1=\cdots=\Psi_k$. Prove the condition in (b) holds whenever the 1-resolvent kernel of X^j has a density u^1 in $L^k(\mathbb{R}^d)$, viz.

$$\int_{\mathbb{R}^d}\left(u^1(x)\right)^k dx<\infty.$$

6. *(A simple criterion for transience)* The purpose of this exercise is to give a test for transience which is simpler than that of Theorem I.17. The assumption that 0 is regular for itself is still in force; but we stress that the result also holds in the general case, though this is much harder to establish.
(a) Prove that X is transient whenever

$$\int_{-r}^r\Re\left(\frac{1}{\Psi(\xi)}\right)d\xi<\infty$$

for some $r>0$. [Hint: use Lemma 11 to show that if X were recurrent, then the local time at level 0, $L(0,\cdot)$, would be bounded.]
(b) Using Theorem I.17, establish the converse of (a).

5. Comments

We refer to Geman and Horowitz (1980) for a survey on occupation measures, local times and their applications. Theorem 1 is from Hawkes (1985) and Proposition 4 from Getoor and Kesten (1972). Barlow, Perkins and Taylor (1986, 1986-a) present intrinsic constructions of the local times of a Lévy process. One can also represent the local times using stochastic calculus and the Meyer-Tanaka formula, see Dellacherie and Meyer (1980) and Protter (1990). Typically, when the Gaussian coefficient is positive, the local time is proportional to the so-called 'semimartingale local time'; nonetheless, when the Gaussian coefficient is zero, the semimartingale local times are always identically zero. The integral representation of a positive continuous additive functional is due to Griego (1967), see also Bally and Stoica (1987).

The Hilbert transform of the local times was studied first in the Brownian case by Yor (1982) and Yamada (1985). Yamada (1986) pointed out that H appears in certain limit theorems for Brownian occupation times, this has been recently extended by Fitzsimmons and Getoor (1992-a) to stable processes. Biane and Yor (1987) obtained remarkable formulas and identities involving the Hilbert transform of the Brownian local times, see also Bertoin (1990) and Yor (1995) for further developments. Fitzsimmons and Getoor (1992) discovered that formulas of Biane and Yor can be extended to symmetric Lévy processes; the present approach based on the Feynman-Kac formula follows Bertoin (1995-c). Carmona, Master and Simon (1990) gave interesting applications of the Feynman-Kac formula for relativistic Schrödinger operators.

Finding conditions for the continuity of the local times of Lévy processes has motivated much work over the last 30 years; see Boylan (1964), Getoor and Kesten (1972), Barlow (1985), Barlow and Hawkes (1985). The end of the story is in Barlow (1988). In the symmetric case, Marcus and Rosen (1992) give a different approach to the results of Barlow and Hawkes, based on Dynkin's isomorphism theorem. In subsequent work, Marcus and Rosen (1992-a,b, 1993, 1994, 1994-a,b) derive from Dynkin's formula numerous sample path properties of the local times of symmetric Lévy processes. As pointed out by Barlow (1985), results on the spatial continuity of local times can be used to decide whether the range of a process is nowhere dense. This question originates with Kesten (1976), see also Barlow (1981), Barlow, Perkins and Taylor (1986-a) and Mountford and Port (1991).

Exercise 1 is from Evans (1989) and Exercise 2 from Bertoin (1995-c); the latter extends the result of Fitzsimmons and Getoor (1992) in the symmetric case. More sophisticated versions of Exercise 3 appear in Barlow and Hawkes (1985) and Bass and Khoshnevisan (1992). Exercise 4 is from Bertoin (1995), extending Marcus and Rosen (1994) in the symmetric case. See also Khoshnevisan (1995). We refer to Marcus and Rosen (1994-a) and Bertoin and Caballero (1995) for the limit case $\alpha = 1$. Wee (1991) proved a law of the iterated logarithm for the supremum of the local times. Exercise 5 is a key step in investigating multiple points in the sample paths of a Lévy process; it originates from Evans (1987) and Le Gall, Rosen and Shieh (1989), see also Rogers (1989). The necessary and sufficient condition for the existence of k-multiple points has been obtained by Evans (1987) and Fitzsimmons and Salisbury (1989), solving a conjecture of Hendricks and Taylor. Earlier results in that field are due to Taylor (1966), Fristedt (1967-a), Hawkes (1978) and Le Gall (1987).

VI

Fluctuation Theory

Fluctuation theory for real-valued Lévy processes is the analogue in continuous time of that developed for random walks by Feller and Spitzer. It provides a number of important formulas for distributions related to a Lévy process and its extrema. It also enables us to investigate some sample path properties. Excursion theory plays a major rôle in the approach.

1. The reflected process and the ladder process

Fluctuation theory in discrete times concerns the joint behaviour of a real-valued random walk and its extrema, see Spitzer (1964), Feller (1971) and Borovkov (1970, 1976). Originally, it relied heavily on analytic methods, in particular the Wiener-Hopf factorization. Then an alternative path-wise approach was proposed, which enables one to circumvent the use of deep analytic techniques. The key ingredient in the developing of the theory is the regenerative property at the successive indices when the random walk reaches a new supremum. A problem that arises when one tries to adapt the arguments to continuous times is that, in general, the instants when a Lévy process attains its supremum do not form a discrete set. This difficulty can be overcome using the notion of local time for Markov processes which was introduced in Chapter IV. One is then able to obtain directly results analogous to those known for random walks, and also to get new information on the infinitesimal behaviour of Lévy processes.

In this section, we observe first that reflecting a real-valued Lévy process at the level of its previous supremum yields a nice Markov process, and then we develop some material on the reflected process using results of Chapter IV.

We will assume henceforth that the dimension of the state space is $d = 1$, and we introduce the so-called supremum and infimum processes

$$S_t = \sup\{0 \vee X_s : 0 \le s \le t\}, \quad I_t = \inf\{0 \wedge X_s : 0 \le s \le t\}.$$

We see that S and $-I$ are two nonnegative increasing right-continuous processes, which are both adapted to the filtration $(\mathcal{F}_t)$. We call $S - X$ and $X - I$ the *reflected* processes, respectively at the supremum and at the infimum. Plainly, $X - I$ can also be viewed as the dual process reflected at the supremum, and we shall therefore restrict our attention to $S - X$. Note that for every $x \ge 0$, $S - X$ starts from x under $\mathbb{P}_{-x}$ and has the same law as $(x \vee S) - X$ under $\mathbb{P}_0$.

Proposition 1 *The reflected process $S - X$ is a Markov process in the filtration $(\mathcal{F}_t)$ and its semigroup has the Feller property.*

Proof First, we prove the Markov property. Let T be a finite stopping time and $s \ge 0$. Note the identity

$$\begin{aligned} S_{T+s} &= S_T \vee \sup\{X_{T+u} : 0 \le u \le s\} \\ &= X_T + (S_T - X_T) \vee \sup\{X_{T+u} - X_T : 0 \le u \le s\}. \end{aligned}$$

We can thus rewrite $S_{T+s} - X_{T+s}$ as

$$(S_T - X_T) \vee \sup\{X_{T+u} - X_T : 0 \le u \le s\} - (X_{T+s} - X_T).$$

By the independence and homogeneity of the increments of X, we see that the law of $S_{T+s} - X_{T+s}$ conditionally on $\mathcal{F}_T$ is the same as that of $(x \vee S_s) - X_s$ under $\mathbb{P}_0$ with $x = S_T - X_T$, and this is the same as that of $S_s - X_s$ under $\mathbb{P}_x$.

Next, if f is a continuous function on $(-\infty, 0]$ which tends to 0 at ∞, then for every fixed $t > 0$

$$\mathbb{E}_x(f(S_t - X_t)) = \mathbb{E}_0(f((x \vee S_t) - X_t)).$$

It follows by dominated convergence that this quantity varies continuously in the variable $x \ge 0$ and tends to 0 as x tends to ∞. □

The reflected process is a 'nice' Markov process in the sense of chapter IV: we already know that the filtration $(\mathcal{F}_t)$ is right-continuous and we have just proved that the reflected process has the Markov property. We point out that in general, $(\mathcal{F}_t)$ is strictly larger than the natural filtration of the reflected process. More precisely, when the Lévy process jumps across its previous supremum, the information on the overshoot (the quantity by which X exceeds its previous supremum) is lost in the reflected process.

We also stress a simple path property that will be useful in the sequel. Roughly, we claim that almost surely, a Lévy process does not jump downwards at a time at which it is about to reach the level of its previous supremum. Specifically, fix $s > 0$ and consider for every $\epsilon > 0$ the stopping time $T_\epsilon = \inf\{t \geq s : X_s > S_s - \epsilon\}$. Then, conditionally on $\mathrm{T} = \lim_{\epsilon \downarrow 0} \mathrm{T}_\epsilon < \infty$, one has $X_{\mathrm{T}} \geq S_s$ a.s. Indeed, this follows immediately from the quasi left-continuity stated in Proposition I.7.

We denote by $L = (L(t), t \geq 0)$ a local time of $S - X$ at 0 in the sense of sections IV.2-4, and by $L^{-1}(\cdot) = \inf\{s > 0 : L(s) > \cdot\}$ its right-continuous inverse. The inverse local time will be called the (ascending) *ladder time process* in the sequel. Recall from chapter IV that the closure of its range coincides with the closure of the zero set of the reflected process, $\mathscr{Z} = \{t : X_t = S_t\}$, which is sometimes referred to as the ladder time set.

The next step consists in introducing the so-called *ladder height process* H, using the inverse local time to time-change the supremum process. Specifically we put

$$H(t) = S_{L^{-1}(t)} \quad \text{if } L^{-1}(t) < \infty\,, \quad H(t) = \infty \quad \text{otherwise.}$$

Note that according to Proposition IV.7(iii), one has $H(t) = X_{L^{-1}(t)}$ whenever $L^{-1}(t) < \infty$. The pair (L^{-1}, H) is known as the *ladder process* . The next lemma sheds light on the probabilistic structure of the ladder process. The result is slightly more complex when 0 is transient for the reflected process, as it involves killing at a certain rate (recall that, if Y is a Lévy process and τ an independent time with an exponential distribution of parameter $q > 0$, then the process with lifetime τ which coincides with Y on the time interval $[0, \tau)$ is called a Lévy process killed at rate q).

Lemma 2 (i) *Suppose that* 0 *is recurrent for the reflected process* $S - X$. *Then the ladder process* (L^{-1}, H) *is a Lévy process.*

(ii) *Suppose that* 0 *is transient for the reflected process* $S - X$. *Then* $L(\infty)$ *has an exponential distribution, say with parameter* q, *and the process* $((L^{-1}(t), H(t)), t < L(\infty))$ *is a Lévy process killed at rate* q.

Proof (i) The proof essentially mimics the first part of that of Theorem IV.8 where it has already been shown that the ladder time process is a subordinator.

According to Proposition IV.7, for each $t \geq 0$, $L^{-1}(t)$ is a finite stopping time, and by the Markov property, the shifted process

$$\widetilde{X}_s = X_{L^{-1}(t)+s} - X_{L^{-1}(t)}, \qquad s \geq 0,$$

is independent of $\mathscr{F}_{L^{-1}(t)}$ and has the law $\mathbb{P}$. Then, in the obvious notation, we have $\widetilde{L}^{-1}(s) = L^{-1}(t+s) - L^{-1}(t)$ (see e.g. the proof of Theorem IV.8), and

$$\widetilde{S}_s = \sup\{X_{L^{-1}(t)+u} : 0 \leq u \leq s\} - X_{L^{-1}(t)} = S_{L^{-1}(t)+s} - S_{L^{-1}(t)},$$

where the last equality comes from the identity $X_{L^{-1}(t)} = S_{L^{-1}(t)}$. This shows that the shifted process $(\widetilde{L}^{-1}, \widetilde{H})$ is independent of $\mathscr{F}_{L^{-1}(t)}$ and has the same law as (L^{-1}, H). So the ladder process has independent homogeneous increments, and since it clearly has right-continuous paths with left limits, it is a Lévy process.

(ii) The proof is similar to the second part of that of Theorem IV.8, we omit the details. □

In particular, the law of the ladder process is characterized by the bivariate Laplace exponent κ,

$$\exp\{-\kappa(\alpha,\beta)\} = \mathbb{E}\left(\exp -\{\alpha L^{-1}(1) + \beta H(1)\}\right), \qquad \alpha, \beta > 0$$

(recall the convention $e^{-\infty} = 0$). It is therefore important to evaluate this quantity explicitly, which will be one of the main purposes of the next section.

We conclude this section with two easy consequences of the duality lemma for the extrema of X. First, we have a useful identity in law (recall that I denotes the infimum process).

Proposition 3 *For each fixed $t > 0$, the pairs of variables $(S_t, S_t - X_t)$ and $(X_t - I_t, -I_t)$ have the same distribution under $\mathbb{P}$.*

Proof Put $\widetilde{X}_t = X_t$ and $\widetilde{X}_s = X_t - X_{(t-s)-}$ for $0 \leq s < t$. Denote the infimum of $\widetilde{X}$ by $\widetilde{I}_t = \inf\{\widetilde{X}_s : 0 \leq s \leq t\}$, and observe (using the right-continuity of the sample paths) that $(S_t, S_t - X_t) = (\widetilde{X}_t - \widetilde{I}_t, -\widetilde{I}_t)$ a.s. We know from Lemma II.2 that the processes $(X_s, 0 \leq s \leq t)$ and $(\widetilde{X}_s, 0 \leq s \leq t)$ have the same law under $\mathbb{P}$, which proves the proposition.

□

As a first example of an application of Proposition 3, we point out that $\mathbb{P}(S_t = X_t) = 0$ for all $t > 0$ if and only if 0 is regular for $(-\infty, 0)$. Second, we observe that the local extrema of X are all distinct, except in the compound Poisson case.

Proposition 4 *Assume that X is not a compound Poisson process. Then a.s. for all $0 \le s < t$, if $X_u = S_u$ for some $u \in (s, t)$, then $S_s < S_t$.*

Proof Plainly, we only need to prove the assertion when s and t range over some countable dense subset of $[0, \infty)$. So take s and t fixed and consider the stopping time $T = \inf\{u > s : X_u = S_u\}$.

First, suppose that 0 is regular for $(0, \infty)$, that is X visits $(0, \infty)$ at arbitrarily small times a.s. On the event $\{T < t\}$, we can apply the Markov property and deduce that $S_s \le S_T < S_t$ a.s.

Next, suppose that 0 is regular for $(-\infty, 0)$, so the dual process $\widehat{X} = -X$ visits $(0, \infty)$ at arbitrarily small times a.s. The preceding argument applied to X time-reversed at t and the duality lemma show again that $S_s < S_t$ a.s. provided that $X_u = S_u$ for some $u \in (s, t)$.

Finally, the zero-one law of Blumenthal implies that 0 is not regular for $(-\infty, 0)$ or for $(0, \infty)$ only in the compound Poisson case. This finishes the proof. □

2. Fluctuation identities

The fluctuation identities provide a number of formulas for the distributions of certain random variables related to extrema of a Lévy process. They were deduced first, from analogous identities for random walks due to Spitzer and Borovkov, by Rogozin (1966) and other authors using approximations based on discrete time skeletons. We will use here a direct approach due to Greenwood and Pitman (1980), which relies on excursion theory for the reflected process.

The random variables of interest in fluctuation theory are the following. Let $\tau = \tau(q)$ be an exponentially distributed random time with parameter $q > 0$, which is independent of the Lévy process. We will be mostly concerned with the value $(S_\tau, S_\tau - X_\tau)$ of the supremum and of the reflected process at time τ, and by $G_\tau = \sup\{t < \tau : X_t = S_t\}$, the last zero of the reflected process before τ. Note that, according to Proposition 4, G_τ is actually the unique instant $t \in [0, \tau]$ such that $X_t = S_\tau$ or $X_{t-} = S_\tau$, except possibly in the compound Poisson case.

Theorem 5 *The joint distribution of the variables G_τ, $\tau - G_\tau$, S_τ and $S_\tau - X_\tau$ is specified by:*

(i) *The pairs (G_τ, S_τ) and $(\tau - G_\tau, S_\tau - X_\tau)$ are independent.*

(ii) *For every $\alpha, \beta > 0$,*

$$\mathbb{E}\left(\exp\{-\alpha G_\tau - \beta S_\tau\}\right) = \exp\left(\int_0^\infty dt \int_{[0,\infty)} (\mathrm{e}^{-\alpha t - \beta x} - 1) t^{-1} \mathrm{e}^{-qt} \mathbb{P}(X_t \in dx)\right),$$

and

$$\mathbb{E}\left(\exp\{-\alpha(\tau - G_\tau) - \beta(S_\tau - X_\tau)\}\right) = \exp\left(\int_0^\infty dt \int_{(-\infty,0)} (\mathrm{e}^{-\alpha t + \beta x} - 1) t^{-1} \mathrm{e}^{-qt} \mathbb{P}(X_t \in dx)\right).$$

We will prove the two parts of Theorem 5 separately. The proof of the first relies on a path decomposition of the Lévy process, which can be thought of as a substitute for the Markov property at a random time that is not a stopping time. Specifically, we split the path of the Lévy process on the time interval $[0, \tau]$ at the instant of the last supremum G_τ, and we observe that the two resulting pieces are independent. The independence is not really surprising if we take the point of view of excursion theory, see in particular Exercise IV.4. But it is perhaps just as simple to follow an elementary approach.

Lemma 6 (i) *If 0 is irregular for the process $S - X$, then the processes $(X_t, 0 \le t \le G_\tau)$ and $(X_{G_\tau + t} - X_{G_\tau}, 0 \le t < \tau - G_\tau)$ are independent.*

(ii) *If 0 is regular for the process $S - X$, then the processes $(X_t, 0 \le t < G_\tau)$ and $(X_{G_\tau + t} - X_{G_\tau -}, 0 \le t < \tau - G_\tau)$ are independent.*

We stress that in general one cannot replace $X_{G_\tau -}$ by X_{G_τ} in (ii); more precisely one can check that these quantities coincide a.s. if and only if 0 is regular for $(-\infty, 0)$. When 0 is transient for the reflected process, a similar result holds when one replaces τ by ∞ and G_τ by the last zero of the reflected process.

Proof (i) The successive passage times of the reflected process at 0, $T_0 = 0, \cdots, T_{n+1} = \inf\{t > T_n : X_t = S_t\}, \cdots$, are stopping times. For each fixed n, put $\widetilde{X}_t = X_{T_n + t} - X_{T_n}$ on $\{T_n < \infty\}$. Applying the Markov property, we get that for any bounded functionals F and K, in the obvious notation

$$\begin{aligned}
&\mathbb{E}\left(F(X_t, 0 \le t \le G_\tau)K(X_{G_\tau+t} - X_{G_\tau}, 0 \le t < \tau - G_\tau), T_n < \tau \le T_{n+1}\right) \\
&= \mathbb{E}\left(\int_{T_n}^{T_{n+1}} q\mathrm{e}^{-qs} F(X_t, 0 \le t \le T_n)K(\tilde{X}_t, 0 \le t < s - T_n)\, ds\right) \\
&= \mathbb{E}(F(X_t, 0 \le t \le T_n)\mathrm{e}^{-qT_n})\, \mathbb{E}\left(\int_0^{\tilde{T}_1} q\mathrm{e}^{-qs} K(\tilde{X}_t, 0 \le t < s)\, ds\right).
\end{aligned}$$

Observe that the second term in the foregoing product does not depend on n. Taking the sum over n shows that the processes $(X_t, 0 \le t \le G_\tau)$ and $(X_{G_\tau+t} - X_{G_\tau}, 0 \le t < \tau - G_\tau)$ are independent.

(ii) For each fixed $\varepsilon > 0$, denote the integer part of $\varepsilon^{-1}L(\tau)$ by $[\varepsilon^{-1}L(\tau)] = n$. Because $L(\tau)$ has an exponential distribution, we have $\varepsilon n < L(\tau) < \varepsilon(n+1)$ a.s. We deduce $L^{-1}(n\varepsilon) \le \tau \le L^{-1}((n+1)\varepsilon)$ a.s. The additive property of the local time and the same argument as in (i) show that the processes $(X_t, 0 \le t \le L^{-1}(n\varepsilon))$ and $(X_{L^{-1}(n\varepsilon)+t} - X_{L^{-1}(n\varepsilon)}, 0 \le t < \tau - L^{-1}(n\varepsilon))$ are independent. According to Proposition IV.7(iii), $L^{-1}(n\varepsilon)$ increases to G_τ as ε tends to zero, which proves Lemma 6. □

We now prove the first part of Theorem 5.

Proof of Theorem 5(i) Suppose first that 0 is irregular for the reflected process, so that the ladder time set is discrete. Then obviously G_τ is a zero of the reflected process $S - X$, $X_{G_\tau} = S_\tau$, and the statement follows from Lemma 6.

Next, suppose that 0 is regular for the reflected process. Again by Lemma 6, all that is needed is to check the identity $X_{G_\tau-} = S_\tau$ a.s. This obviously holds on the event $\{G_\tau = \tau\}$ because X is continuous at time τ a.s., so we suppose henceforth that $G_\tau < \tau$. Since the supremum process stays constant over the excursion intervals of the reflected process, we have $S_\tau = S_{G_\tau}$. If X is continuous at time G_τ, then $X_{G_\tau-} = X_{G_\tau} = S_{G_\tau-} = S_{G_\tau}$. If X makes a negative jump at time G_τ, then $X_{G_\tau-} = S_{G_\tau-} = S_{G_\tau} > X_{G_\tau}$. Finally, we prove that X cannot make a positive jump at time G_τ. Fix $\varepsilon > 0$ and consider the instant $\varepsilon(n)$ when X accomplishes its n-th jump with height at least ε. It is a stopping time and since 0 is regular for the reflected process, the Markov property implies that $S - X$ returns to 0 immediately after $\varepsilon(n)$ provided that $X_{\varepsilon(n)} = S_{\varepsilon(n)}$. Thus X cannot make a jump with height greater than ε at time G_τ, and the proof is complete. □

Theorem 5(i) shows that (τ, X_τ), which has a known distribution, can be expressed as the sum of two independent variables, (G_τ, S_τ) and

$(\tau - G_\tau, X_\tau - S_\tau)$. In the next three lemmas, we will see that these three pairs have infinitely divisible laws, and this will actually be sufficient to characterize their respective distributions. It is easy to see that the law of a random walk evaluated at some independent time with a geometric distribution is infinitely divisible. Here is the analogue in continuous times.

Lemma 7 *Let $Z = (Z_t, t \geq 0)$ be a d-dimensional Lévy process and τ a random time independent of Z, which has an exponential distribution with parameter $q > 0$. Then (τ, Z_τ) has an infinitely divisible law, and its Lévy measure is $\mu(dt, dz) = t^{-1}\mathrm{e}^{-qt}\mathbb{P}(Z_t \in dz)dt$ $(t > 0, z \in \mathbb{R}^d)$.*

Proof Denote the characteristic exponent of Z by ψ. For every $\alpha > 0$ and $\beta \in \mathbb{R}^d$, we have

$$\begin{aligned}
&\mathbb{E}\left(\exp\{-\alpha\tau + \mathrm{i}\langle\beta, Z_\tau\rangle\}\right) \\
&= \int_0^\infty q\mathrm{e}^{-qt}\exp\{-(\alpha + \psi(\beta))t\}dt \\
&= \exp\left(\int_0^\infty \left(\exp\{-(\alpha+\psi(\beta))t\} - 1\right) t^{-1}\mathrm{e}^{-qt}dt\right),
\end{aligned}$$

where, in the last equality, we used the Frullani integral. Finally, the foregoing quantity can also be written as

$$\exp\left(\int_0^\infty \int_{\mathbb{R}^d} \left(\exp\{-\alpha t + \mathrm{i}\langle\beta, z\rangle\} - 1\right)\mathbb{P}(Z_t \in dz)t^{-1}\mathrm{e}^{-qt}dt\right),$$

which entails our assertion. More precisely, it is easily checked that $\mathbb{E}(|Z_t| \wedge 1) = O(\sqrt{t})$ as $t \to 0+$, which ensures that the integral

$$\int_0^\infty \int_{\mathbb{R}^d} (1 \wedge (t + |z|))\mathbb{P}(Z_t \in dz)t^{-1}\mathrm{e}^{-qt}dt$$

converges, and then we can identify the Lévy measure in the preceding Lévy-Khintchine formula. □

The ladder process lies at the heart of the proof of the next lemma, which underlines again the importance of excursion theory in this chapter.

Lemma 8 *The law of (G_τ, S_τ) is infinitely divisible.*

Proof We shall focus on the case when 0 is regular for the reflected process; the case when 0 is irregular is quite similar and even simpler. We saw in the proof of Theorem 5(i) that $S_\tau = S_{G_\tau-}$ a.s., and we can thus express $\mathbb{E}(\exp\{-\alpha G_\tau - \beta S_\tau\})$ as

$$\mathbb{E}(\exp\{-\alpha G_\tau - \beta S_{G_\tau}\}) = \mathbb{E}\left(\int_0^\infty dt\, q e^{-qt} \exp\{-\alpha G_t - \beta S_{G_t}\}\right)$$

$$= \mathbb{E}\left(\int_0^\infty dt\, q e^{-qt} \mathbf{1}_{\{X_t = S_t\}} \exp\{-\alpha t - \beta S_{t-}\}\right)$$

$$+ \mathbb{E}\left(\sum_g \exp\{-\alpha g - \beta S_{g-}\} e^{-qg} \int_0^{d-g} q e^{-qt} dt\right)$$

where Σ_g refers to the summation taken over all the excursion intervals (g, d) of the reflected process. Let $\mathrm{d} \geq 0$ be the drift coefficient of the ladder time process, so by Corollary IV.6, the first term in the sum can be rewritten as

$$q\mathrm{d}\mathbb{E}\left(\int_0^\infty dL(t) e^{-qt} \exp\{-\alpha t - \beta S_t\}\right)$$

$$= q\mathrm{d}\mathbb{E}\left(\int_0^\infty dt \exp\{-(\alpha + q)L^{-1}(t) - \beta H(t)\}\right)$$

$$= q\mathrm{d}/\kappa(\alpha + q, \beta),$$

where κ is the bivariate Laplace exponent of (L^{-1}, H).

On the other hand, the compensation formula of excursion theory enables us to re-express the second term in the sum as

$$\mathbb{E}\left(\int_0^\infty dL(t) \exp\{-(\alpha + q)t - \beta S_t\}\right) n(1 - e^{-q\zeta})$$

$$= \mathbb{E}\left(\int_0^\infty dt \exp\{-(\alpha + q)L^{-1}(t) - \beta H(t)\}\right) n(1 - e^{-q\zeta})$$

$$= n(1 - e^{-q\zeta})/\kappa(\alpha + q, \beta),$$

where n denotes the excursion measure of the reflected process and ζ the lifetime of the generic excursion.

Finally, recall from Theorem IV.8 that the Laplace exponent of L^{-1} is given by $\kappa(q, 0) = q\mathrm{d} + n(1 - e^{-q\zeta})$, so putting the pieces together, we find

$$\mathbb{E}(\exp\{-\alpha G_\tau - \beta S_\tau\}) = \kappa(q, 0)/\kappa(\alpha + q, \beta). \tag{1}$$

It is straightforward to check from the Lévy-Khintchine formula that $\tilde{\kappa}(\alpha, \beta) = \kappa(\alpha + q, \beta) - \kappa(q, 0)$ is the Laplace exponent of a Lévy process $\tilde{Z}$ taking values in $[0, \infty) \times [0, \infty)$, and (1) shows that $(L^{-1}(\tau-), H(\tau-))$ has the same law as $\tilde{Z}$ evaluated at an independent exponential time with parameter $\kappa(q, 0)$. It follows from Lemma 7 that this law is infinitely divisible.

The final lemma of this group is readily derived from the preceding by time reversal.

Lemma 9 *The law of* $(\tau - G_\tau, X_\tau - S_\tau)$ *is infinitely divisible.*

Proof First assume that X is not a compound Poisson process, and consider the reversed process $\widetilde{X}_t = X_{(\tau-t)-} - X_\tau$, $t \in [0,\tau)$, which has the same law as $(-X_t, 0 \le t < \tau)$. Proposition 4 guarantees that, in the obvious notation, $\widetilde{G_\tau} = \tau - G_\tau$ and $\widetilde{S}_\tau = S_\tau - X_\tau$, and it then follows from Lemma 8 applied to the dual Lévy process that the law of $(\tau - G_\tau, X_\tau - S_\tau)$ is infinitely divisible.

Finally, in the compound Poisson case, put $X_t^{(\varepsilon)} = X_t + \varepsilon t$, so that $X^{(\varepsilon)}$ is a Lévy process but not a compound Poisson process. It is immediate that in the obvious notation, $G_\tau^{(\varepsilon)}$ and $S_\tau^{(\varepsilon)}$ converge to G_τ and S_τ, respectively. We deduce from the foregoing that the law of $(\tau - G_\tau, X_\tau - S_\tau)$ is infinitely divisible. □

We are finally able to prove the second part of Theorem 5.

Proof of Theorem 5(ii) Denote the Lévy measures of (τ, X_τ), (G_τ, S_τ) and $(\tau - G_\tau, X_\tau - S_\tau)$ by μ, μ^+ and μ^-, respectively. Since $(G_\tau, S_\tau) \in [0,\infty) \times [0,\infty)$ a.s., the support of μ^+ is contained in $[0,\infty) \times [0,\infty)$, and similarly, that of μ^- is contained in $[0,\infty) \times (-\infty, 0]$. Observe also that, since an exponential variable can take arbitrarily small values, so do G_τ, $\tau - G_\tau$, S_τ and $X_\tau - S_\tau$. This implies that the distributions of $(\tau - G_\tau, X_\tau - S_\tau)$ and (G_τ, S_τ) contain no drift component. Moreover, they obviously have no Gaussian component and are therefore determined by their Lévy measures.

Then we use Theorem 5(i) to express (τ, X_τ) as the sum of two independent variables, namely $(\tau, X_\tau) = (G_\tau, S_\tau) + (\tau - G_\tau, X_\tau - S_\tau)$. It follows that $\mu = \mu^+ + \mu^-$. Recall from Lemma 7 that

$$\mu(dt, dx) = t^{-1} e^{-qt} \mathbb{P}(X_t \in dx) dt \qquad (t > 0, x \in \mathbb{R}),$$

and assume first that X is not a compound Poisson process, so that $\mathbb{P}(X_t = 0) = 0$ for almost every $t > 0$ (see Proposition I.15). We deduce that

$$\mu^+(dt, dx) = t^{-1} e^{-qt} \mathbb{P}(X_t \in dx) dt \qquad (t > 0, x > 0),$$

$$\mu^-(dt, dx) = t^{-1} e^{-qt} \mathbb{P}(X_t \in dx) dt \qquad (t > 0, x < 0).$$

This establishes (ii) in this case.

In the compound Poisson case, we consider $X_t^{(\varepsilon)} = X_t + \varepsilon t$, $t \ge 0$, which is a Lévy process but not a compound Poisson process. When ε decreases to 0, then in the obvious notation, $G_\tau^{(\varepsilon)}$ and $S_\tau^{(\varepsilon)}$ converge to G_τ and S_τ, and the measures

$$\mathbf{1}_{\{x>0\}}\mathbb{P}(X_t^{(\varepsilon)} \in dx) \text{ and } \mathbf{1}_{\{x<0\}}\mathbb{P}(X_t^{(\varepsilon)} \in dx)$$

converge to

$$\mathbf{1}_{\{x\geq 0\}}\mathbb{P}(X_t \in dx) \text{ and } \mathbf{1}_{\{x<0\}}\mathbb{P}(X_t \in dx),$$

respectively. Thus (ii) follows by approximation. □

We conclude this section with two important by-products of the fluctuation identities.

First, recall that Ψ stands for the characteristic exponent of X, so that $q/(q+\Psi)$ is the characteristic function of X_τ. Theorem 5 thus yields the following remarkable factorization:

$$\frac{q}{q+\Psi} = \Psi_q^+ \, \Psi_q^- , \qquad (2)$$

where Ψ_q^+ (respectively, Ψ_q^-) is the characteristic function of the infinitely divisible nonnegative variable S_τ (respectively, nonpositive variable $X_\tau - S_\tau$) with zero drift. The identity (2) is known as the *Wiener-Hopf factorization* of the Lévy process, and Ψ_q^+ and Ψ_q^- as the right and left Wiener-Hopf factors, respectively. Such a factorization is unique (by the argument in the proof of Theorem 5(ii)), which sometimes allows explicit calculation of the factors. For instance, in the Brownian case, one has $\Psi(\lambda) = (1/2)\lambda^2$ and obviously

$$\frac{q}{q+\frac{1}{2}\lambda^2} = \frac{\sqrt{2q}}{\sqrt{2q}-\mathrm{i}\lambda} \times \frac{\sqrt{2q}}{\sqrt{2q}+\mathrm{i}\lambda} \qquad (\lambda \in \mathbb{R}).$$

The first factor in the right-hand side is the characteristic function of an infinitely divisible distribution on $[0,\infty)$ with no drift, namely the exponential law with parameter $\sqrt{2q}$. The second is the characteristic function of the negative of an exponential variable with parameter $\sqrt{2q}$. Since the Wiener-Hopf factorization is unique, this shows that in the Brownian case, $S_\tau - X_\tau$ and S_τ are two independent exponential variables with parameter $\sqrt{2q}$. Of course, it is also easy to check the latter assertion directly. In general, the Wiener-Hopf factors are known explicitly only in some special cases such as when the Lévy measure gives no mass to the positive (or the negative) half-line, see chapter VII, and in some instances when the process is stable, see Doney (1987).

Second, we turn our attention back to the bivariate Laplace exponent κ of the ladder process (L^{-1}, H).

Corollary 10 *There is a constant $k > 0$ (which depends on the normal-*

ization of the local time L) such that

$$\kappa(\alpha,\beta) = k \exp\left(\int_0^\infty dt \int_{[0,\infty)} (\mathrm{e}^{-t} - \mathrm{e}^{-\alpha t - \beta x}) t^{-1} \mathbb{P}(X_t \in dx)\right).$$

Proof For $\alpha > 1$, the formula follows from Theorem 5(ii) and (1) applied for $q = 1$. Both sides are analytic functions in the variable $\alpha > 0$, and the identity can thus be extended. □

Corollary 10 yields interesting relations between the bivariate Laplace exponents κ of the ladder process and $\widehat{\kappa}$, the exponent of the ladder process in the dual Lévy process $\widehat{X} = -X$. The latter is given by

$$\widehat{\kappa}(\alpha,\beta) = \widehat{k} \exp\left(\int_0^\infty dt \int_{(-\infty,0]} (\mathrm{e}^{-t} - \mathrm{e}^{-\alpha t + \beta x}) t^{-1} \mathbb{P}(X_t \in dx)\right).$$

In particular, we have

$$\kappa(\lambda,0) = k \exp\left(\int_0^\infty (\mathrm{e}^{-t} - \mathrm{e}^{-\lambda t}) t^{-1} \mathbb{P}(X_t \geq 0) dt\right),$$

$$\widehat{\kappa}(\lambda,0) = \widehat{k} \exp\left(\int_0^\infty (\mathrm{e}^{-t} - \mathrm{e}^{-\lambda t}) t^{-1} \mathbb{P}(X_t \leq 0) dt\right)$$

for every $\lambda > 0$. Assume now that X is not a compound Poisson process, so that $\mathbb{P}(X_t = 0) = 0$ for a.e. t. Using the Frullani identity, we deduce that for some constant $k' > 0$

$$\kappa(\lambda,0)\widehat{\kappa}(\lambda,0) = k'\lambda \qquad (\lambda > 0). \tag{3}$$

Observe also from (1) that the Wiener-Hopf factors (2) are given by

$$\Psi_q^+(\lambda) = \kappa(q,0)/\kappa(q,-\mathrm{i}\lambda) \text{ and } \Psi_q^-(\lambda) = \widehat{\kappa}(q,0)/\widehat{\kappa}(q,\mathrm{i}\lambda) \qquad (\lambda \in \mathbb{R}).$$

Letting q tend to $0+$ and using (3), we deduce

$$k'\Psi(\lambda) = \kappa(0,-\mathrm{i}\lambda)\widehat{\kappa}(0,\mathrm{i}\lambda). \tag{4}$$

One often also refers to (4) as the Wiener-Hopf factorization of the characteristic exponent.

3. Some applications of the ladder time process

When one specializes the fluctuation identities to the single variable $G_\tau = G_{\tau(q)} = \sup\{t < \tau(q) : X_t = S_t\}$, one obtains the remarkably simple formula

$$\mathbb{E}(\exp\{-\lambda G_{\tau(q)}\}) = \exp\left(\int_0^\infty (\mathrm{e}^{-\lambda t} - 1) t^{-1} \mathrm{e}^{-qt} \mathbb{P}(X_t \geq 0) dt\right). \tag{5}$$

This identity enables us to determine the asymptotic behaviour of the Lévy process, for both small and large times.

First, recall that 0 is regular for $[0,\infty)$ if the Lévy process visits $[0,\infty)$ at arbitrarily small times a.s., and this is equivalent to 0 being regular for the reflected process $S - X$. Observe also that, except in the compound Poisson case, 0 is regular for $(0,\infty)$ whenever it is regular for $[0,\infty)$, since otherwise the local suprema of X would not be distinct (see Proposition 4). We have the following criterion.

Proposition 11 (Rogozin's criterion) (i) *The origin is regular for* $[0,\infty)$ *if and only if*

$$\int_0^1 t^{-1}\mathbb{P}(X_t \geq 0)dt = \infty.$$

(ii) *When X has bounded variation, denote the drift coefficient by* d. *Then* $\lim_{t\to 0+} t^{-1}X_t = \mathrm{d}$ *a.s., in particular* 0 *is regular for* $[0,\infty)$ *if* $\mathrm{d} > 0$, *and* 0 *is irregular for* $[0,\infty)$ *if* $\mathrm{d} < 0$.

Proof (i) The origin is irregular for the reflected process if and only if the first return time of $S - X$ to 0 is positive a.s. This holds if and only if $\mathbb{P}(G_{\tau(1)} = 0) > 0$, that is if and only if

$$\lim_{\lambda\to\infty} \mathbb{E}(\exp\{-\lambda G_{\tau(1)}\}) > 0.$$

It follows from (5) that the latter condition is equivalent to that stated in the proposition.

(ii) When X has bounded variation, it can be expressed as the difference of two subordinators, and one simply invokes Proposition III.8. □

We mention for completeness that Rogozin (1968) proved that, when X has unbounded variation,

$$\limsup_{t\to 0+} t^{-1}X_t = \infty \qquad \text{a.s.};$$

and as a consequence, 0 is then regular for $[0,\infty)$. In the remaining case when X has bounded variation and zero drift, there is also an explicit test in terms of the Lévy measure to decide whether the origin is regular for $(0,\infty)$; see Bertoin (1997).

Next, we turn our attention to the asymptotic behaviour for large times.

Theorem 12 *Suppose that X is not identically zero.*

(i) *If* $\int_1^\infty t^{-1}\mathbb{P}(X_t \geq 0)dt < \infty$, *then* $\lim_{t\to\infty} X_t = -\infty$ *a.s. and we say that X drifts to* $-\infty$.

(ii) *If* $\int_1^\infty t^{-1}\mathbb{P}(X_t \le 0)dt < \infty$, *then* $\lim_{t\to\infty} X_t = \infty$ *a.s. and we say that X drifts to* ∞.

(iii) *If both (i) and (ii) fail, then* $\limsup_{t\to\infty} X_t = \limsup_{t\to\infty} \widehat{X}_t = \infty$ *a.s. and we say that X oscillates.*

Observe that 0 is a transient point for the reflected process if and only if (i) holds. Note also that the Lévy process is transient whenever it drifts to ∞ or to $-\infty$, but the converse is false (for instance, the symmetric stable process with index $\alpha \in (0,1)$ is transient (see Theorem I.17), and oscillates since then $\mathbb{P}(X_t \ge 0) = 1/2$ for all $t > 0$). On the other hand, recall that the Chung and Fuchs criterion is available in the case when $\mathbb{E}(|X_1|) < \infty$ (see Exercises I.9-10, and also Exercise 3).

Proof (i) Letting q tend to 0+ in (5), we see by monotone convergence that the last passage time of the reflected process at 0, $G_\infty = \sup\{t : X_t = S_t\}$, has Laplace transform

$$\mathbb{E}(\exp\{-\lambda G_\infty\}) = \exp\left(\int_0^\infty (e^{-\lambda t} - 1)t^{-1}\mathbb{P}(X_t \ge 0)dt\right).$$

Then letting λ tend to 0+, we deduce by dominated convergence that $G_\infty < \infty$ a.s. That is 0 is transient for the reflected process and as a consequence, the overall supremum S_∞ is finite a.s.

Plainly, the assumption implies that $\int_1^\infty t^{-1}\mathbb{P}(X_t < 0)dt = \infty$, so applying (5) to the dual process $\widehat{X} = -X$ shows that 0 is recurrent for the reflected dual process $\widehat{S} - \widehat{X} = X - I$. By Lemma 2, the dual ladder height process $\widehat{H}$ is a subordinator (there is no killing). It follows that $\widehat{S}_\infty = -I_\infty = \infty$ a.s.

Now we know that X visits $(-\infty, -x]$ for every $x > 0$, and the first passage time $T_{(-\infty,-x]}$ is finite a.s. Because $S_\infty < \infty$ a.s., we see that for any $\varepsilon > 0$, we can pick x_ε large enough such that

$$\mathbb{P}(X_t > x/2 \text{ for some } t \ge 0) < \varepsilon \qquad \text{for all } x \ge x_\varepsilon.$$

We deduce from the Markov property applied at time $T_{(-\infty,-x]}$ that

$$\mathbb{P}\left(X_t > -x/2 \text{ for some } t > T_{(-\infty,-x]}\right) \le \varepsilon.$$

This shows that X drifts to $-\infty$.

(ii) follows from (i) replacing X by $-X$.

(iii) The argument in (i) shows that both $S_\infty = \infty$ and $I_\infty = -\infty$ a.s. □

We now arrive at one of the most striking parts of the theory, known as the arcsine laws for Lévy processes. It was discovered by Lévy (1939) in the Brownian case, and then by Sparre Andersen (1953) and Spitzer (1956) for random walks. Consider the Lévy process on a fixed time interval $[0,t]$, and introduce the time it spends in $[0,\infty)$,

$$A_t = \int_0^t \mathbf{1}_{\{X_s \geq 0\}} ds \qquad (t \geq 0),$$

and the instant of its last supremum $G_t = \sup\{s < t : X_s = S_s\}$, $t \geq 0$. Recall that for $\rho \in (0,1)$, the generalized arcsine law with parameter ρ is the probability distribution on $[0,1]$ given by

$$\frac{s^{\rho-1}(1-s)^{-\rho}}{\Gamma(\rho)\Gamma(1-\rho)} ds = \frac{\sin \rho\pi}{\pi} s^{\rho-1}(1-s)^{-\rho} ds \qquad (0 < s < 1).$$

The first result concerns the special case when the probability that $X_t \geq 0$ is the same for all $t > 0$. Note that the latter holds when X is a stable process, and when X is symmetric but not a compound Poisson process (however, these are not the only examples, for instance a stable process time-changed by an independent subordinator fulfills this condition).

Theorem 13 *Assume that $\mathbb{P}(X_t \geq 0) = \rho \in (0,1)$ for all $t > 0$. Then for every $t > 0$, the random variables $t^{-1}A_t$ and $t^{-1}G_t$ are both distributed according to the generalized arcsine law with parameter ρ.*

The second result is a limit theorem in the same vein.

Theorem 14 *The following assertions are equivalent:*

(i) *$t^{-1}\int_0^t \mathbb{P}(X_s \geq 0)ds$ converges to some $\rho \in [0,1]$ as $t \to \infty$ (respectively, as $t \to 0+$).*

(ii) *$t^{-1}A_t$ converges in distribution as $t \to \infty$ (respectively, as $t \to 0+$).*

(iii) *$t^{-1}G_t$ converges in distribution as $t \to \infty$ (respectively, as $t \to 0+$).*

(iv) *The Laplace exponent $\kappa(\cdot, 0)$ of the ladder time process L^{-1} is regularly varying at 0+ (respectively, at ∞) with index ρ*

Moreover, when (i) holds, the limit law which appears in (ii) or (iii) is the generalized arcsine law with parameter ρ if $\rho \in (0,1)$, and the Dirac point mass at 0 (respectively, at 1) if $\rho = 0$ (respectively, $\rho = 1$).

Hypothesis (i) is known as *Spitzer's condition* (at 0+ or at ∞), and will appear in further important results. We point out that there exist Lévy processes for which Spitzer's condition fails, and the interested reader can easily build a counter-example making use of the forthcoming Proposition VII.6. Bertoin and Doney (1997) (extending Doney (1995) for random walks) showed recently that Spitzer's condition is equivalent to the apparently stronger condition $\lim \mathbb{P}(X_t \geq 0) = \rho$ (as t goes to 0+ or ∞).

The key to Theorems 13-14 is the following extension to continuous times of a celebrated combinatorial identity due to Sparre Andersen.

Lemma 15 (Sparre Andersen's identity) *For every $t > 0$, A_t and G_t have the same law.*

Proof For the sake of simplicity, we focus on the case $t = 1$. Consider for every integer $n > 0$, the random walk with n steps $X_k^{(n)} = X_{k/n}$, $k = 0, 1, \cdots, n$. Denote by

$$A^{(n)} = \sum_{k=1}^{n} \mathbf{1}_{\{X_{k/n} \geq 0\}}$$

the total number of indices when $X^{(n)}$ is nonnegative, and by

$$G^{(n)} = \max\{k : X_k^{(n)} \geq X_j^{(n)} \text{ for all } j = 0, \cdots, n\},$$

the index of the last maximum of $X^{(n)}$. According to Sparre Andersen's identity (see Theorem 2 of section XII.8 in Feller (1971)), the preceding two variables have the same distribution.

Assume first that X is not a compound Poisson process, so according to Proposition I.15, $\mathbb{P}(X_t = 0) = 0$ for a.e. t. Using the right-continuity of the path and the theorem of dominated convergence, we see that $n^{-1}A^{(n)}$ converges to A_1 a.s. On the other hand, it is easy to check from the fact that the paths are right-continuous with left limits that $\limsup n^{-1}G^{(n)} \leq G_1$ a.s. Similarly, using the fact that the local suprema of X are distinct (cf. Proposition 4), we have $\liminf n^{-1}G^{(n)} \geq G_1$ a.s. Hence A_1 and G_1 have also the same distribution. Finally the compound Poisson case follows by approximations, using the process $(X_t + \varepsilon t, t \in [0, 1])$ and letting ε tend to 0+. □

Proof of Theorem 13 By Lemma 15, we simply have to check that $t^{-1}G_t$ has the generalized arcsine distribution. We first rewrite (5) as

$$\int_0^\infty q\mathrm{e}^{-qt}\mathbb{E}(\exp\{-\lambda G_t\})dt = \exp\left(\rho \int_0^\infty (\mathrm{e}^{-\lambda t} - 1)t^{-1}\mathrm{e}^{-qt}dt\right)$$

$$= \left(\frac{q}{q+\lambda}\right)^\rho \qquad \text{[by the Frullani integral].}$$

Then we invert the double Laplace transform and get that for almost every $t > 0$,

$$\mathbb{P}(G_t \in ds) = s^{\rho-1}(t-s)^{-\rho}[\Gamma(\rho)\Gamma(1-\rho)]^{-1}ds \qquad (s \in [0, t]).$$

Finally, the process $t \to G_t$ is right-continuous, so the foregoing identity holds for all $t > 0$. □

Proof of Theorem 14 We see from Lemma 15 that (ii) and (iii) are equivalent, and an immediate application of Fubini's theorem shows that (ii) implies (i). The ladder time process L^{-1} is a (possibly killed) subordinator, and we know from Proposition IV.7 that G_t coincides with the left limit of L^{-1} at the instant $L(t)$ when it crosses the level t. Theorem III.6 shows that (iii) and (iv) are equivalent. Moreover (i) readily yields (ii) in the cases $\rho = 0$ and $\rho = 1$.

We assume henceforth that (i) holds with $\rho \in (0, 1)$ and aim at checking (iii). For simplicity, we will only consider the case of small time, the argument for large time being the same. Moreover (i) and Lemma 15 show that $\mathbb{E}[L^{-1}(L(t)-)]/t$ converges to ρ, so (iii) now follows from the arcsine laws for subordinators (Theorem III.6). □

Finally, we present an easy application of Theorem 14 to the ascending ladder time set, that is the zero set of the reflected process.

Corollary 16 *Assume that*

$$\lim_{t\to 0+} t^{-1} \int_0^t \mathbb{P}(X_s \geq 0)ds = \rho \in [0, 1].$$

Then the Hausdorff dimension of $\{t : X_t = S_t\}$ *is* ρ *a.s.*

Proof We know from Theorem 14 that the characteristic exponent of the ladder time process is regularly varying at ∞ with index ρ. Then Theorem III.15 implies that the Hausdorff dimension of the range of the ladder time process is ρ a.s. Finally we know from Proposition IV.7 that the range of L^{-1} differs from $\{t : X_t = S_t\}$ by a countable set. This finishes the proof. □

4. Some applications of the ladder height process

Lemma 2 exhibits two subordinators (killed at a certain rate in the transient case) which are naturally related to the supremum of a Lévy process: the ladder time and the ladder height processes. Sections 2 and 3 emphasize the rôle of the first, and we now present some typical applications of the latter.

The *renewal function* $\mathcal{V}$ associated with the ladder height process H appears in every single result of this section. It is given by

$$\mathcal{V}(x) = \int_0^\infty \mathbb{P}(H(t) \leq x)dt = \mathbb{E}\left(\int_0^\infty \mathbf{1}_{\{S_t \leq x\}} dL(t)\right), \qquad x \in [0, \infty).$$

When 0 is recurrent for the reflected process $S - X$, H is a subordinator and the definition agrees with that in section III.1. In the transient case,

H has the law of a subordinator σ killed at rate $q > 0$, and $\mathscr{V}$ is the distribution function of the q-resolvent measure of σ. In both cases, $\mathscr{V}$ is increasing and right-continuous, and its Laplace transform is given for every $\lambda > 0$ by

$$\lambda \int_0^\infty e^{-\lambda x} \mathscr{V}(x) dx = 1/\kappa(0, \lambda) \tag{6}$$

where κ is the bivariate Laplace exponent of the ladder process.

The renewal function $\mathscr{V}$ is closely related to the first instant when X crosses a positive level. Here is a simple expression for $\mathscr{V}$ in the case when the Lévy process does not oscillate.

Proposition 17 (i) *If X drifts to $-\infty$, then there exists a constant $k > 0$ such that for all $x \geq 0$*

$$\mathscr{V}(x) = k\mathbb{P}(S_\infty \leq x) = k\mathbb{P}(T_{(x,\infty)} = \infty).$$

(ii) *If X drifts to ∞, then there exists a constant $k > 0$ such that for all $x \geq 0$*

$$\mathscr{V}(x) = k\mathbb{E}(T_{(x,\infty)}).$$

(iii) *If X oscillates, then we have for every $x > 0$*

$$\mathbb{P}(S_\infty < \infty) = 0 \quad \textit{and} \quad \mathbb{E}(T_{(x,\infty)}) = \infty.$$

Proof (i) We know from Lemma 2(ii) that $L(\infty)$ has an exponential distribution, say with parameter $k > 0$, and that there is a subordinator $\sigma = (\sigma_t, t \geq 0)$ independent of $L(\infty)$ such that $H(t) = \sigma_t$ for all $t < L(\infty)$. In particular, $\mathbb{P}(H(t) \leq x) = e^{-kt}\mathbb{P}(\sigma_t \leq x)$ for all $t \geq 0$, and since σ is continuous at $L(\infty)$ a.s., we have

$$\begin{aligned}
\mathbb{P}(S_\infty \leq x) &= \mathbb{P}(H(L(\infty)-) \leq x) \\
&= \mathbb{P}(\sigma_{L(\infty)} \leq x) \\
&= \int_0^\infty k e^{-kt} \mathbb{P}(\sigma_t \leq x) dt \\
&= k \int_0^\infty \mathbb{P}(H(t) \leq x) dt = k\mathscr{V}(x).
\end{aligned}$$

(ii) For every $q > 0$, let $\tau(q)$ be an independent exponential time with parameter q. Using (1) in the second equality below, we have for every $\lambda > 0$

$$\lambda \int_0^\infty dx e^{-\lambda x} \mathbb{P}(T_{(x,\infty)} > \tau(q)) = \mathbb{E}(\exp -\lambda S_{\tau(q)}) = \kappa(q, 0)/\kappa(q, \lambda). \tag{7}$$

Next denote the bivariate Laplace exponent of the dual ladder process by $\hat{\kappa}$, and note that $\lim_{q\to 0+} \hat{\kappa}(q,0) > 0$, because the dual Lévy process drifts to $-\infty$. Then assume that X is not a compound Poisson process, so that according to (3),

$$\lim_{q\to 0+} q^{-1}\kappa(q,0) = k$$

for some constant $k > 0$. We deduce by monotone convergence

$$\lambda \int_0^\infty dxe^{-\lambda x}\mathbb{E}(T_{(x,\infty)}) = \lambda \lim_{q\to 0+} q^{-1} \int_0^\infty dxe^{-\lambda x}\mathbb{P}(T_{(x,\infty)} > \tau(q))$$
$$= k/\kappa(0,\lambda).$$

The comparison with (6) shows that $\mathbb{E}(T_{(\cdot,\infty)}) = k\mathscr{V}(\cdot)$ almost everywhere. Note that the mapping $x \to \mathbb{E}(T_{(x,\infty)})$ is right-continuous, since $T_{(x,\infty)}$ decreases to $T_{(x',\infty)}$ as $x \downarrow x'$. But $\mathscr{V}$ is right-continuous as well, so $\mathbb{E}(T_{(\cdot,\infty)}) = k\mathscr{V}(\cdot)$ everywhere.

The compound Poisson case follows readily by approximations, introducing a small drift (alternatively, one can also use the analogue for random walks, see e.g. in Feller (1971) on page 396).

(iii) We already know that the overall supremum is infinite a.s. in the oscillating case. If $\mathbb{E}(T_{(x,\infty)})$ were finite for some $x > 0$, an immediate induction based on the Markov property would then imply that $\mathbb{E}(T_{(nx,\infty)}) \leq n\mathbb{E}(T_{(x,\infty)})$ for every integer $n > 0$. As a consequence, the integral

$$\int_0^\infty e^{-y}\mathbb{E}(T_{(y,\infty)})dy$$

would be finite. But the same argument as in (ii) gives

$$\int_0^\infty e^{-y}\mathbb{E}(T_{(y,\infty)})dy = \frac{k'}{\hat{\kappa}(0,0)\kappa(0,1)}$$

and now $\hat{\kappa}(0,0) = 0$, because X (and therefore $\hat{X}$) oscillates. Thus $\mathbb{E}(T_{(x,\infty)}) = \infty$ for all $x > 0$. □

In the oscillating case, we have the following connection between the renewal function, the tail of the distribution of $T_{(x,\infty)}$ and Spitzer's condition. The result is the analogue in continuous time of a theorem due to Rogozin (1971).

Theorem 18 *The following assertions are equivalent for each $\rho \in (0,1)$:*

(i)
$$\lim_{t\to\infty} t^{-1} \int_0^t \mathbb{P}(X_s \geq 0)ds = \rho\ .$$

(ii) *There exists $x > 0$ such that the tail of the distribution function $\mathbb{P}(T_{(x,\infty)} > \cdot)$ is regularly varying at ∞ with index $-\rho$.*

Moreover, if (i) holds, then (ii) holds for every $x > 0$, and more precisely, we have for every $x, y > 0$

$$\lim_{t\to\infty} \left(\mathbb{P}(T_{(x,\infty)} > t)/\mathbb{P}(T_{(y,\infty)} > t)\right) = \mathscr{V}(x)/\mathscr{V}(y).$$

Proof For every $q > 0$, consider the function given by

$$\mathscr{V}^q(x) = \int_0^\infty \mathbb{E}(\exp\{-qL^{-1}(t)\}H_t \le x)dt, \qquad x \ge 0.$$

In particular, its Laplace transform is

$$\lambda \int_0^\infty e^{-\lambda x}\mathscr{V}^q(x)dx = 1/\kappa(q,\lambda) \qquad (\lambda > 0).$$

Inverting the Laplace transform in (7), we thus get

$$q \int_0^\infty e^{-qt}\mathbb{P}(T_{(x,\infty)} > t)dt = \kappa(q,0)\mathscr{V}^q(x) \tag{8}$$

for almost every $x \ge 0$. Recall that a.s., $T_{(x,\infty)}$ decreases to $T_{(x',\infty)}$ as $x \downarrow x'$, so both sides are right-continuous in the variable x and (8) holds for every $x \ge 0$.

By the Tauberian theorem and the monotone density theorem (cf. section O.7), we see that (ii) is equivalent to the assertion that the left-hand side of (8) is regularly varying at 0+ with index $\rho \in (0,1)$ in the variable q. Clearly $\mathscr{V}^q(x)$ increases to $\mathscr{V}(x)$ as $q \to 0+$ for each $x > 0$, so that (ii) is satisfied if and only if $\kappa(q,0)$ is regularly varying at 0+ with index $\rho \in (0,1)$. According to Theorems III.6 and 14, the latter holds if and only if $\frac{1}{t}\int_0^t \mathbb{P}(X_s \ge 0)ds$ converges to ρ as $t \to \infty$. Finally, in that case, the Tauberian theorem and (8) imply

$$\begin{aligned}&\lim_{t\to\infty} \left(\mathbb{P}(T_{(x,\infty)} > t)/\mathbb{P}(T_{(y,\infty)} > t)\right)\\ &= \lim_{q\to 0+} \mathbb{E}(1-\exp\{-qT_{(x,\infty)}\})/\mathbb{E}(1-\exp\{-qT_{(y,\infty)}\})\\ &= \mathscr{V}(x)/\mathscr{V}(y).\end{aligned}$$

□

Another connection between the renewal function and the first passage time across a positive level appears in the solution of the following problem. We say that a path $\omega \in \Omega$ started at 0 *creeps across* $x > 0$ if it enters (x,∞) continuously, that is if the first passage time in (x,∞) is not the instant of a jump of the path. The probability that the Lévy process creeps across a given level has been evaluated by Millar (1973). Here is his result.

Theorem 19 *The following assertions are equivalent:*

(i) *There exists $x > 0$ such that* $\mathbb{P}(X$ creeps across $x) > 0$.

(ii) *The ladder height process H has a positive drift coefficient.*
(iii) *The renewal measure $d\mathcal{V}$ is absolutely continuous with respect to the Lebesgue measure on $[0, \infty)$ and has a bounded density.*

Moreover, when these assertions hold, there is a version v of the density of the renewal measure $d\mathcal{V}$ which is continuous and positive on $(0, \infty)$. Finally, $\lim_{x \to 0+} v(x) = v(0+) > 0$ and $\mathbb{P}(X$ creeps across $x) = v(x)/v(0+)$ *for all $x > 0$.*

Proof It is obvious that assertions (i-iii) fail when 0 is irregular for $S - X$, because then the sample paths of the supremum process are step functions a.s. The same argument shows that (i-iii) also fail in the compound Poisson case. So we assume henceforth that 0 is regular for $(0, \infty)$.

The stopping time $T_{(x,\infty)} = \inf\{t \geq 0 : X_t > x\}$ is clearly a zero of the reflected process whenever $T_{(x,\infty)} < \infty$. By the Markov property, it cannot be a left-end point of an excursion interval of $S - X$. So according to Proposition IV.7, we have

$$T_{(x,\infty)} = L^{-1}(\ell_x), \quad \text{where } \ell_x = L(T_{(x,\infty)}).$$

We then check that $\ell_x = \inf\{t \geq 0 : H(t) > x\}$ a.s. This identity is plain when $H(\ell_x) > x$ because $X_t \leq x$ for all $t < T_{(x,\infty)}$. If $H(\ell_x) = x$, then the Markov property and the assumption that 0 is regular for $(0, \infty)$ ensure that X visits (x, ∞) immediately after $T_{(x,\infty)}$, and thus $H(t) > x$ for all $t > \ell_x$. In conclusion, we have

$$\mathbb{P}(X \text{ creeps across } x) = \mathbb{P}(H(\ell_x) = x),$$

where ℓ_x is the first passage time of the subordinator H strictly above x. All that is needed now is to invoke Theorems III.4 and III.5 (see also the discussion after Theorem III.5 when H is a killed subordinator). □

It is interesting to observe that the Gaussian coefficient of X is positive if and only if

$$\mathbb{P}(X \text{ creeps across } x) > 0 \text{ and } \mathbb{P}(\widehat{X} \text{ creeps across } y) > 0$$

for some (and then all) $x, y > 0$. This assertion follows immediately from the Wiener-Hopf factorization (4) and Theorem 19, since, according to Proposition I.2, the quantities $\Psi(\lambda)/\lambda^2$, $\kappa(0, i\lambda)/i\lambda$ and $\widehat{\kappa}(0, i\lambda)/i\lambda$ converge as $\lambda \to \infty$ respectively to the Gaussian coefficient of X, the drift coefficient of H and the drift coefficient of $\widehat{H}$.

We now conclude this section with a nice formula for the potential kernel of the Lévy process killed as it enters $(-\infty, 0)$. It was discovered by Spitzer (1964) for random walks, and we shall therefore exclude the compound Poisson case. Denote by $\widehat{\mathcal{V}}$ the renewal function of the dual ladder height process.

Theorem 20 *Suppose that X is not a compound Poisson process. There exists a constant $k > 0$ such that for every measurable function $f : [0,\infty) \to [0,\infty)$ and $x \geq 0$, one has*

$$\mathbb{E}_x\left(\int_0^{T_{(-x,0)}} f(X_t)dt\right) = k\int_{[0,\infty)} d\mathcal{V}(y)\int_{[0,x]} d\widehat{\mathcal{V}}(z)f(x+y-z).$$

Proof With no loss of generality, we may suppose that f is continuous with compact support. Introduce an independent exponential time $\tau(q)$ with parameter $q > 0$. By Theorem 5(i) and Proposition 3, we get

$$\begin{aligned}\mathbb{E}_x&\left(\int_0^{T_{(-x,0)}} f(X_t)e^{-qt}dt\right)\\ &= q^{-1}\mathbb{E}_x\left(f(X_{\tau(q)}), \tau(q) < T(-\infty,0)\right)\\ &= q^{-1}\int_{[0,\infty)} \mathbb{P}(S_{\tau(q)} \in dy)\int_{[0,x]} \widehat{\mathbb{P}}(S_{\tau(q)} \in dz)f(x+y-z).\end{aligned}$$

Then by (3), we can rewrite the foregoing as

$$k\int_{[0,\infty)} \frac{1}{\kappa(q,0)}\mathbb{P}(S_{\tau(q)} \in dy)\int_{[0,x]} \frac{1}{\widehat{\kappa}(q,0)}\widehat{\mathbb{P}}(S_{\tau(q)} \in dz)f(x+y-z).$$

On the other hand, we know by (1) and (6) that

$$\lim_{q\to 0+} \frac{1}{\kappa(q,0)}\mathbb{P}(S_{\tau(q)} \in dy) = d\mathcal{V}(y)$$

in the sense of vague convergence, and we have a similar result for the dual. This entails the theorem. □

5. Increase times

This final section, which is essentially taken from Doney (1995-a), is devoted to a problem which was first studied by Dvoretzky, Erdös and Kakutani (1961) in the Brownian case, and in which fluctuation theory has a key rôle. We say that an instant $t > 0$ is an *increase time* for the path ω if for some $\varepsilon > 0$

$$\omega(t') \leq \omega(t) \leq \omega(t'') \qquad \text{for all } t' \in [t-\varepsilon, t] \text{ and } t'' \in [t, t+\varepsilon].$$

We are concerned with the existence of increase times on the sample path of a Lévy process. Plainly, the probability that there exist increase times necessarily equals 0 or 1.

There are two simple cases for which one can immediately settle the alternative that applies. First, when 0 is irregular for $(-\infty, 0)$, the Markov property and the duality lemma show that any fixed time $t > 0$ is an increase time a.s. Second, when 0 is irregular for $(0, \infty)$ and is not a compound Poisson process (so that 0 is regular for $(-\infty, 0)$), one readily deduces from the Markov property applied at the (discrete) ladder times that there exist no increase times a.s. These two trivial cases will be implicitly excluded in the sequel, that is we assume from now on that *0 is regular for $(-\infty, 0)$ and for $(0, \infty)$.*

We point out that the Markov property implies that any given stopping time is not an increase time a.s. In particular, a fixed time t, or the instant of a positive jump, is not an increase time a.s. Here, a.s. refers to an event of zero probability which depends on t and one should not conclude that there exist no increase times a.s.

We now introduce notation relevant for the study. As usual, let $\tau = \tau(1)$ be an exponential time with parameter 1 which is independent of X. Denote by F and $\widehat{F}$ the distribution functions of S_τ and $\widehat{S}_\tau$,

$$F(x) = \mathbb{P}(S_\tau \le x)\,,\ \widehat{F}(x) = \mathbb{P}(\widehat{S}_\tau \le x) \qquad (x \ge 0).$$

To start with, we note simple bounds connecting F and the renewal function $\mathcal{V}$ of the ladder height process H. Of course, similar bounds hold for $\widehat{F}$ and $\widehat{\mathcal{V}}$, the renewal function of the dual ladder height process $\widehat{H}$.

Lemma 21 *The functions F and $\mathcal{V}$ are continuous, and there exists a constant $c > 0$ such that $c\mathcal{V}(x) \le F(x) \le c^{-1}\mathcal{V}(x)$ for all $x > 0$ small enough.*

Proof The calculation in the proof of Lemma 8 shows that

$$\begin{aligned}\int_{[0,\infty)} e^{-\beta x} dF(x) &= \mathbb{E}\left(\exp\{-\beta S_\tau\}\right) \\ &= \kappa(1,0)\mathbb{E}\left(\int_0^\infty e^{-t}\exp\{-\beta S_t\} dL(t)\right)\end{aligned}$$

so that inverting the Laplace transform gives

$$F(x) = \kappa(1,0)\mathbb{E}\left(\int_0^\infty \mathbf{1}_{\{S_s \le x\}} e^{-s} dL(s)\right).$$

This entails immediately $F(x) \le \kappa(1,0)\mathscr{V}(x)$ and also that

$$F(x) \ge e^{-1}\kappa(1,0)\mathbb{E}\left(\int_0^1 \mathbf{1}_{\{S_s\le x\}} dL(s)\right).$$

On the other hand, we have

$$\begin{aligned}\mathscr{V}(x) &= \mathbb{E}\left(\int_0^1 \mathbf{1}_{\{S_s\le x\}} dL(s)\right) + \mathbb{E}\left(\int_1^\infty \mathbf{1}_{\{S_s\le x\}} dL(s)\right)\\ &\le \frac{eF(x)}{\kappa(1,0)} + \mathbb{P}(S_1 \le x)\mathscr{V}(x),\end{aligned}$$

where the inequality stems from the Markov property. This completes the proof of the lemma, provided for instance that x is such that $\mathbb{P}(S_1 \le x) \le 1/2$.

Finally, the continuity assertion follows from Proposition I.15 and the fact that H is not a compound Poisson process (because 0 is regular for $(0,\infty)$). □

Then recall that $T_x = \inf\{t \ge 0 : X_t > x\}$ stands for the first passage time above x, and put $\widehat{T}_x = \inf\{t \ge 0 : -X_t > x\}$. It will be convenient to write indifferently Y_t and $Y(t)$ in the sequel. The key to the study lies in the following characterization.

Lemma 22 *The probability that there exist increase times equals* 1 *if*

$$\lim_{\varepsilon\to 0+} \widehat{F}(\varepsilon)^{-1} \int_{[0,\infty)} \mathbb{P}\left(y < S(\widehat{T}_\varepsilon) - X(\widehat{T}_\varepsilon), \widehat{T}_\varepsilon < \tau\right) F(dy) < \infty$$

and 0 *otherwise.*

Using Lemma 21, we see that we may replace $\widehat{F}$ by $\widehat{\mathscr{V}}$ in Lemma 22. An integration by parts also shows that we may replace F by $\mathscr{V}$.

Proof Consider the set of paths having global increase times on $[0,\tau]$,

$$\mathscr{I} = \{\exists t \in [0,\tau] : X_t = S_t \text{ and } X_s \ge X_t \text{ for all } s \in [t,\tau]\}.$$

Since 0 is regular for $(-\infty,0)$, $I_\tau < 0$ and $X_\tau < S_\tau$ a.s. (by Proposition 3) and we may replace the closed interval $[0,\tau]$ in the definition of $\mathscr{I}$ by the open interval $(0,\tau)$. It can then be immediately seen that $\mathbb{P}(\mathscr{I}) = 0$ if and only if there are no increase times a.s. For every $\varepsilon > 0$, introduce the set of paths having approximate increase times,

$$\mathscr{I}_\varepsilon = \{\exists t \in [0,\tau] : X_t = S_t \text{ and } X_s \ge X_t - \varepsilon \text{ for all } s \in [t,\tau]\}.$$

The family $\mathscr{I}_\varepsilon$ decreases as $\varepsilon \downarrow 0$ to some set which plainly contains $\mathscr{I}$. Conversely, take any path $\omega \in \mathscr{I}_\varepsilon$ for every $\varepsilon > 0$, and consider the début

$$\begin{aligned}d(\varepsilon) = \inf\{t \in [0,\tau] :\ & X_t(\omega) = S_t(\omega)\\ & \text{and } X_s(\omega) \ge X_t(\omega) - \varepsilon \text{ for all } s \in [t,\tau]\}.\end{aligned}$$

Then $d(\varepsilon)$ increases to, say, $d \in [0, \tau]$ as $\varepsilon \downarrow 0$ and one has

$$X_{d-}(\omega) = S_{d-}(\omega) \text{ and } X_s(\omega) \geq X_{d-}(\omega) \text{ for all } s \in [d, \tau].$$

It follows from the assumption that 0 is regular for $(-\infty, 0)$ that a.s., $X_{j-} < S_{j-}$ when j is the instant of a jump (just apply the Markov property to the time-reversed path). Thus $\omega \in \mathscr{I}$ except for a set of paths of zero probability, and in conclusion we have

$$\mathbb{P}(\mathscr{I}) = \lim_{\varepsilon \to 0+} \mathbb{P}(\mathscr{I}_\varepsilon). \tag{9}$$

To calculate $\mathbb{P}(\mathscr{I}_\varepsilon)$, we put $\sigma(0) = 0$ and then consider the first passage time below $X_{\sigma(0)} - \varepsilon$, $\rho(1) = \widehat{T}_\varepsilon$, then the first instant $\sigma(1)$ after $\rho(1)$ at which X reaches a new supremum, and more generally we introduce by induction

$$\sigma(k) = \inf\{t > \rho(k) : X_t > S_{\rho(k)}\},$$
$$\rho(k+1) = \inf\{t > \sigma(k) : X_t - X_{\sigma(k)} < -\varepsilon\}.$$

The motivation for this definition is that $\mathscr{I}_\varepsilon \neq \emptyset$ if and only if there exists an integer k with $\sigma(k) \leq \tau \leq \rho(k+1)$; more precisely $\sigma(k)$ then coincides with the début $d(\varepsilon)$. We thus note the identity

$$\mathscr{I}_\varepsilon = \{\sigma(k) \leq \tau \leq \rho(k+1) \text{ for some integer } k \geq 0\},$$

so that by the Markov property and the lack of memory of the exponential law

$$\mathbb{P}(\mathscr{I}_\varepsilon) = \sum_k \mathbb{P}(\sigma(k) \leq \tau \leq \rho(k+1)) = \mathbb{P}(\rho(1) \geq \tau) \sum_k \mathbb{P}(\sigma(k) \leq \tau).$$

The Markov property at the $\sigma(k)$'s and the lack of memory of the exponential law also show that $\mathbb{P}(\sigma(k) \leq \tau) = \mathbb{P}(\sigma(1) \leq \tau)^k$, and hence

$$\mathbb{P}(\mathscr{I}_\varepsilon) = \mathbb{P}(\rho(1) \geq \tau) / \mathbb{P}(\sigma(1) > \tau). \tag{10}$$

On the one hand, $\mathbb{P}(\rho(1) \geq \tau) = \mathbb{P}(\widehat{T}_\varepsilon \geq \tau) = \widehat{F}(\varepsilon)$. On the other hand, the Markov property at $\rho(1) = \widehat{T}_\varepsilon$ gives

$$\begin{aligned}\mathbb{P}(\sigma(1) > \tau) &= \mathbb{P}(\rho(1) \geq \tau) + \mathbb{P}(\rho(1) < \tau < \sigma(1)) \\ &= \widehat{F}(\varepsilon) + \mathbb{E}\left(F(S(\widehat{T}_\varepsilon) - X(\widehat{T}_\varepsilon)), \widehat{T}_\varepsilon < \tau\right) \\ &= \widehat{F}(\varepsilon) + \int_{[0,\infty)} \mathbb{P}\left(y < S(\widehat{T}_\varepsilon) - X(\widehat{T}_\varepsilon), \widehat{T}_\varepsilon < \tau\right) F(dy).\end{aligned}$$

By (9-10), this establishes the lemma. □

Lemma 22 is nice because it provides a necessary and sufficient condition for the existence of increase times, but is not really satisfactory since it involves the distribution of $S(\widehat{T}_\varepsilon) - X(\widehat{T}_\varepsilon)$ which is not known explicitly. Our next purpose is to deduce from Lemma 22 simpler criteria in terms of the renewal functions $\mathscr{V}$ and $\widehat{\mathscr{V}}$ for either the existence or the absence of increase times. Further explicit results will be given when X is stable (Proposition VIII.7) and when X has no positive jumps (Corollary VII.9 and Proposition VII.10). First, one has a simple condition that guarantees the absence of increase times.

Corollary 23 *Suppose that* $\liminf_{\varepsilon\to 0+}\left(\widehat{\mathscr{V}}(\varepsilon)/\mathscr{V}(\varepsilon)\right) = 0$. *Then there are no increase times a.s.*

Proof For every $y \in [0,\varepsilon]$, we have

$$\mathbb{P}\left(y < S(\widehat{T}_\varepsilon) - X(\widehat{T}_\varepsilon), \widehat{T}_\varepsilon < \tau\right) = \mathbb{P}\left(\widehat{T}_\varepsilon < \tau\right) = 1 - \widehat{F}(\varepsilon)$$

so the integral in Lemma 22 is at least $(1 - \widehat{F}(\varepsilon))F(\varepsilon)$. Our assertion follows now from Lemma 21. □

It is interesting to observe that the hypothesis of Corollary 23 is fulfilled when $\widehat{X}$ can creep across a positive level (see Theorem 19) and the Gaussian coefficient is zero. Indeed, then X does not creep across a positive level a.s. (by the discussion at the end of section 4), and it follows from Theorems 19 and II.16(iv) that

$$\lim_{\varepsilon\to 0+} \varepsilon^{-1}\widehat{\mathscr{V}}(\varepsilon) < \infty \quad , \quad \lim_{\varepsilon\to 0+} \varepsilon^{-1}\mathscr{V}(\varepsilon) = \infty.$$

We mention that more generally, any Lévy process which can creep downwards (even with a positive Gaussian coefficient) has no increase times a.s., see Bertoin (1995-b).

We next present a more precise condition for the absence of increase times.

Corollary 24 *Suppose that* $\widehat{\mathscr{V}}$ *is regularly varying at* $0+$ *with index* $\alpha \in (0,1)$ *and that the integral* $\int_{0+}\left(\mathscr{V}(dx)/\widehat{\mathscr{V}}(x)\right)$ *diverges. Then there are no increase times a.s.*

Proof Denote by $\widehat{\Phi} = \widehat{\kappa}(0,\cdot)$ the Laplace exponent of the dual ladder height process $\widehat{H}$ and by $\widehat{\mu}$ its Lévy measure. The assumption that $\widehat{\mathscr{V}}$ is regularly varying with index $\alpha < 1$ implies that $\widehat{H}$ has zero drift coefficient, see e.g. Theorem III.5. Moreover, since $1/\widehat{\Phi}$ is the Laplace transform of the renewal measure $\widehat{\mathscr{V}}$, an Abelian theorem shows that $\widehat{\mathscr{V}}(x) \sim c\widehat{\Phi}(1/x)$ for some constant $c > 0$, and in particular $\widehat{\Phi}$ is regularly varying at ∞. It follows then from the Tauberian theorem and the

Lévy-Khintchine formula that

$$1/\widehat{\mathcal{V}}(x) \sim c'\widehat{\mu}(x,\infty) \qquad (x \to 0+) \tag{11}$$

for some constant $c' > 0$.

Then denote the first passage time of $\widehat{H}$ above ε by $\sigma = \inf\{t \geq 0 : \widehat{H}_t > \varepsilon\}$ and deduce from Proposition III.2 that for every $y > \varepsilon$

$$\mathbb{P}(\widehat{H}_\sigma > y) = \int_{[0,\varepsilon]} \widehat{\mathcal{V}}(dx)\widehat{\mu}(y-x,\infty) \geq \widehat{\mathcal{V}}(\varepsilon)\widehat{\mu}(y,\infty). \tag{12}$$

On the other hand, we clearly have $\widehat{H}_\sigma = \widehat{X}(\widehat{T}_\varepsilon)$ and thus

$$\begin{aligned}\mathbb{P}\left(y < S(\widehat{T}_\varepsilon) - X(\widehat{T}_\varepsilon), \widehat{T}_\varepsilon < \tau\right) &\geq \mathbb{P}\left(y < S(\widehat{T}_\varepsilon) - X(\widehat{T}_\varepsilon)\right) - \widehat{F}(\varepsilon)\\ &\geq \mathbb{P}\left(y < -X(\widehat{T}_\varepsilon)\right) - \widehat{F}(\varepsilon)\\ &= \mathbb{P}(\widehat{H}_\sigma > y) - \widehat{F}(\varepsilon).\end{aligned}$$

We now deduce from (11-12) and Lemma 21 that for y and ε small enough, the foregoing quantity is greater than $(c''\widehat{\mathcal{V}}(\varepsilon)/\widehat{\mathcal{V}}(y)) - \mathcal{V}(\varepsilon)$ for some constant $c'' > 0$. The assumption that $\int_{0+}\left(\mathcal{V}(dx)/\widehat{\mathcal{V}}(x)\right) = \infty$ and Lemma 21 thus show that the limit in Lemma 22 is infinite, and there are no increase times a.s. □

In the converse direction, the following condition guarantees the existence of increase times.

Corollary 25 *Suppose that the integral $\int_{0+}\left(\mathcal{V}(dx)/\widehat{\mathcal{V}}(x)\right)$ converges. Then there exist increase times a.s.*

Proof We first point out the following inequality for the renewal measure:

$$\widehat{\mathcal{V}}(2x) \leq 2\widehat{\mathcal{V}}(x) \qquad (x \geq 0) \tag{13}$$

which is plain from the Markov property applied at the first instant when $\widehat{H}$ exceeds $x > 0$. Then we observe that, since the integral converges, $\mathcal{V} = o(\widehat{\mathcal{V}})$ at $0+$, so by Lemma 21, all that is needed is to check

$$\lim_{\varepsilon\to 0+} \widehat{\mathcal{V}}(\varepsilon)^{-1} \int_{[4\varepsilon,\infty)} \mathbb{P}\left(y < S(\widehat{T}_\varepsilon) - X(\widehat{T}_\varepsilon), \widehat{T}_\varepsilon < \tau\right) F(dy) < \infty. \tag{14}$$

Now, for $y \geq 4\varepsilon$, denote by $T_{y/2}$ the first instant when X exceeds $y/2$ and observe that the integrand in (14) is bounded from above by

$$\mathbb{P}\left(T_{y/2} \leq \widehat{T}_\varepsilon \wedge \tau\right) + \mathbb{P}\left(y/2 < -X(\widehat{T}_\varepsilon)\right) = A + B.$$

To get an upper bound for A, we note the inequality

$$\mathbb{P}_x\left(\widehat{T}_\varepsilon > \tau\right) = \widehat{F}(x+\varepsilon) \geq \widehat{F}(y/2) \qquad \text{for all } x \geq y/2$$

and use the Markov property at $T_{y/2}$ in the second inequality below:

$$\widehat{F}(\varepsilon) = \mathbb{P}(\widehat{T}_\varepsilon > \tau) \geq \mathbb{P}\left(T_{y/2} \leq \widehat{T}_\varepsilon \wedge \tau, \widehat{T}_\varepsilon > \tau\right)$$
$$\geq \mathbb{P}\left(T_{y/2} \leq \widehat{T}_\varepsilon \wedge \tau\right) \widehat{F}(y/2).$$

So we find $A \leq \widehat{F}(\varepsilon)/\widehat{F}(y/2)$.

Then we re-express B in terms of the dual ladder height process $\widehat{H}$ and its first passage time σ above ε, as $B = \mathbb{P}(H_\sigma > y/2)$. Applying Proposition III.2 to evaluate this quantity in terms of the renewal function $\widehat{\mathcal{V}}$ and the Lévy measure $\widehat{\mu}$ of $\widehat{H}$, we get (recall that $y \geq 4\varepsilon$)

$$B = \int_0^\varepsilon \widehat{\mathcal{V}}(dx)\widehat{\mu}(y/2 - x, \infty) \leq \widehat{\mathcal{V}}(\varepsilon)\widehat{\mu}(y/2 - \varepsilon, \infty) \leq \widehat{\mathcal{V}}(\varepsilon)\widehat{\mu}(y/4, \infty\sigma).$$

On the other hand, note that for all $r > 0$

$$\widehat{\mathcal{V}}(r)\widehat{\mu}(r, \infty) \leq \int_0^r \widehat{\mathcal{V}}(dx)\widehat{\mu}(r - x, \infty) \leq 1$$

where the second inequality stems again from Proposition III.2. In conclusion, we find $B \leq \widehat{\mathcal{V}}(\varepsilon)/\widehat{\mathcal{V}}(y/4)$.

Invoking Lemma 21, we thus see that (14) holds provided that

$$\int_{0+} \frac{d\mathcal{V}(y)}{\widehat{\mathcal{V}}(y/4)} < \infty.$$

But using (13), we get

$$\int_{0+} \frac{d\mathcal{V}(y)}{\widehat{\mathcal{V}}(y/4)} \leq 4 \int_{0+} \frac{d\mathcal{V}(y)}{\widehat{\mathcal{V}}(y)} < \infty$$

and the proof is now complete. □

Very recently, Fourati (1998) proposed a different method based on the general theory of processes to study the existence of increase times. She proved that $\int_{0+} \left(d\mathcal{V}(y)/\widehat{\mathcal{V}}(y)\right) < \infty$ is a necessary and sufficient condition for the existence of increase times. In other words, the hypothesis that $\widehat{\mathcal{V}}$ is regularly varying in Corollary 24 is superfluous.

6. Exercises

1. *(Upwards passage times)* For every $x > 0$, denote the first passage time above x by $T(x) = \inf\{t \geq 0 : X_t > x\}$, and the so-called overshoot by $K(x) = X_{T(x)} - x$.
(a) Check that $(T(x), K(x)) = (L^{-1}(\tau_x), H(\tau_x) - x)$ a.s., where $\tau_x = \inf\{t \geq 0 : H(t) > x\}$.

(b) Deduce the following formula: For every $\alpha, \beta, q > 0$

$$\int_0^\infty e^{-qx}\mathbb{E}\left(\exp\{-\alpha T(x) - \beta K(x)\}\right) dx = \frac{\kappa(\alpha, q) - \kappa(\alpha, \beta)}{(q-\beta)\kappa(\alpha, q)}.$$

2. *(A limit theorem for the reflected process)* Suppose that X drifts to ∞. Check that $X_t - I_t$ converges in law as $t \to \infty$ and that the repartition function of the limit is proportional to the renewal function $\mathscr{V}$ of the ladder height process H.

3. *(Moment conditions)* Suppose that the first moment of X_1 exists, $\mathbb{E}(X_1) = \mu \in (-\infty, \infty)$.
(a) Prove that $T_{(1,\infty)} = \inf\{t \geq 0 : X_t > 1\}$ has a finite expectation if and only if $\mu > 0$. [Hint: assume first that the support of the Lévy measure is bounded from above and apply Doob's optional sampling theorem to the martingale $X_t - \mu t$.]
(b) Deduce that X drifts to ∞, oscillates, or drifts to $-\infty$ according as $\mu > 0$, $\mu = 0$ or $\mu < 0$. [Actually, a stronger statement is true: X is recurrent when $\mu = 0$, see Exercise I.10, and also the strong law of large numbers in Exercise I.9.]
(c) Prove that $\mathbb{E}(H(1)) < \infty$ and $\hat{\mathbb{E}}(H(1)) < \infty$ if and only if $\mu = 0$ and $\mathbb{E}(X_1^2) < \infty$.

4. *(Cramér's estimate)* Suppose that there exists $\nu > 0$ such that

$$\mathbb{E}(\exp \nu X_1) = 1.$$

(a) Check that the characteristic exponent Ψ has an analytic extension to the complex strip $\Im(z) \in [-\nu, 0]$ and that $\Psi^\natural(\lambda) = \Psi(\lambda - i\nu)$ $(\lambda \in \mathbb{R})$ is the characteristic exponent of a Lévy process, which will be denoted by $X^\natural$ in the sequel. Deduce from Exercise 3 that X drifts to $-\infty$ and $X^\natural$ drifts to ∞.
(b) Check that the renewal function $\mathscr{V}^\natural$ of the ladder height process of $X^\natural$ has $\mathscr{V}^\natural(dx) = ke^{\nu x}\mathscr{V}(dx)$ for some constant $k > 0$. Then deduce from the renewal theorem that

$$\lim_{x\to\infty} e^{\nu x}\mathbb{P}(S_\infty > x) = C$$

for some constant $C < \infty$, and that $C > 0$ if and only if $\mathbb{E}(|X_1^\natural|) < \infty$.

5. *(Occupation measure of the excursion law)* Let $\hat{n}$ stand for the excursion measure of the dual reflected process $\hat{S} - \hat{X} = X - I$. Prove that for every Borel function $f : (0, \infty) \to [0, \infty)$

$$\hat{n}\left(\int_0^\zeta f(\epsilon(t))dt\right) = \int_{(0,\infty)} f(x)d\mathscr{V}(x),$$

where $\epsilon = (\epsilon(t), 0 \leq t < \zeta)$ denotes the generic excursion with lifetime ζ. [Hint: use the expression for the entrance law under the excursion measure given in section IV.4.] This identity shows that the measure $d\mathcal{V}$ is invariant for the dual reflected process $X - I$ whenever the latter is recurrent, that is whenever X does not drift to ∞. See Dellacherie, Maisonneuve and Meyer (1992).

6. *(Heavy supremum)* One says that a Lévy process has a heavy supremum if $\int_0^\infty \mathbb{P}(X_t = S_t)dt > 0$.
(a) Prove that X has a heavy supremum if and only if the ladder height process of $\hat{X} = -X$ is a (possibly killed) compound Poisson process.
(b) Shows that the condition
$$\int_0^1 t^{-1}\mathbb{P}(X_t < 0)dt < \infty$$
is necessary and sufficient for heaviness of the supremum.
(c) Prove that the supremum is heavy for both a Lévy process and its dual only in the compound Poisson case.

7. *(An invariant function for the dual process)* For every $q > 0$, denote by $\mathcal{V}^q$ the distribution function of the q-resolvent measure of the ladder height process. Using the identity (8), check that the process
$$e^{-qt}\mathcal{V}^q(X_t)\mathbf{1}_{\{t<T_{(-\infty,0)}\}} - \frac{1-\exp\{-q(t \wedge T_{(-\infty,0)})\}}{\kappa(q,0)} \qquad (t \geq 0)$$
is a $\widehat{\mathbb{P}}_x$-martingale for every $x > 0$. Deduce that
$$\mathcal{V}(X_t)\mathbf{1}_{\{t<T_{(-\infty,0)}\}} \qquad (t \geq 0)$$
is also a $\widehat{\mathbb{P}}_x$-martingale for every $x > 0$ whenever X does not drift to $-\infty$.

8. *(Another fluctuation identity)* Prove the following identity between measures on $(0,\infty) \times (0,\infty)$:
$$\mathbb{P}(X_t \in dx)dt = t\int_0^\infty \mathbb{P}(L^{-1}(u) \in dt, H(u) \in dx)u^{-1}du\,.$$
[Hint: calculate the bivariate Laplace transform of each term and use Corollary 10.]

9. *(Creeping with bounded variation)* Suppose that X has bounded variation and drift coefficient d. Prove that the probability that X creeps across $x > 0$ is zero if and only if $\mathrm{d} \leq 0$. [Hint: use the Wiener-Hopf factorization (4), Proposition I.2 and Proposition 11.]

7. Comments

The principle of reflection goes back to Désiré André, it can be viewed as a variation of Kelvin's method of images. It was further developed by Lévy (1965) and Skorohod (1965) in the Brownian case. The Markov property of the reflected process for Lévy processes has been implicitly used by many authors including Fristedt (1974). It seems that the first rigorous proof is in Bingham (1975) to whom we refer for a survey of applications of fluctuation theory in applied probability, see also Borovkov (1976). The ladder process is a key concept for fluctuation theory in discrete time, see Spitzer (1964) and Feller (1971); it has been introduced in continuous time by Rubinovitch (1971). Millar (1981) observed an interesting connection between the rate of growth of a Lévy process at a local minimum and that of the ladder process at the origin.

The first results on the distribution of suprema of Lévy processes are in Baxter and Donsker (1957), Rogozin (1966) and Shtatland (1966). The fluctuation identities of Theorem 5 were derived from an analogous identity for random walks by Pecherskii and Rogozin (1969), see also Gusak and Korolyuk (1969). Exercise 1 presents a closely related identity due to Gusak (1969), which is sometimes referred to as the second factorization identity. See also Pecherskii (1974) and McGill (1989, 1989-a) for an alternative approach and applications to Brownian motion. The fluctuation identity of Exercise 8 is taken from Bertoin and Doney (1995). Lemma 6 is due to Millar (1977, 1978), Lemma 7 to Huff (1969) and Corollary 10 to Fristedt (1974); see also Rubinovitch (1971), Prabhu and Rubinovitch (1973). We refer to Prabhu (1972) and Greenwood (1975, 1975-a, 1976) for more on the Wiener-Hopf factorization for Lévy processes.

Proposition 11 and Theorem 12 are from Rogozin (1966). A possible drawback of these results is that they are given in terms of the probabilities $\mathbb{P}(X_t \geq 0)$ which are usually not known explicitly. Erickson (1973) obtained a characterization of random walks that drift to ∞ in terms of their step-distributions; and its immediate continuous-time analogue provides a characterization of Lévy processes that drift to ∞ in terms of the tails of their Lévy measure. The arcsine laws are due to Spitzer (1956) in discrete time, a different proof in continuous time has been recently given by Getoor and Sharpe (1994). Recent developments in connection with the arcsine laws appear in Fitzsimmons and Getoor (1995), Knight (1996), Getoor and Sharpe (1994-a) and Bertoin and Yor (1996). The key Lemma 15 originates from Sparre Andersen (1953); Bertoin (1993) presents a closely related identity involving processes, see also Doney (1993). Wendel (1960) proved another remarkable identity involving the time spent above some fixed level, see also Port (1963).

We refer to Doney (1982-a, 1989) and Greenwood and Novikov (1986) for further results in connection with Theorem 18. See also Berman (1986), Braverman (1997), Braverman and Samorodnitsky (1995) and Willekens (1987) for estimates of the distribution function of the supremum S_1 in terms of the Lévy measure. Theorem 19 is due to Millar (1973), after early work of Rogozin (1965), and Millar also gives more explicit criteria for creeping. Rogers (1984) has an alternative approach. Theorem 20 and Exercises 5 and 7 are from Silverstein (1980), Exercise 4 from Bertoin and Doney (1994-a). We refer to Doney (1982) and Chow (1986) for developments (in discrete time) in connection with Exercise 3.

The notion of increase time was introduced by Dvoretzky, Erdös and Kakutani (1961), who proved that the Brownian motion never increases a.s. This result was extended to certain Lévy processes by Bertoin (1991, 1994, 1995-b). The present approach to studying the existence of increase times for a general Lévy process follows Doney (1995-a) who adapts ideas of Adelman (1985) and Burdzy (1990).

VII

Lévy Processes with no Positive Jumps

The (continuous) supremum process S of a Lévy process X with no positive jumps is a local time for the reflected process $S - X$. Fluctuation theory can then be developed in a remarkably simple and complete form. In particular, the probability that X makes its first exit from an interval at the upper boundary is evaluated explicitly in terms of the so-called scale function. This enables us to construct the law of a new process which can be thought of as X conditioned to stay positive, and which displays many interesting connections with X.

1. Fluctuation theory with no positive jumps

Lévy processes with no positive jumps form a remarkable class of real-valued Lévy processes, which were first studied in applied probability as models for queuing, insurance risk, and dam theory. See in particular Borovkov (1976) and Prabhu (1981). They also have a great interest from a more theoretical point of view as well, because they are the processes for which fluctuation theory takes the nicest form and can be developed explicitly to its full extent.

Throughout this chapter, we suppose that X is a real-valued Lévy process with no positive jumps, that is its Lévy measure has support in $(-\infty, 0]$. Some authors say that X is spectrally negative. The degenerate case when X is either the negative of a subordinator or a deterministic drift has no interest and will be implicitly excluded in the sequel.

To start with, we observe that, although X_t may take values of both signs, its exponential moments are finite, i.e.

$$\mathbb{E}(\exp\{\lambda X_t\}) < \infty \qquad \text{for all } \lambda > 0. \tag{1}$$

To see this, note first that due to the absence of positive jumps, for all $a \geq 0$,

$$X_{T_{[a,\infty)}} = a, \quad \mathbb{P}\text{-a.s. on } \{T_{[a,\infty)} < \infty\},$$

where $T_{[a,\infty)}$ denote the first passage time above a. Next, consider an independent exponential time $\tau(q)$ with parameter $q > 0$. Using the lack of memory of the exponential law, we deduce from the Markov property applied at time $T_{[a,\infty)}$ that for every $a, b > 0$

$$\mathbb{P}(T_{[a+b,\infty)} < \tau(q)) = \mathbb{P}(T_{[a,\infty)} < \tau(q))\mathbb{P}(T_{[b,\infty)} < \tau(q)),$$

so that the supremum $S_{\tau(q)}$ of X on the time interval $[0, \tau(q)]$ has an exponential distribution. Denote its parameter by $\Phi(q)$ (this quantity will be specified in the sequel). Since $S_{\tau(q)}$ converges in probability to 0 as q tends to ∞, we have $\Phi(q) > \lambda$ provided that q is chosen large enough. This implies (1).

The characteristic function $\lambda \to \mathbb{E}(\exp\{i\lambda X_t\})$ ($\lambda \in \mathbb{R}$) can therefore be extended to define an analytic function in the complex lower half-plane $\{\Im(\lambda) \leq 0\}$. On the other hand, the Lévy-Khintchine formula and the fact that the Lévy measure vanishes on the positive semi-axis show that the characteristic exponent $\Psi(\lambda)$ is well defined and analytic on $\{\Im(\lambda) \leq 0\}$. Hence, if we introduce the notation

$$\psi(\lambda) = -\Psi(-i\lambda) = a\lambda + \frac{1}{2}Q\lambda^2 + \int_{(-\infty,0)} (e^{\lambda x} - 1 - \lambda x \mathbf{1}_{\{x > -1\}})\Pi(dx)\,,$$

we then see that the identity

$$\mathbb{E}(\exp\{\lambda X_t\}) = \exp\{t\psi(\lambda)\}$$

holds whenever $\Re(\lambda) \geq 0$. We call ψ the Laplace exponent of X. The mapping $\psi : [0, \infty) \to (-\infty, \infty)$ is strictly convex (by Hölder's inequality) and $\lim_{\lambda\to\infty} \psi(\lambda) = \infty$, because $\mathbb{P}(X_1 > 0) > 0$ since the case of the negative of a subordinator has been excluded.

The main weakness of the fluctuation identities for general Lévy processes is that they involve the one-dimensional distributions of the process which are seldom explicitly known. In the case of the absence of positive jumps, these identities can be expressed directly in terms of the basic data, the Laplace exponent ψ and its inverse Φ which we now introduce. We denote by $\Phi(0)$ the largest solution of the equation $\psi(\lambda) = 0$. Observe that 0 is always a solution and that if $\Phi(0) > 0$, then by strict convexity, 0 and $\Phi(0)$ are the only solutions. In all cases, the mapping

$\psi : [\Phi(0), \infty) \to [0, \infty)$ is continuous and increasing, it is a bijection and we denote the inverse bijection by $\Phi : [0, \infty) \to [\Phi(0), \infty)$,

$$\psi \circ \Phi(\lambda) = \lambda \qquad (\lambda > 0).$$

The key to the simplification of the fluctuation identities is provided by the following theorem.

Theorem 1 *The point 0 is regular for $(0, \infty)$ and the continuous increasing process $S_t = \sup\{X_s : 0 \le s \le t\}$ is a local time at 0 for the reflected process $S - X$. Its right-continuous inverse,*

$$T(x) = \inf\{s \ge 0 : S_s > x\} = \inf\{s \ge 0 : X_s > x\} \qquad (x \ge 0)$$

is a subordinator, killed at an independent exponential time if X drifts to $-\infty$; its Laplace exponent is Φ.

Proof We have already observed that S evaluated at an independent exponential time τ has an exponential distribution, in particular $\mathbb{P}(S_\tau = 0) = 0$. This implies that X visits $(0, \infty)$ at arbitrarily small times a.s. Plainly, S is a continuous increasing process that increases only when $S - X = 0$, and is not identically zero. Moreover, the identity

$$S_{t+s} = S_t + \sup\{0 \vee (X_{t+u} - S_t) : 0 \le u \le s\}$$

shows that the hypotheses of Proposition IV.5 are fulfilled, and S is therefore a local time at 0 for the reflected process.

We know from Lemma VI.2 that its inverse is a subordinator (killed at some rate in the case when X drifts to $-\infty$). To determine its Laplace exponent, we observe that $\mathbb{E}(\exp\{\Phi(\lambda)X_t\}) = e^{t\lambda}$ for all $\lambda > 0, t \ge 0$. It follows from the independence and stationarity of the increments that the process

$$\exp\{\Phi(\lambda)X_t - \lambda t\}, \qquad t \ge 0,$$

is a martingale. Applying the optional sampling theorem at the bounded stopping time $T(x) \wedge t$, we get

$$\mathbb{E}(\exp\{\Phi(\lambda)X(T(x) \wedge t) - \lambda(T(x) \wedge t)\}) = 1.$$

The absence of positive jumps implies that $\Phi(\lambda)X(T(x) \wedge t) - \lambda(T(x) \wedge t)$ is bounded from above by $x\Phi(\lambda)$, and converges as t tends to ∞ to $\Phi(\lambda)x - \lambda T(x)$ on $\{T(x) < \infty\}$ and to $-\infty$ on $\{T(x) = \infty\}$. We deduce by dominated convergence that

$$\mathbb{E}(\exp\{-\lambda T(x)\}, T(x) < \infty) = \exp\{-x\Phi(\lambda)\}. \qquad \square$$

In particular, Theorem 1 shows that S can be recovered from the reflected process $S - X$, for instance by approximations involving numbers of excursion intervals of length greater than a certain quantity (see section IV.2), while in general this feature is lost for Lévy processes with positive jumps.

As a first application, we calculate the distribution of the supremum evaluated at an independent exponential time, and specify the asymptotic behaviour at infinity of X in terms of that of its Laplace exponent at the origin. To this end, note that points are necessarily recurrent when X oscillates, because X cannot cross a level by a positive jump.

Corollary 2 (i) *Let $\tau(q)$ be an independent exponential time with parameter $q > 0$. Then $S_{\tau(q)}$ has an exponential distribution with parameter $\Phi(q)$.*

(ii) *X drifts to ∞, oscillates or drifts to $-\infty$ according as the right derivative of the Laplace exponent at the origin, $\psi'(0+)$, is positive, zero, or negative. In the latter case, the overall supremum S_∞ has an exponential distribution with parameter $\Phi(0) > 0$.*

Proof (i) We have for every $x > 0$

$$\mathbb{P}(S_{\tau(q)} > x) = \mathbb{P}(T(x) \leq \tau(q)) = \mathbb{E}(\exp\{-qT(x)\}) = \exp\{-\Phi(q)x\}.$$

(ii) According to Proposition VI.17, X drifts to ∞ if and only if the expectation of $T(x) = T_{(x,\infty)}$ is finite. We deduce from Theorem 1 that the latter holds if and only if $\Phi(0) = 0$ and the right derivative of Φ at 0 has $\Phi'(0+) < \infty$. Since Φ is the inverse of ψ, this is equivalent to $\psi'(0+) > 0$. On the other hand, letting q tend to 0+ in (i), we see that $S_\infty < \infty$ a.s., that is X drifts to $-\infty$, if and only if $\Phi(0) > 0$, which is equivalent to $\psi'(0+) < 0$ by the convexity of ψ. Alternatively, this result can be deduced from Exercise VI.3. □

The feature that the first passage process T is a (possibly killed) subordinator has many interesting consequences, some are presented in Exercises 7–8.

The comparison between the fluctuation identities of section VI.2 and Corollary 2 yields a remarkable identity involving the one-dimensional distributions of X and the first passage process $T = (T(x), x \geq 0)$ (which can also be viewed as a special case of Exercise VI.8).

Corollary 3 *The measures $t\mathbb{P}(T(x) \in dt)dx$ and $x\mathbb{P}(X_t \in dx)dt$ coincide on $[0,\infty) \times [0,\infty)$.*

Proof On the one hand, we have according to Corollary 2 and Theorem VI.5(ii)

$$\frac{\Phi(q)}{\Phi(q)+\lambda} = \mathbb{E}(\exp\{-\lambda S_{\tau(q)}\})$$
$$= \exp\left(\int_0^\infty dt \int_{[0,\infty)} (e^{-\lambda x}-1)t^{-1}e^{-qt}\mathbb{P}(X_t \in dx)\right).$$

On the other hand, the Frullani integral yields

$$\frac{\Phi(q)}{\Phi(q)+\lambda} = \exp\left(\int_0^\infty (e^{-\lambda x}-1)x^{-1}e^{-\Phi(q)x}dx\right)$$
$$= \exp\left(\int_0^\infty dx \int_{[0,\infty)} (e^{-\lambda x}-1)x^{-1}e^{-qt}\mathbb{P}(T(x) \in dt)\right),$$

where in the last equality, we used the fact that Φ is the Laplace exponent of the first passage process T. The comparison between the two identities establishes the formula of Corollary 3. □

The absence of positive jumps ensures that the ladder height process H has $H(t) = S(T(t)) = t$ on $T(t) < \infty$. In other words, H is a unit drift, killed at rate $\Phi(0)$ when X drifts to $-\infty$. In particular, its renewal function is $\mathscr{V}(x) = (1-\exp\{-\Phi(0)x\})/\Phi(0)$ if X drifts to $-\infty$, and $\mathscr{V}(x) = x$ otherwise. The bivariate Laplace exponent of the ladder process κ is given by

$$\kappa(\alpha,\beta) = \Phi(\alpha)+\beta \qquad (\alpha,\beta \geq 0). \tag{2}$$

The fluctuation formulas are now especially simple.

Theorem 4 (i) *Let $\tau = \tau(q)$ be an independent exponential time with parameter $q > 0$ and $G_\tau = \sup\{t < \tau : X_t = S_t\}$. We have for every $\alpha, \beta > 0$*

$$\mathbb{E}(\exp\{-\alpha G_\tau - \beta S_\tau\}) = \frac{\Phi(q)}{\Phi(\alpha+q)+\beta},$$
$$\mathbb{E}(\exp\{-\alpha(\tau - G_\tau) - \beta(S_\tau - X_\tau)\}) = \frac{q\,(\Phi(\alpha+q)-\beta)}{\Phi(q)\,(\alpha+q-\psi(\beta))}$$

where the right-hand side in the latter equality is $q\Phi'(\beta)/\Phi(\beta)$ for $\beta = \Phi(\alpha+q)$.

(ii) *The bivariate Laplace exponent $\hat{\kappa}$ of the dual ladder process is given by*

$$\hat{\kappa}(\alpha,\beta) = c\frac{\alpha-\psi(\beta)}{\Phi(\alpha)-\beta}, \qquad \alpha,\beta > 0,$$

where $c > 0$ is some meaningless constant.

Proof (i) The first identity follows from (2) and equation (1) of chapter VI. The second is derived from the first and Theorem VI.5(i) whenever $\beta < \Phi(\alpha + q)$, and the formula is extended analytically for $\beta \geq \Phi(\alpha + q)$.

(ii) The formula follows from (i) and (VI.1). □

Another interesting by-product of Theorem 4 and Proposition VI.3 is the following formula for the characteristic function of the infimum evaluated at $\tau(q)$, an independent exponential time with parameter $q > 0$:

$$\mathbb{E}(\exp\{\lambda I_{\tau(q)}\}) = \frac{q\,(\Phi(q) - \lambda)}{\Phi(q)\,(q - \psi(\lambda))} \qquad (\lambda > 0). \tag{3}$$

This enables us to determine completely the initial path behaviour of X at the origin.

Corollary 5 *The following assertions are equivalent:*

(i) 0 *is irregular for* $\{0\}$*;*
(ii) 0 *is irregular for* $(-\infty, 0)$*;*
(iii) $\lim_{\lambda\to\infty} \lambda^{-1}\psi(\lambda) < \infty$*;*
(iv) X *has bounded variation.*

Proof (i)⇔(ii) Since 0 is always regular for $(0, \infty)$ and X has no positive jumps, 0 is necessarily regular for $\{0\}$ whenever it is regular for $(-\infty, 0)$. The converse is obvious because X is not a compound Poisson process, so 0 is irregular for $(-\infty, 0)$ if and only if it is irregular for $(-\infty, 0]$.

(ii)⇔(iii) Letting λ tend to ∞ in (3), we see that the probability that $I_{\tau(q)} = 0$ is positive if and only if (iii) holds.

(iii)⇔(iv) This follows readily from the Lévy-Khintchine formula. □

Next, we turn our attention to Spitzer's condition, which is the key both to the arcsine laws and to the asymptotic behaviour of the tail distribution of first passage times (see Theorems VI.14 and VI.18).

Proposition 6 *For every* $\rho \in (0, 1]$*, the following conditions are equivalent:*

(i) $\lim \frac{1}{t}\int_0^t \mathbb{P}(X_s \geq 0)ds = \rho$ *as* $t \to 0+$ *(respectively, as* $t \to \infty$*).*
(ii) *The Laplace exponent* ψ *is regularly varying at* ∞ *(respectively, at* $0+$*) with index* $1/\rho$.

Observe from the Lévy-Khintchine formula that if (ii) holds, then necessarily $\rho \geq 1/2$.

Proof Recall from Theorem VI.14 that (i) holds if and only if the Laplace exponent of the ladder time process, that is Φ, is regularly varying with index ρ. The latter is equivalent to (ii), because ψ is the inverse of Φ. □

We conclude this section with a relation between Lévy processes with no positive jumps which drift to $-\infty$, and others which drift to ∞. This connection will enable us to reduce the study of the former to that of the latter. More precisely, recall that $\Phi(0)$ is the largest root of the Laplace exponent ψ, so the process $\exp\{\Phi(0)X_t\}$ $(t \geq 0)$ is an exponential martingale. This leads us to consider the probability measure $\mathbb{P}^\natural$ on $(\Omega, \mathscr{F})$ given by

$$\mathbb{P}^\natural(\Lambda) = \mathbb{E}(\exp\{\Phi(0)X_t\}, \Lambda), \qquad \Lambda \in \mathscr{F}_t.$$

Of course, only the case when $\Phi(0) > 0$ (that is when X drifts to $-\infty$) is relevant, and we will henceforth focus on it. It is easy to check that X is still a Lévy process (obviously with no positive jumps) under $\mathbb{P}^\natural$. Specifically, for every $\lambda > 0$ and $\Lambda \in \mathscr{F}_t$, we have

$$\begin{aligned}
&\mathbb{E}^\natural(\exp\{\lambda(X_{t+s} - X_t)\}, \Lambda) \\
&\quad = \mathbb{E}(\exp\{\Phi(0)(X_{t+s})\} \exp\{\lambda(X_{t+s} - X_t)\}, \Lambda) \\
&\quad = \mathbb{E}(\exp\{(\Phi(0) + \lambda)(X_{t+s} - X_t)\})\mathbb{E}(\exp\{\Phi(0)X_t\}, \Lambda) \\
&\quad = \exp\{s\psi(\Phi(0) + \lambda)\}\mathbb{P}^\natural(\Lambda).
\end{aligned}$$

This shows our assertion, and more precisely, the Laplace exponent under $\mathbb{P}^\natural$ is $\psi^\natural(\cdot) = \psi(\Phi(0) + \cdot)$.

Observe that the derivative of $\psi^\natural$ at $0+$ is $\psi'(\Phi(0)) > 0$, so according to Corollary 2(ii), X drifts to ∞ under $\mathbb{P}^\natural$. Moreover, the first passage process $T = (T(x), x \geq 0)$ is a subordinator under $\mathbb{P}^\natural$ with Laplace exponent $\Phi^\natural(\cdot) = \Phi(\cdot) - \Phi(0)$. This shows that the law of the supremum process S under $\mathbb{P}$ is the same as that of $S \wedge \tau$ under $\mathbb{P}^\natural$, where τ is an independent exponential time with parameter $\Phi(0)$.

The next lemma suggests that $\mathbb{P}^\natural$ can be thought of as the law of the initial Lévy process conditioned to drift to ∞.

Lemma 7 (i) *For every $t > 0$ and $\Lambda \in \mathscr{F}_t$,*

$$\lim_{x \to \infty} \mathbb{P}(\Lambda \mid S_\infty > x) = \mathbb{P}^\natural(\Lambda).$$

(ii) *For every $x > 0$, the law of $(X_t, 0 \leq t < T(x))$ is the same under $\mathbb{P}^\natural$ as under $\mathbb{P}(\cdot \mid T(x) < \infty)$.*

Proof (i) Recall from Corollary 2 that the overall supremum S_∞ has an exponential distribution with parameter $\Phi(0)$ under $\mathbb{P}$. By the Markov property, we have

$$\begin{aligned}
&\mathbb{P}(\Lambda \mid S_\infty > x)\\
&= e^{\Phi(0)x}(\mathbb{P}(\Lambda, T(x) < t) + \mathbb{P}(\Lambda, t \le T(x) < \infty))\\
&= e^{\Phi(0)x}\left(\mathbb{P}(\Lambda, T(x) < t) + \mathbb{E}(\exp\{\Phi(0)(X_t - x)\}, \Lambda, t \le T(x))\right)\\
&= e^{\Phi(0)x}\mathbb{P}(\Lambda, T(x) < t) + \mathbb{P}^{\natural}(\Lambda, t \le T(x)).
\end{aligned}$$

On the one hand, the first term in the foregoing sum converges to 0 as $x \to \infty$, because by Corollary 2,

$$\mathbb{P}(S_{\tau(1)} > x) = \exp\{-\Phi(1)x\} = o(\exp\{-\Phi(0)x\}) \qquad (x \to \infty),$$

where $\tau(1)$ stands for an independent exponential time with parameter 1. On the other hand, the second term converges to $\mathbb{P}^{\natural}(\Lambda)$.

(ii) We deduce from the Markov property that the distribution of $(X_t, 0 \le t < T(x))$ is the same under $\mathbb{P}(\cdot \mid T(x) < \infty)$ as under $\mathbb{P}(\cdot \mid T(x') < \infty)$ for any $x' > x$. The assertion (ii) follows from (i) by letting x' tend to ∞. □

2. The scale function

The absence of positive jumps not only confers a great simplicity for fluctuation theory, but also allows us to tackle problems which have no known solution for general Lévy processes. Here, we first consider the so-called two-sided-exit problem, which consists in determining the probability that X makes its first exit from an interval $[-x, y]$ $(x, y > 0)$ at the upper boundary point. This problem has a remarkable simple solution, which is essentially due to Takács (1966).

Theorem 8 *For every $x, y > 0$, the probability that X makes its first exit from $[-x, y]$ at y is*

$$\mathbb{P}(I_{T(y)} \ge -x) = W(x)/W(x+y),$$

where $W : [0, \infty) \to [0, \infty)$ is the unique absolutely continuous increasing function with Laplace transform

$$\int_0^\infty e^{-\lambda x} W(x) dx = \frac{1}{\psi(\lambda)} \qquad (\lambda > \Phi(0)).$$

Proof First, suppose that X does not drift to $-\infty$, so 0 is a regular recurrent point for the reflected process $S - X$. Recall from Theorem

IV.10 that the excursion process of $S-X$ away from 0 is then a Poisson point process. It follows that the process of the height of the excursion,

$$h_t = \sup\{(S-X)_{T(t-)+s} : 0 \le s < T(t)-T(t-)\}\,, \qquad t \ge 0,$$

is a Poisson point process; we denote its characteristic measure by ν. Because the events $\{I_{T(y)} \ge -x\}$ and $\{h_t \le x+t \text{ for all } t \in [0,y]\}$ coincide, we have

$$\begin{aligned}\mathbb{P}(I_{T(y)} \ge -x) &= \exp\left\{-\int_0^y \nu(x+t,\infty)dt\right\}\\ &= \exp\left\{-\int_x^\infty \nu(t,\infty)dt + \int_{x+y}^\infty \nu(t,\infty)dt\right\}\\ &= W(x)/W(x+y)\,,\end{aligned}$$

where W is the absolutely continuous increasing function given by

$$W(x) = c\exp\left\{-\int_x^\infty \nu(t,\infty)dt\right\}$$

and $c>0$ an arbitrary constant which will be chosen later on.

In order to determine W in terms of the Laplace exponent ψ, we suppose first that X drifts to ∞. Taking the limit as $y \to \infty$, we find

$$\mathbb{P}(I_\infty \ge -x) = W(x)/W(\infty),$$

so that W is proportional to the distribution function of $-I_\infty$. It follows from (3) that the Laplace-Stieltjes transform of the latter is proportional to $\lambda/\psi(\lambda)$ (recall that $\Phi'(0+) < \infty$ when X drifts to ∞), and therefore

$$\int_0^\infty e^{-\lambda x} W(x)dx = c'/\psi(\lambda) \qquad (\lambda > 0)$$

For simplicity, we now choose c in such a way that $c' = 1$, and the theorem is proved when X drifts to ∞. The case when X oscillates follows by approximation, adding a small positive drift.

Finally, the case of the Lévy process drifting to $-\infty$ is reduced to the case of drifting to ∞, by using Lemma 7(ii). More precisely, we have

$$\begin{aligned}\mathbb{P}(I_{T(y)} \ge -x) &= \mathbb{P}(T(y)<\infty)\mathbb{P}(I_{T(y)} \ge -x \mid T(y) < \infty)\\ &= e^{-\Phi(0)y}\mathbb{P}^\natural(I_{T(y)} \ge -x)\\ &= e^{-\Phi(0)y} W^\natural(x)/W^\natural(x+y)\\ &= W(x)/W(x+y),\end{aligned}$$

where $W^\natural$ is the continuous increasing function with Laplace transform $1/\psi^\natural(\cdot) = 1/\psi(\Phi(0)+\cdot)$ and $W(x) = \exp\{\Phi(0)x\}W^\natural(x)$.

The function W is called the *scale function* of the Lévy process, by analogy with Feller's theory of real-valued diffusions (see e.g. Rogers and

Williams (1987)). Identifying the Laplace transforms in Theorem 4(ii) and (VI.6), we observe the identity

$$W(dx) = c\exp\{\Phi(0)x\}\widehat{\mathcal{V}}(dx), \tag{4}$$

where $\widehat{\mathcal{V}}$ denotes the renewal function of the dual ladder height process $\widehat{H}$ and $c > 0$ some meaningless constant. Compare also with Proposition VI.17(i) when X drifts to ∞.

We present now a striking application of the scale function to the sample path behaviour of the Lévy process. Recall from section VI.5 that an instant $t > 0$ is an increase time for a path ω if ω stays above $\omega(t)$ immediately after time t and below $\omega(t)$ immediately before time t. First, we have the following characterization of the class of Lévy processes with no positive jumps that possess increase times in terms of their scale functions.

Corollary 9 *The probability that there exist increase times is* 0 *or* 1 *according as the integral* $\int_{0+}(dx/W(x))$ *diverges or converges.*

Proof The case when X drifts to $-\infty$ reduces to that when X drifts to ∞, by invoking Lemma 7 and the identity $W(x) = \exp\{\Phi(0)x\}W^{\natural}(x)$. So we may suppose henceforth that X does not drift to $-\infty$, that is $\Phi(0) = 0$.

Recall the notation of section VI.5. By Lemma VI.22, all that we need is to express the condition

$$\lim_{\varepsilon\to 0+} \widehat{F}(\varepsilon)^{-1}\int_{[0,\infty)} \mathbb{P}\left(y < S(\widehat{T}_\varepsilon) - X(\widehat{T}_\varepsilon), \widehat{T}_\varepsilon < \tau\right) F(dy) < \infty \tag{5}$$

in terms of the scale function. To this end, note that

$$\mathbb{P}\left(y < S(\widehat{T}_\varepsilon) - X(\widehat{T}_\varepsilon)\right) - \mathbb{P}\left(y < S(\widehat{T}_\varepsilon) - X(\widehat{T}_\varepsilon), \widehat{T}_\varepsilon < \tau\right)$$
$$\leq \mathbb{P}(\widehat{T}_\varepsilon \geq \tau) = \widehat{F}(\varepsilon)$$

so in (5), we may replace the integrand by

$$\mathbb{P}\left(y < S(\widehat{T}_\varepsilon) - X(\widehat{T}_\varepsilon)\right) = W(\varepsilon)/W(y+\varepsilon) \qquad \text{[by Theorem 8]}.$$

On the one hand, we know from Corollary 2 that $F(dy) = ae^{-ay}dy$, where $a = \Phi(1)$. On the other hand, recall that W is proportional to the renewal function $\widehat{\mathcal{V}}$ of the dual ladder height process. Invoking Lemma VI.21, we now see that (5) holds if and only if

$$\lim_{\varepsilon\to 0+} \frac{1}{W(\varepsilon)}\int_{[0,\infty)} \frac{W(\varepsilon)}{W(y+\varepsilon)} e^{-ay}dy < \infty,$$

and the latter condition is equivalent to that in Corollary 9. □

Corollary 9 is attractively simple, but its main drawback is that it involves the scale function which is only known through its Laplace transform. We now re-express the integral test in terms of the Laplace exponent ψ, the Lévy measure Π and the Gaussian coefficient Q.

Proposition 10 *The following assertions are equivalent:*

(i) $$\int_{0+} \frac{dx}{W(x)} < \infty,$$

(ii) $$\int^{\infty} s^{-3}\psi(s)ds < \infty,$$

(iii) $$Q = 0 \quad and \quad \int_{(-1,0)} x^2 \log|1/x| \, \Pi(dx) < \infty.$$

Proof (i)$\Leftrightarrow$(ii) It is easy to verify that the case when $\Phi(0) > 0$ reduces to that when $\Phi(0) = 0$ by considering the Laplace exponent $\psi^{\natural}(\cdot) = \psi(\Phi(0) + \cdot)$. So we assume henceforth that $\Phi(0) = 0$ and then (4) shows that W is the renewal function of the dual ladder height process, which has Laplace exponent $\psi(\lambda)/\lambda$. An application of Proposition III.1 gives the estimate

$$W(x) \asymp \frac{1}{x\psi(1/x)}$$

which in turn entails that the integrals (i) and (ii) converge or diverge simultaneously.

(ii)$\Leftrightarrow$(iii) Assume first that $Q > 0$. It then follows from the Lévy-Khintchine formula that $\psi(\lambda) \sim \frac{1}{2}Q\lambda^2$ as $\lambda \to 0+$ and *a fortiori* (ii) fails.

Assume henceforth that $Q = 0$. By the Lévy-Khintchine formula

$$\int_1^{\infty} s^{-3}\psi(s)ds = \int_1^{\infty} ds\, s^{-3}\left(as + \int_{(-\infty,-1]} (e^{sx} - 1)\Pi(dx)\right)$$
$$+ \int_{(-1,0)} \Pi(dx) \int_1^{\infty} s^{-3}(e^{sx} - 1 - sx)ds.$$

On the one hand, the first term in the sum is always finite, and on the other hand, one has

$$\int_1^{\infty} s^{-3}(e^{sx} - 1 - sx)ds = x^2 \int_{-x}^{\infty} s^{-3}(1 - e^{-s} + s)ds$$
$$\sim \frac{1}{2}x^2 \log|1/x| \qquad (x \to 0-).$$

It follows that (ii) and (iii) are equivalent. □

It is interesting to point out that 'most' Lévy process with no positive jumps and zero Gaussian component possess increase times (this holds for instance in the stable case with index $\alpha \in (1, 2)$), but there are Lévy processes with no positive jumps and zero Gaussian which have no increase times.

3. The process conditioned to stay positive

It is well known that one can use the scale function of a real-valued diffusion process for conditioning the latter to stay positive. Specifically, the law of the diffusion started at $x > 0$ and conditioned to exit an interval $[0, y]$ at the upper boundary point is absolutely continuous with respect to the law of the original diffusion killed when it quits $[0, y]$; and the density for a finite time interval $[0, t]$ can be expressed in terms of the scale function evaluated at the value of the diffusion at time t. Then it is easy to verify that the density converges as y tends to infinity, and the limit law is often very useful for studying the original process. For instance, this procedure shows that the three-dimensional Bessel process can be thought of as a one-dimensional Brownian motion conditioned to stay positive, see McKean (1963) and Williams (1974) for details.

The purpose of this section is to show that this approach applies to Lévy processes with no positive jumps as well, and to present some connections between the Lévy process and the process conditioned to stay positive. Further relations will be developed in section 4. The solution of the two-sided-exit problem in terms of the scale function W readily entails the following lemma.

Lemma 11 *The process*

$$\mathbf{1}_{\{t<T_{(-\infty,0)}\}} W(X_t), \qquad t \geq 0,$$

is a $\mathbb{P}_x$-martingale for every $x > 0$.

Proof Fix some level $y > x$ and denote by $\sigma(y)$ the first exit time of $(0, y)$. We deduce from the Markov property and Theorem 8 that for every $t \geq 0$,

$$\mathbb{P}_x(X_{\sigma(y)} = y \mid \mathscr{F}_t) = \mathbf{1}_{\{t<\sigma(y)\}} \frac{W(X_t)}{W(y)} + \mathbf{1}_{\{t \geq T(y)=\sigma(y)\}}.$$

Therefore, the process

$$\mathbf{1}_{\{t<\sigma(y)\}} W(X_t) + W(y) \mathbf{1}_{\{t \geq T(y)=\sigma(y)\}}, \qquad t \geq 0,$$

is a $\mathbb{P}_x$-martingale for all $x > 0$. On the one hand, the first term in the sum increases to

$$\mathbf{1}_{\{t<T_{(-\infty,0)}\}} W(X_t)$$

as y goes to ∞. On the other hand, the expectation of the second term is bounded by $\mathbb{E}(W(S_t), S_t > y)$. But S_t has finite exponential moments of every order (by Corollary 2), and

$$\int_0^\infty \mathrm{e}^{-\lambda x} W(x) dx < \infty$$

for λ large enough, so $\mathbb{E}(W(S_t), S_t > y)$ tends to 0 as $y \to \infty$. This shows our assertion. □

Now we introduce a new probability measure $\mathbb{P}_x^\uparrow$, which is given on $(\Omega, \mathscr{F}_t)$ by

$$\mathbb{P}_x^\uparrow(\Lambda) = \frac{1}{W(x)} \mathbb{E}_x\left(W(X_t), \Lambda, t < T_{(-\infty,0)}\right), \qquad \Lambda \in \mathscr{F}_t. \tag{6}$$

The martingale property stated in Lemma 11 ensures the consistency of this definition, namely this family of probability measures on $(\Omega, \mathscr{F}_t)$, $t \geq 0$, is projective and induces a unique probability measure on $(\Omega, \mathscr{F})$.

When the Lévy process drifts to ∞, $W(x) = c\mathbb{P}_x(T_{(-\infty,0)} = \infty)$ for some constant $c > 0$ and it follows readily that $\mathbb{P}_x^\uparrow$ can be identified as the conditional law $\mathbb{P}_x(\cdot \mid T_{(-\infty,0)} = \infty)$. More precisely, the Markov property entails that for every $\Lambda \in \mathscr{F}_t$,

$$\begin{aligned}
&\mathbb{P}_x(\Lambda \mid T_{(-\infty,0)} = \infty) \\
&= \mathbb{P}_x(\Lambda, T_{(-\infty,0)} = \infty)/\mathbb{P}_x(T_{(-\infty,0)} = \infty) \\
&= \mathbb{P}_x(\Lambda, t < T_{(-\infty,0)}, T_{(-\infty,0)} \circ \theta_t = \infty)/\mathbb{P}_x(T_{(-\infty,0)} = \infty) \\
&= \mathbb{E}_x\left(\mathbb{P}_{X_t}(T_{(-\infty,0)} = \infty)/\mathbb{P}_x(T_{(-\infty,0)} = \infty), \Lambda, t < T_{(-\infty,0)}\right) \\
&= \frac{1}{W(x)} \mathbb{E}_x\left(W(X_t), \Lambda, t < T_{(-\infty,0)}\right).
\end{aligned}$$

This feature is lost when the Lévy process does not drift to ∞, because then $\mathbb{P}_x(T_{(-\infty,0)} = \infty) = 0$. Nonetheless, in every case, an easy variation of this argument and that in the proof of Lemma 11 show that $\mathbb{P}_x^\uparrow$ is the limit of the conditional law $\mathbb{P}_x(\cdot \mid T(y) < T_{(-\infty,0)})$ as $y \to \infty$. To this end, we also observe from Lemma 7 that, when the Lévy process drifts to $-\infty$, the law of the process killed as it exceeds y is the same under $\mathbb{P}_x(\cdot \mid T(y) < T_{(-\infty,0)})$ as under $\mathbb{P}_x^\natural(\cdot \mid T(y) < T_{(-\infty,0)})$, where $\mathbb{P}_x^\natural$ is the law of the Lévy process started at x and conditioned to drift to ∞. As a consequence,

$$\mathbb{P}_x^{\uparrow} = \mathbb{P}_x^{\natural\uparrow} \qquad \text{for all } x > 0,$$

which is also easy to check directly from (6).

It is plain from (6) and Lemma 11 that under $\mathbb{P}_x^{\uparrow}$, X is a process taking values in $[0, \infty)$. It will be referred to as the Lévy process started at x and conditioned to stay positive. Recall from Proposition II.4 that the Lévy process killed as it enters $(-\infty, 0)$ is a Markov process. We deduce from (6) that X is a Markov process under $(\mathbb{P}_x^{\uparrow}, x > 0)$ as well, and its semigroup is given by

$$\mathbb{P}_x^{\uparrow}(X_t \in dy) = \frac{W(y)}{W(x)} \mathbb{P}_x(X_t \in dy, t < T_{(-\infty,0)}).$$

Referring to the pioneer work of Doob (1957), specialists say that $\mathbb{P}_x^{\uparrow}$ is obtained by an h-transform of the law of the Lévy process killed at time $T_{(-\infty,0)}$, with the harmonic function W. Here is a first elementary result.

Lemma 12 *For every* $0 < x < y$,

$$\mathbb{P}_y^{\uparrow}(X_t \geq x \text{ for all } t \geq 0) = \exp\{\Phi(0)x\} W(y-x)/W(y).$$

As a consequence, $\lim_{t\to\infty} X_t = \infty$ $\mathbb{P}_x^{\uparrow}$*-a.s. for every* $x > 0$.

Proof We deduce readily from (6) and Theorem 8 that for every $0 < x < y < z$,

$$\mathbb{P}_y^{\uparrow}(T_{(z,\infty)} < T_{(0,x)}) = \frac{W(z)W(y-x)}{W(y)W(z-x)}.$$

By Theorem 4, we have

$$\lim_{z\to\infty} \left(W(z)/W(z+x)\right) = \mathbb{P}(T(x) < \infty) = \exp\{-\Phi(0)x\},$$

and our first assertion follows.

Next, we deduce from the foregoing that for every $x > 0$

$$\lim_{y\to\infty} \mathbb{P}_y^{\uparrow}(X_t \geq x \text{ for all } t \geq 0) = 1,$$

and the second assertion follows from the Markov property applied at the first instant when the process exceeds y. □

The potential operator of $\mathbb{P}_x^{\uparrow}$ is easy to calculate.

Proposition 13 *Let* $f \geq 0$ *be a Borel function. Then for every* $x > 0$*, we have*

$$\begin{aligned} U^{\uparrow} f(x) &= \mathbb{E}_x^{\uparrow}\left(\int_0^{\infty} f(X_t)dt\right) \\ &= \frac{1}{W(x)} \int_0^{\infty} f(u)W(u)\left(e^{-u\Phi(0)}W(x) - \mathbf{1}_{\{u<x\}}W(x-u)\right) du. \end{aligned}$$

Proof Let τ be an independent exponential time with parameter $q > 0$. According to (6), we have

$$\mathbb{E}_x^{\uparrow}(f(X_\tau)) = \frac{1}{W(x)}\mathbb{E}\left(f((X-I)_\tau + I_\tau + x)W((X-I)_\tau + I_\tau + x), \tau < T_{(-\infty,-x)}\right).$$

Recall from Theorem VI.5(i) that the variables $(X-I)_\tau$ and I_τ are independent, and that the first has an exponential distribution with parameter $\Phi(q)$ (by Corollary 2 and Proposition VI.3). We deduce

$$\mathbb{E}_x^{\uparrow}\left(\int_0^\infty dt f(X_t)e^{-qt}\right) = \frac{\Phi(q)}{qW(x)}\int_0^\infty f(u)W(u) \times \left(\int_{[x-u,x]} \mathbb{P}(-I_\tau \in dy)\exp\{-(u+y-x)\Phi(q)\}\right) du. \quad (7)$$

According to (3) the characteristic function of the measure

$$\Phi(q)q^{-1}\mathbb{P}(-I_\tau \in dy)$$

tends to $(\lambda - \Phi(0))/\psi(\lambda)$ as q goes to $0+$. By Theorem 8, the latter is the Laplace transform of the measure $W(dy) - \Phi(0)W(y)dy$, and the proposition follows by letting q tend to $0+$ in (7). □

It is interesting to note that the mapping $x \to U^{\uparrow}f(x)$ is continuous if f has compact support, thanks to the continuity of the scale function. In other words, the potential operator of the Lévy process conditioned to stay positive has the strong Feller property.

Next, we study the behaviour of $\mathbb{P}_x^{\uparrow}$ as $x \to 0+$.

Proposition 14 *The probability measures $\mathbb{P}_x^{\uparrow}$ converge as $x \to 0+$ in the sense of finite-dimensional distributions. The limit is denoted by $\mathbb{P}_0^{\uparrow} = \mathbb{P}^{\uparrow}$ and the canonical process X is a Feller process for the family $(\mathbb{P}_x^{\uparrow}, x \geq 0)$.*

We point out that the argument below shows that the convergence also holds weakly on the space $\mathcal{D}([0,\infty), \mathbb{R}_+)$ endowed with Skorohod's topology.

Proof For every $x > 0$, consider $\sigma(x) = \inf\{t \geq 0 : X_t - I_t > x\}$, the first passage time above x for the dual reflected process, and $G_{\sigma(x)} = \sup\{t < \sigma(x) : X_t = I_t\}$, the instant of its last zero before $\sigma(x)$. It can be immediately seen that both $\sigma(x)$ and $G_{\sigma(x)}$ are finite $\mathbb{P}$-a.s., no matter whether X drifts to ∞, to $-\infty$ or oscillates. Denote by $Q^{(x)}$

the law under $\mathbb{P}$ of the piece of path on the time interval $[G_{\sigma(x)}, \sigma(x))$, $((X-I)_{G_{\sigma(x)}+t}, 0 \le t < \sigma(x) - G_{\sigma(x)})$.

Recall that we are using the canonical notation, that θ and $\mathbf{k}$ stand for the shift and the killing operators, respectively, and that $T(x) = \inf\{t \ge 0 : X_t > x\}$. It follows from the Markov property applied at the stopping time $\sigma(x)$ for the dual reflected process $X - I$ that for every $y > x$, the processes $X \circ \mathbf{k}_{T(x)}$ and $X \circ \theta_{T(x)}$ are independent under $Q^{(y)}$. Moreover, the former has the law $Q^{(x)}$ and the latter the law of the Lévy process started from x, killed as it leaves $[0, y]$, and conditioned to exit at the upper boundary point.

Next, let y tend to ∞, so that the law of $X \circ \theta_{T(x)}$ under $Q^{(y)}$ converges to $\mathbb{P}^{\uparrow}_x$. Thus, if we denote by $\mathbb{P}^{\uparrow}$ the unique probability measure on Ω under which the processes $X \circ \mathbf{k}_{T(x)}$ and $X \circ \theta_{T(x)}$ are independent and their respective laws are $Q^{(x)}$ and $\mathbb{P}^{\uparrow}_x$, then $Q^{(y)}$ converges to $\mathbb{P}^{\uparrow}$ as y goes to ∞. In particular, the law $\mathbb{P}^{\uparrow}$ does not depend on x, and the first passage time above x, $T(x)$, decreases to 0 $\mathbb{P}^{\uparrow}$-a.s. as x tends to $0+$. Recall that the paths are right-continuous. Splitting the process at $T(x)$ under $\mathbb{P}^{\uparrow}$ and then letting x tend to $0+$ show that $\mathbb{P}^{\uparrow}_x$ converges in the sense of finite-dimensional distributions (and also in the sense of Skorohod) to $\mathbb{P}^{\uparrow}$.

Finally, the canonical process X has the (simple) Markov property under $\mathbb{P}^{\uparrow}_x$, and this extends to $\mathbb{P}^{\uparrow}$ by approximation. The Feller property is a consequence of (6), of the continuity of the scale function W and of the easily shown fact that the distribution of $-I_t$ under $\mathbb{P}$ has no atom on $(0, \infty)$. □

The argument in the proof of Proposition 14 also yields an interesting connection between $\mathbb{P}^{\uparrow}$ and the excursion measure $\widehat{n}$ of the dual reflected process $\widehat{S} - \widehat{X} = X - I$.

Proposition 15 *There is a constant $k > 0$ (which only depends on the normalization of the local time at 0 of $X - I$), such that for every $t > 0$ and $\Lambda \in \mathscr{F}_t$,*

$$\widehat{n}(\Lambda, t < \zeta) = k\mathbb{E}^{\uparrow}(W(X_t)^{-1}, \Lambda).$$

In particular, the law under $\widehat{n}$ of the height of the generic excursion, $h(\epsilon) = \sup\{\epsilon(t) : 0 \le t < \zeta\}$, is given by

$$\widehat{n}(h > x) = k/W(x), \qquad x > 0.$$

Proof Fix $x > 0$. We deduce from Proposition O.2 and Theorem IV.10 that the conditional law $\hat{n}(\cdot \mid h > x)$ is the law of the first excursion of $X - I$ away from 0 with height $h > x$. The Markov property shows that under $\hat{n}(\cdot \mid h > x)$, the process $X \circ \theta_{T(x)}$ has the law of the Lévy process started at x and killed as its enters $(-\infty, 0)$. We now deduce from Theorem 8 that

$$\hat{n}(h > y \mid h > x) = W(x)/W(y) \qquad \text{for every } y > x,$$

which entails the second assertion of Proposition 15, and then the first follows from (6) and Proposition 14, by letting x tend to $0+$. □

One should observe that when the Lévy process drifts to ∞, the dual reflected process $X - I$ is transient and $W(\infty) < \infty$. We deduce by letting t tend to ∞ in Proposition 15 that $\mathbb{P}^{\uparrow}$ can be viewed in that case as the conditional law $\hat{n}(\cdot \mid \zeta = \infty)$, or equivalently, that the law of $X - I$ shifted at its last passage time at the origin is $\mathbb{P}^{\uparrow}$.

An interesting application of this result is the following formula for the entrance law $\mathbb{P}^{\uparrow}(X_t \in dx)$.

Corollary 16 *We have for every* $x, t > 0$

$$\mathbb{P}^{\uparrow}(X_t \in dx) = \frac{xW(x)}{t}\mathbb{P}(X_t \in dx).$$

Proof. First, we determine the entrance law under the excursion measure $\hat{n}$, relying on the calculations at the end of section IV.4. To this end, we know on the one hand from Corollary 2 and Proposition VI.3 that if $\tau(q)$ is an independent exponential time with parameter $q > 0$, then

$$\mathbb{P}(X_{\tau(q)} - I_{\tau(q)} \in dx) = \Phi(q)\exp\{-\Phi(q)x\}dx.$$

On the other hand, we know from equation (3) of section VI.2 and Theorem 1 that the Laplace exponent of the inverse local time at 0 of $X - I$ is given by

$$\hat{\kappa}(q, 0) = k'q/\Phi(q),$$

for some constant $k' > 0$. It follows now from equation (7) of Chapter IV (but beware of the difference in notation) that the entrance law under $\hat{n}$ has

$$\int_0^{\infty} e^{-qt}\hat{n}(\epsilon(t) \in dx, t < \zeta)dt = k' \exp\{-\Phi(q)x\}dx.$$

We then use Theorem 1 to invert the Laplace transform and obtain

$$\hat{n}(\epsilon(t) \in dx, t < \zeta)dt = k'\mathbb{P}(T(x) \in dt)dx. \tag{8}$$

We now deduce from Proposition 15 the following identity between measures:

$$\mathbb{P}^{\uparrow}(X_t \in dx)dt = k'' W(x)\mathbb{P}(T(x) \in dt)dx,$$

for some constant $k'' > 0$. Integrating the function $(t, x) \to e^{-qt}$ with respect to these measures yields $1/q = k''/q$, so $k'' = 1$. Finally we use Corollary 3 and the right-continuity of the paths; this finishes the proof. □

We conclude this section with a useful description, in terms of $\mathbb{P}^{\uparrow}$, of the Lévy process taken up to its first passage time below the origin. Of course, only the case when 0 is irregular for $(-\infty, 0)$ is relevant, so we will assume that the Lévy process has bounded variation (see Corollary 5). As a consequence, its Laplace exponent has the form

$$\psi(\lambda) = \lambda\left(\mathrm{d} - \int_0^\infty e^{-\lambda x}\Pi(-\infty, -x)dx\right), \tag{9}$$

where $\mathrm{d} > 0$ is the drift coefficient and Π the Lévy measure.

Theorem 17 *Suppose that* 0 *is irregular for* $(-\infty, 0)$, *and put* $\xi = T_{(-\infty,0)}$.

(i) $\mathbb{P}(\xi = \infty) = 0$ *if* $\psi'(0+) \leq 0$, *and* $\mathbb{P}(\xi = \infty) = \psi'(0+)/\mathrm{d}$ *otherwise. Moreover, in the latter case,* $\mathbb{P}^{\uparrow} = \mathbb{P}(\cdot \mid \xi = \infty)$.

(ii) *The law of the pair* $(X_{\xi-}, X_{\xi})$ *is given on* $\{0 \leq x \leq -y\}$ *by*

$$\mathbb{P}(X_{\xi-} \in dx, X_{\xi} - X_{\xi-} \in dy \mid \xi < \infty) = c^{-1}\exp\{-\Phi(0)x\}dx\Pi(dy),$$

where $c = \mathrm{d}$ *if* $\psi'(0+) \leq 0$ *and* $c = \mathrm{d} - \psi'(0+)$ *if* $\psi'(0+) > 0$.

(iii) *Let* $\varsigma(x) = \sup\{t : X_t \leq x\}$ *stand for the last passage time of the canonical process X below* $x \geq 0$. *Then the law of* $(X_t, t < \varsigma(x))$ *under* $\mathbb{P}^{\uparrow}$ *is a version of the conditional law of* $(X_t, t < \xi)$ *under* $\mathbb{P}(\cdot \mid X_{\xi-} = x)$.

Proof (i) We deduce from Theorem 8 that $\mathbb{P}(\xi = \infty) = W(0)/W(\infty)$, that $W(0) = 1/\mathrm{d}$, and that

$$W(\infty) = \begin{cases} 1/\psi'(0+) & \text{if } \psi'(0+) > 0, \\ \infty & \text{otherwise.} \end{cases}$$

This proves the first assertion. Moreover, in the case when $\psi'(0+) > 0$, we have already noted that $\mathbb{P}_x^{\uparrow} = \mathbb{P}_x(\cdot \mid \xi = \infty)$ for all $x > 0$. Letting x tend to 0+ yields the second assertion by Proposition 14.

(ii) Since 0 is irregular for $(-\infty, 0)$, we see from the Markov property applied at time ξ that $X_\xi < 0$ whenever $\xi < \infty$, and in particular, ξ is then the instant of a jump. Recall that the jump process ΔX is a Poisson point process with characteristic measure Π, so that by the compensation formula of section O.5, we have

$$\mathbb{P}(X_{\xi-}\in dx, \Delta X_\xi \in dy, \xi < \infty) = \mathbb{E}\left(\sum_{t\geq 0} \mathbf{1}_{\{t<\xi\}}\mathbf{1}_{\{X_{t-}\in dx\}}\mathbf{1}_{\{\Delta X_t\in dy\}}\right)$$
$$= \mathbb{E}\left(\int_0^\xi \mathbf{1}_{\{X_t\in dx\}}dt\right)\Pi(dy).$$

On the other hand, the excursion measure $\widehat{n}$ of the dual reflected process $\widehat{S}-\widehat{X} = X - I$ is proportional to the law of $(X_t, t<\xi)$ under $\mathbb{P}$, and we deduce from (8) that

$$\mathbb{E}\left(\int_0^\xi \mathbf{1}_{\{X_t\in dx\}}dt\right) = k\mathbb{P}(T(x)<\infty)dx = k\exp\{-\Phi(0)x\}dx,$$

for some constant $k>0$. This establishes the second assertion, the normalizing constant c being calculated using (9).

(iii) First, we develop some information about the process $(X_t, t<\varsigma(x))$ under $\mathbb{P}^\uparrow$. We derive from the Markov property and Lemma 12 that for every $\Lambda\in\mathcal{F}_t$,

$$\begin{aligned}\mathbb{P}^\uparrow(\Lambda, \varsigma(x)>t) &= \mathbb{P}^\uparrow(\Lambda) - \mathbb{E}^\uparrow\left(\exp\{\Phi(0)x\}W(X_t-x)/W(X_t), \Lambda, X_t\geq x\right)\\ &= k\mathbb{E}\left(W(X_t) - \mathrm{e}^{x\Phi(0)}W(X_t-x)\mathbf{1}_{\{X_t>x\}}, \Lambda, t<\xi\right), \qquad (10)\end{aligned}$$

where the last identity follows from Proposition 15 and the fact that the excursion measure $\widehat{n}$ is proportional to the law of $(X_t, t<\xi)$ under $\mathbb{P}$.

All that is needed now is to integrate (10) with respect to the law

$$\mathbb{P}(X_{\xi-}\in dx \mid \xi<\infty) = c^{-1}\exp\{-\Phi(0)x\}\Pi(-\infty,-x)dx$$

(according to (ii)). In this direction, we invert the Laplace transform in the obvious identity

$$\frac{\mathrm{d}}{\psi(\lambda)} - \frac{1}{\psi(\lambda)}\left(\mathrm{d}-\frac{\psi(\lambda)}{\lambda}\right) = \frac{1}{\lambda}$$

and get

$$\mathrm{d}W(y) - \int_0^x \exp\{\Phi(0)x\}W(y-x)\exp\{-\Phi(0)x\}\Pi(-\infty,-x)dx = 1. \quad (11)$$

Assume first that $\psi'(0+)\leq 0$, so $\mathbb{P}(\xi=\infty)=0$ and (from (9))

$$\mathrm{d} = \int_0^\infty \exp\{-\Phi(0)x\}\Pi(-\infty,-x)dx.$$

Using this in (10-11) shows that

$$\int_0^\infty \mathbb{P}^\uparrow(\Lambda, \varsigma(x)>t)\mathbb{P}(X_{\xi-}\in dx, \xi<\infty) = \mathbb{P}(\Lambda, t<\xi),$$

and our assertion is proved. Finally, assume that $\psi'(0+)>0$, so (9) yields

$$\psi'(0+) = \mathrm{d} - \int_0^\infty \Pi(-\infty,-x)dx.$$

Using this in (9-10) gives

$$\frac{\psi'(0+)}{\mathrm{d}}\mathbb{P}^{\uparrow}(\Lambda) + \int_0^{\infty} \mathbb{P}^{\uparrow}(\Lambda, \varsigma(x) > t)\mathbb{P}(X_{\xi-} \in dx, \xi < \infty) = \mathbb{P}(\Lambda, t < \xi),$$

and the proof is now complete. □

4. Some path transformations

The purpose of this section is to present two important path transformations connecting $\mathbb{P}$ and $\mathbb{P}^{\uparrow}$. They were established first in the Brownian case by Williams (1974) and Pitman (1975), respectively. The first involves time reversal. For each $x > 0$, recall that the last passage time below the level x is denoted by $\varsigma(x) = \sup\{t : X_t \leq x\}$.

Theorem 18 *For every $x > 0$, the law of the process killed at its first passage time above x, $(X_t, 0 \leq t < T(x))$ under $\mathbb{P}(\cdot \mid T(x) < \infty)$, is the same as that of the process time-reversed at its last passage time below x, $(x - X_{(\varsigma(x)-t)-}, 0 \leq t < \varsigma(x))$, under $\mathbb{P}^{\uparrow}$.*

Proof One can prove Theorem 18 by tedious bare-hands calculations, but it would not be really instructive as this approach essentially consists of checking in our setting a celebrated theorem on time reversal for Markov processes, due to Nagasawa (1964). Therefore, we shall take Nagasawa's result for granted, and refer to section XVIII.2 in Dellacherie, Maisonneuve and Meyer (1992) for a detailed account.

Assume first that $\Phi(0) = 0$, so that the Lévy process does not drift to $-\infty$ and there is no need to condition $\mathbb{P}$ on $\{T(x) < \infty\}$ in the statement. Observe that for every $t > 0$, $\sup\{s : X_{s+t} < x\} = \varsigma(x) - t$ if $t < \varsigma(x)$ and $= 0$ otherwise (one says that $\varsigma(x)$ is a co-optional time). The Lévy process 'conditioned to stay positive' and the dual Lévy process killed as it enters $(-\infty, 0)$ are both Feller processes. By the fundamental theorem on time reversal for Markov processes, all that we need is to check a relation of duality between their potential measures.

We deduce from Corollaries 3 and 16 that the potential measure at the origin of $\mathbb{P}^{\uparrow}$ is given by

$$U^{\uparrow}(0, dy) = \int_0^{\infty} \mathbb{P}^{\uparrow}(X_t \in dy)dt = W(y)dy \qquad (y \geq 0).$$

Then, from Proposition 13, the potential measure at $x > 0$, $U^{\uparrow}(x, dy)$, is absolutely continuous with respect to $U^{\uparrow}(0, dy)$, with density

$$u^{\uparrow}(x, y) = 1 - \mathbf{1}_{\{y<x\}}\frac{W(x-y)}{W(x)} \qquad (y \geq 0).$$

Next, we calculate the potential operator of the dual Lévy process killed as it enters $(-\infty, 0)$. For every continuous function $f \geq 0$ with compact support and for every real number $x > 0$, we have

$$\widehat{\mathbb{E}}_x\left(\int_0^{T_{(-\infty,0)}} f(X_t)dt\right) = \lim_{q\to 0+} \widehat{\mathbb{E}}_x\left(\int_0^{\infty} \mathrm{e}^{-qt} f(X_t)\mathbf{1}_{\{t<T_{(-\infty,0)}\}}dt\right)$$
$$= \lim_{q\to 0+} q^{-1}\mathbb{E}\left(f(S_{\tau(q)} - X_{\tau(q)} + x - S_{\tau(q)}), S_{\tau(q)} \leq x\right),$$

where $\tau(q)$ denotes an independent exponential time with parameter $q > 0$. Recall that $S_{\tau(q)}$ and $S_{\tau(q)} - X_{\tau(q)}$ are independent, that the first has an exponential distribution with parameter $\Phi(q)$ and that the latter has the same law as $-I_{\tau(q)}$. Moreover, it follows from (3) that the measure $\Phi(q)q^{-1}\mathbb{P}(-I_{\tau(q)} \in dy)$ converges weakly to $W(dy)$ as $q \to 0+$. We deduce

$$\widehat{\mathbb{E}}_x\left(\int_0^{T_{(-\infty,0)}} f(X_t)dt\right) = \int_0^{\infty} f(y)\left(W(y) - \mathbf{1}_{\{x<y\}}W(y-x)\right)dy.$$

In other words, the potential measure of the dual Lévy process started at $x > 0$ and killed as it enters $(-\infty, 0)$ is absolutely continuous with respect to the measure $U^{\uparrow}(0, dy)$ with density $u^{\uparrow}(y, x)$.

The fundamental theorem of time reversal for Markov processes applies. By the absence of positive jumps and Lemma 12, $\mathbb{P}^{\uparrow}(X_{\varsigma(x)-} = x, \varsigma(x) < \infty) = 1$ and we get that the law of the time-reversed process $(X_{(\varsigma(x)-t)-}, 0 \leq t < \varsigma(x))$ under $\mathbb{P}^{\uparrow}$ is the same as that of the dual Lévy process started at x and killed as it enters $(-\infty, 0)$. This establishes Theorem 18 when the Lévy process does not drift to $-\infty$.

Now assume that the Lévy process drifts to $-\infty$, and recall that $\mathbb{P}^{\natural}$ stands for the law of the Lévy process 'conditioned to drift to ∞'. By Lemma 7(ii) and Theorem 18 applied to $\mathbb{P}^{\natural}$, we see that the law of $(X_t, 0 \leq t < T(x))$ under $\mathbb{P}(\cdot \mid T(x) < \infty)$ is the same as that of $(x - X_{(\varsigma(x)-t)-}, 0 \leq t < \varsigma(x))$ under $\mathbb{P}^{\natural\uparrow}$. But we know that $\mathbb{P}^{\natural\uparrow} = \mathbb{P}^{\uparrow}$, and the proof is now complete. □

One readily deduces from Theorem 18 the following path decomposition of the Lévy process conditioned to stay positive, killed at its last passage time at x.

Corollary 19 *For every $x > 0$, the processes $X \circ k_{\varsigma(x)} = (X_t, t < \varsigma(x))$ and $X \circ \theta_{\varsigma(x)} - x = (X_{\varsigma(x)+t} - x, t \geq 0)$ are independent under $\mathbb{P}^{\uparrow}$, and the latter again has the law $\mathbb{P}^{\uparrow}$.*

Proof With no loss of generality, we assume that the Lévy process does not drift to $-\infty$ (otherwise, we work with the Lévy process conditioned to drift to ∞ and use the identity $\mathbb{P}^{\uparrow} = \mathbb{P}^{\natural\uparrow}$).

Fix $x > 0$, take any $y > 0$ and observe that $T(x+y) - T(y)$ is the first passage time above x for the shifted process $(X_{T(y)+t} - y, t \geq 0)$. The Markov property of the Lévy process at $T(y)$ and the time reversal identity of Theorem 18 show that under $\mathbb{P}^{\uparrow}$, the processes $X \circ \mathrm{k}_{\varsigma(x)}$ and $(X_{\varsigma(x)+t} - x, 0 \leq t < \varsigma(x+y) - \varsigma(x))$ are independent and that the latter has the same law as $X \circ \mathrm{k}_{\varsigma(y)}$. Then we let y go to ∞. □

The second path transformation involves the infimum process I and the jumps made by the Lévy process when it reaches a new infimum. Specifically, consider the canonical decomposition of the decreasing process I as the sum of its continuous and discontinuous parts $I_t = I_t^c + I_t^d$, where

$$I_t^d = \sum_{s \leq t} \Delta I_s = \sum_{s \leq t} \mathbf{1}_{\{X_s < I_{s-}\}} (X_s - I_{s-}) \qquad (t \geq 0).$$

Then denote by $J = (J_t, t \geq 0)$ the process of the sum of the jumps of X across its previous infimum,

$$J_t = \sum_{s \leq t} \mathbf{1}_{\{X_s < I_{s-}\}} (X_s - X_{s-}) \qquad (t \geq 0).$$

Theorem 20 *The law under $\mathbb{P}$ of the process $X - 2I^c - J$ is $\mathbb{P}^{\uparrow}$.*

When $I^c \equiv 0$ a.s., the path transformation of Theorem 20 simply consists in discarding the jumps that take the Lévy process to a new infimum. When $I^c \not\equiv 0$, I^c induces a further alteration of the path. It is interesting to note that $I^c \equiv 0$ a.s. if and only if the Lévy process has no Brownian component. Specifically, I^c is not trivial if and only if it is a local time at 0 for the dual reflected process $X - I$, and this is equivalent to the dual ladder height process $\widehat{H}$ having a positive drift. According to Theorem VI.19, the dual Lévy process can then creep across positive levels. But the absence of positive jumps implies that the Lévy process also creeps across positive levels, and the remark after Theorem VI.19 shows that this holds if and only if the Gaussian component of the Lévy process is not zero. (Alternatively, one can also check this assertion by an easy calculation based on (3).)

When one specializes Theorem 20 to the case when X is a Brownian motion, one gets that $R = X - 2I$ is a three-dimensional Bessel process. This is a famous identity due to Pitman (1975), see also Pitman and Rogers (1981) for Brownian motion with drift. It is remarkable that in the Brownian case, X can be recovered from R, because $-I$ of X coincides with the future infimum process $\inf\{R_s : s \geq \cdot\}$ of the Bessel process. This yields an inverse transformation from the Bessel process to the Brownian motion. This feature has been lost in the general case, the sigma-field generated by $X - 2I^c - J$ is usually strictly included in that generated by X. We also mention that when the Lévy process drifts to $-\infty$, there is a path transformation closely related to that of Theorem 20, which connects the Lévy process conditioned to stay *negative* to the Lévy process conditioned to drift to ∞. This provides an analogue of the inverse transformation from the Bessel process to Brownian motion, see Bertoin (1991) for a precise statement.

The rest of this section is devoted to the proof of Theorem 20 which is broken into three lemmas. The case of the Lévy process with bounded variation follows readily from Theorem 17 and Corollary 19. Note that in that case, the infimum process I is pure jump and hence $I^c \equiv 0$.

Lemma 21 *Suppose that X has bounded variation. Then the law of $X - J$ under $\mathbb{P}$ is $\mathbb{P}^\uparrow$.*

Proof Assume first that X does not drift to ∞ and denote by $\{0 = T_0 < T_1 < \cdots\}$ the zero set of $X - I$, that is the sequence of the instants when X reaches a new infimum. The excursions $((X-I)_{T_n+t}, 0 \leq t < T_{n+1} - T_n)$, $n = 0, 1, \cdots$, are independent, and by Theorem 17, each has the law of the Lévy process conditioned to stay positive and killed at its last passage time at some independent variable distributed as $\mathbb{P}(X_{T_1-} \in \cdot)$. On the other hand, the argument in the proof of Corollary 19 shows that when one tacks on a process distributed as $X \circ \mathrm{k}_{\varsigma(x)}$ under $\mathbb{P}^\uparrow$ at the end of an independent process distributed as $X \circ \mathrm{k}_{\varsigma(y)}$ under $\mathbb{P}^\uparrow$, one gets a process distributed as $X \circ \mathrm{k}_{\varsigma(x+y)}$ under $\mathbb{P}^\uparrow$. Therefore, the law under $\mathbb{P}$ of the process $(X_t - J_t, 0 \leq t < T_n)$ obtained after splicing the first n excursions of $X - I$ is that of the Lévy process conditioned to stay positive and killed at its last passage time at an independent variable with distribution $\mathbb{P}(I_{T_n} - J_{T_n} \in \cdot)$. Clearly, $(I_{T_n} - J_{T_n}, n \in \mathbb{N})$ is an increasing random walk under $\mathbb{P}$, in particular it tends to ∞, and letting $n \to \infty$ establishes our assertion.

The only difference in the case where the Lévy process drifts to ∞ is that the number of excursions of $X - I$ is then finite a.s. Again the excursions with finite lifetime are distributed as the Lévy process conditioned to stay positive and killed at its last passage time at an independent variable. According to Theorem 17(i), the law of the excursion with infinite lifetime is that of the Lévy process conditioned to stay positive. The argument is now an immediate modification of that when the Lévy process does not drift to ∞. □

We will deduce Theorem 20 from Lemma 21 by approximation. To this end, we note first that if $B = (B_t, t \geq 0)$ is a real-valued Brownian motion, there exists a sequence $N^{(n)}$ $(n = 1, \cdots)$ of Poisson processes with intensity n^2 such that the compensated Poisson processes $(n^{-1}N_t^{(n)} - nt, t \geq 0)$ converge to B uniformly on every compact time interval a.s. To see this, one can use the convergence in law of the compensated Poisson processes to Brownian motion, so by a representation theorem of Skorohod (1965), on pages 9–10, this convergence can be realized a.s. on the space $\mathscr{D}([0,\infty), \mathbb{R})$ endowed with Skorohod's topology. All that is needed now is to recall that convergence in Skorohod's topology to a continuous limit is equivalent to uniform convergence (e.g. Jacod and Shiryaev (1987) on page 292). Alternatively, one can also construct such a sequence of Poisson processes directly from a Brownian motion.

On the other hand, the proof of Theorem I.1 shows that any Lévy process Y with Lévy measure μ and no Brownian component is the limit, in the sense of a.s. uniform convergence on compact time intervals, of a sequence $(Y^{(n)}, n \in \mathbb{N})$ of compound Poisson processes with drift, where the Lévy measure of $Y^{(n)}$ is $\mu^{(n)}(dx) = \mathbf{1}_{\{|x|>1/n\}}\mu(dx)$.

In conclusion, there is a sequence of Lévy processes $(X^{(n)}, n = 1, \cdots)$ with no positive jumps and bounded variation, which converge to X uniformly on every compact time interval $\mathbb{P}$-a.s., and their respective Lévy measures are

$$\Pi^{(n)}(dx) = \mathbf{1}_{\{|x|>1/n\}}\Pi(dx) + Qn^2\delta_{-1/n}(dx),$$

where $\delta_{-1/n}(dx)$ stands for the Dirac point mass at $-1/n$ and Q is the Gaussian coefficient of the Lévy process.

Next, we turn our attention to the jumps of $X^{(n)}$ across its previous infimum $I^{(n)}$, distinguishing the jumps with length $-1/n$ (which result from the approximation of Brownian motion by the compensated Poisson process) from the others, and the portion of the jump above the previous infimum from that below. Specifically, we decompose the decreasing process $J^{(n)}$ as the sum $J^{(n)} = J^{(n,+)} + J^{(n,-)} + C^{(n,+)} + C^{(n,-)}$, where

$$J_t^{(n,+)} = \sum_{0 \leq s \leq t} \left(I_{s-}^{(n)} - X_{s-}^{(n)}\right) \mathbf{1}_{\{I_s^{(n)} < I_{s-}^{(n)}, \Delta X_s^{(n)} < -1/n\}},$$

$$J_t^{(n,-)} = \sum_{0 \leq s \leq t} \left(X_s^{(n)} - I_{s-}^{(n)}\right) \mathbf{1}_{\{I_s^{(n)} < I_{s-}^{(n)}, \Delta X_s^{(n)} < -1/n\}},$$

$$C_t^{(n,+)} = \sum_{0 \leq s \leq t} \left(I_{s-}^{(n)} - X_{s-}^{(n)}\right) \mathbf{1}_{\{I_s^{(n)} < I_{s-}^{(n)}, \Delta X_s^{(n)} = -1/n\}},$$

$$C_t^{(n,-)} = \sum_{0 \leq s \leq t} \left(X_s^{(n)} - I_{s-}^{(n)}\right) \mathbf{1}_{\{I_s^{(n)} < I_{s-}^{(n)}, \Delta X_s^{(n)} = -1/n\}}.$$

Lemma 22 *The following convergences hold in probability, uniformly over compact intervals of time:*

$$\lim_{n\to\infty} J_t^{(n,-)} = I_t^d\ ,\ \lim_{n\to\infty} J_t^{(n,+)} = J_t - I_t^d\ ,$$
$$\lim_{n\to\infty} C_t^{(n,-)} = I_t^c\ ,\ \lim_{n\to\infty} C_t^{(n,+)} = I_t^c\ ,$$

where $I = I^c + I^d$ is the canonical decomposition of the decreasing process I as the sum of its continuous and its discontinuous part.

Proof This is the technical part in the proof of Theorem 20, we will merely indicate the general lines and refer to section 4 in Bertoin (1992) for the tedious calculations. With no loss of generality, we will focus on the case when the Lévy process does not drift to $-\infty$, since otherwise we simply have to work with the process conditioned to drift to ∞.

First, we fix $\varepsilon > 0$ and we consider the sum of jumps accomplished by $X^{(n)}$ across its previous infimum $I^{(n)}$, where we take into account only the jumps of length $> -\varepsilon$ and $\neq -1/n$ (the latter restriction is made to exclude the jumps of the approximation of the Brownian component). Specifically, we put

$$K_t^{(n,\varepsilon)} = \sum_{0\le s\le t} \Delta X_s^{(n)} \mathbf{1}_{\{X_s^{(n)} < I_{s-}^{(n)}, \Delta X_s^{(n)} \in (-\varepsilon,0)-\{-1/n\}\}}.$$

Applying Theorem 17(iii), one can calculate the expectation of $K^{(n,\varepsilon)}$ evaluated at an independent exponential time τ and check that

$$\lim_{\varepsilon\to 0+} \sup\{\mathbb{E}(K_\tau^{(n,\varepsilon)}) : n = 1, \cdots\} = 0. \tag{12}$$

Next, consider a sample path and suppose that X jumps at time s across I_{s-} and that $\Delta X_s \le -\varepsilon$. Since $X^{(n)}$ and $I^{(n)}$ converge uniformly to X and I, respectively, we see that $X^{(n)}$ jumps at time s across $I_{s-}^{(n)}$ as well and $\Delta X_s^{(n)} = \Delta X_s$, provided that n is large enough. Conversely, for all n large enough, if $X^{(n)}$ jumps at time s across its previous infimum and if the size of this jump is less than $-\varepsilon$, then the same holds for X. This and (12) yield the first line limits of Lemma 22.

The obvious identity $I^{(n)} = I^{(n,-)} + C^{(n,-)}$ and the fact that $I^{(n)}$ converges to I now imply that $C^{(n,-)}$ converges to $I - I^d = I^c$, the continuous part of I.

Finally, we turn our attention to the difference $D^{(n)} = C^{(n,+)} - C^{(n,-)}$. This process is constant on each interval $[T_i^{(n)}, T_{i+1}^{(n)})$, where $0 = T_0^{(n)} < T_1^{(n)} < \cdots$ is the sequence of the instant when $X^{(n)}$ reachs a new infimum. The key point (which follows from Theorem 17(iii)) is that

$$\mathbb{E}\left(D^{(n)}(T_1^{(n)})\right) = 0.$$

We deduce from the Markov property that the chain $D^{(n)}(T_i^{(n)})$, $i = 0, 1, \cdots$, is a $\mathbb{P}$-martingale, and Theorem 17(iii) enables us to calculate its second moment. Then an appeal to Doob's maximal inequality shows that $D^{(n)}$ converges to 0 in $L^2(\mathbb{P})$, uniformly over every compact time interval. As a consequence, $C^{(n,+)}$ converges in probability, uniformly over compact time intervals to I^c. □

Then we denote by $\mathbb{P}^{(n)\uparrow}$ the law of $X^{(n)}$ conditioned to stay positive, that is of $X^{(n)} - J^{(n)}$ under $\mathbb{P}$. Since $X^{(n)} - J^{(n)}$ converges to $X - 2I^c - J$, Theorem 20 is now a consequence of the following lemma.

Lemma 23 *$\mathbb{P}^{(n)\uparrow}$ converges to $\mathbb{P}^{\uparrow}$ in the sense of finite-dimensional distributions.*

Proof With no loss of generality, we may assume that neither the Lévy process nor its approximations $X^{(n)}$ drift to $-\infty$. Denote the first passage of $X^{(n)}$ above x by $T^{(n)}(x) = \inf\{t \geq 0 : X_t^{(n)} > x\}$. Since $X^{(n)}$ converges uniformly over compact time intervals to X, we have for every $x > 0$

$$T(x-) \leq \liminf_{n\to\infty} T^{(n)}(x) \leq \limsup_{n\to\infty} T^{(n)}(x) \leq T(x) \qquad \mathbb{P}\text{-a.s.}$$

But T is $\mathbb{P}$-a.s. continuous at time x (T is a subordinator), and therefore $\lim_{n\to\infty} T^{(n)}(x) = T(x)$. We now deduce from Theorem 18 that the law of $X^{(n)}$ conditioned to stay positive and killed at its last passage time at x converges to that of $X \circ \mathrm{k}_{\varsigma(x)}$ under $\mathbb{P}^{\uparrow}$. To complete the proof, we simply let x tend to ∞. □

5. Exercises

1. *(Occupation of the positive half-line)* Prove that for every $\lambda > 0$,

$$\int_0^\infty e^{-\lambda t}\mathbb{P}(X_t \geq 0)dt = \frac{\Phi'(\lambda)}{\Phi(\lambda)}.$$

2. *(Resolvent densities)* (a) Check that for every $q > 0$, the q-resolvent measure $U^q(0, dx)$ is absolutely continuous with respect to the Lebesgue measure and has a bounded density. Prove that there exists a version of the density which is continuous on $(0, \infty)$, and that there exists a version continuous on $(-\infty, \infty)$ if and only if X has unbounded variation.

(b) Prove that the revolvent density is given on $(0, \infty)$ by

$$u^q(x) = \Phi'(q)\exp\{-\Phi(q)x\}, \qquad x > 0,$$

where Φ' stands for the derivative of Φ. Deduce an alternative solution to Exercise 1.

3. *(Last passage times)* (a) Prove that under $\mathbb{P}^{\uparrow}$, the last-passage-time process $\varsigma = (\varsigma(x), x \geq 0)$ is a subordinator with characteristic exponent $\Phi^{\natural}(\lambda) = \Phi(\lambda) - \Phi(0)$.
(b) Suppose that the Lévy process has bounded variation. Deduce from (a) and Theorem 17 the joint distribution under $\mathbb{P}$ of $(\xi, X_{\xi-}, \Delta X_{\xi})$, where $\xi = T_{(-\infty,0)}$.

4. *(Brownian motion and its supremum)* Suppose that X is a standard Brownian motion, and recall that $S.$ and $T.$ denote the supremum process and the first-passage process, respectively.
(a) Prove that the process $(t - T(t), t \geq 0)$ is a Lévy process with no positive jumps, with Laplace exponent $\psi(\lambda) = \lambda - \sqrt{2\lambda}$.
(b) Check that the events $\{\exists t > 0 : X_t = S_t = t\}$ and $\{\exists x > 0 : T(x) = x\}$ coincide up to a set of zero probability. Then deduce from Theorem 17 that

$$\mathbb{P}(\exists t > 0 : X_t = S_t = t) = 1/2.$$

5. *(Nowhere increase of the dual)* Adapt the proof of Corollary 9 and show that the dual process $\widehat{X} = -X$ has no increase time a.s.

6. *(A fluctuation identity for the process conditioned to stay positive)* Put $K_t = \inf\{X_s : s \geq t\}$ and consider $\tau(q)$, an independent exponential time with parameter $q > 0$. Prove that under $\mathbb{P}^{\uparrow}$, the variables $(X_{\tau(q)} - K_{\tau(q)})$ and $K_{\tau(q)}$ are independent, that the first has Laplace transform

$$\frac{\lambda + \Phi(q) - \Phi(0)}{\psi(\lambda + \Phi(q))} \times \frac{q}{\Phi(q) - \Phi(0)},$$

and that the latter has an exponential distribution with parameter $\Phi(q) - \Phi(0)$. [Hint: use Exercise 3.]

7. *(Hausdorff dimension of the ascending ladder time set)* Prove that the Hausdorff dimension of the ascending ladder time set $\{t : X_t = S_t\}$ is

$$\sup\{\alpha > 0 : \lim_{\lambda \to \infty} \lambda^{-\alpha}\Phi(\lambda) = \infty\} = \sup\{\alpha > 0 : \lim_{\lambda \to \infty} \psi(\lambda)\lambda^{-1/\alpha} = 0\}.$$

Note in particular that this quantity is always greater than or equal to 1/2.

8. *(Rates of growth)* (a) Let $f : [0, \infty) \to [0, \infty)$ be an increasing function such that $s \to f(s)/s$ decreases. Prove that a.s.

$$\liminf_{s \to 0+} \left(S_s/f(s)\right) = 0 \text{ or } \infty$$

according as the integral

$$\int_0^1 \left(\Phi(1/x) - (1/x)\Phi'(1/x)\right) df(x)$$

diverges or converges. [Hint: Apply Theorem III.9.]
(b) Suppose that the Laplace exponent ψ is regularly varying at ∞ with index $\gamma \in (1, \infty)$, and put $\alpha = 1/\gamma$. Check that the inverse Φ is regularly varying at ∞ with index $\alpha = 1/\gamma \in (0, 1)$ and deduce from Theorem III.11 the following extension of Khintchine's law of the iterated logarithm:

$$\limsup_{t \to 0+} \left(X_t \Phi(t^{-1} \log|\log t|)/\log|\log t|\right) = \alpha^{-\alpha}(1-\alpha)^{\alpha-1} \qquad \text{a.s.}$$

6. Comments

There is a remarkable link between Lévy processes with no negative jumps and branching processes, see Bingham (1976) and the references therein. Quite recently, a related connection with random trees and superprocesses was noted by Le Gall and Le Jan (1995). This provides additional motivations for the study of Lévy processes with no positive jumps, if necessary.

Theorem 1 and its companions Corollaries 2-3 are from Zolotarev (1964), see also Borovkov (1965) and Keilson (1963). A special case was established first by Kendall (1957). Theorem 4 is from Fristedt (1974). The concept of the random walk conditioned to drift to ∞ appears in Feller (1971) under the denomination 'associated random walk'. The continuous-time analogue was considered by Doney (1991) and Bertoin (1991); Lemma 7 which motivates the terminology was proved by the latter. Exercise 3 originates from Prabhu (1970) who has a different approach, and Exercise 4 appears in Doney (1991). We refer to Bingham (1975) and Prabhu (1981) for further references and applications of fluctuation theory without positive jumps.

The solution of the two-sided-exit problem was given first by Takács (1966), this was re-discovered by Emery (1973), see also Rogers (1990). We refer to Suprun (1976), Korolyuk, Suprun and Shurenkov (1976) and Suprun and Shurenkov (1980) for sharper results in this field. The application to increase times was pointed out in Bertoin (1991-a) using an argument of random covering, the present proof of Corollary 9 is from Doney (1995-a). We refer to Bertoin (1993-a) for further results on increase times for Lévy processes with no positive jumps.

Sections 3 and 4 are essentially excerpts from Bertoin (1992). More generally, Bertoin (1993) considers Lévy processes (possibly with positive jumps) conditioned to stay positive, and also gives a simpler proof of Theorem 20 based on stochastic calculus. See also Bertoin and Doney (1994) and Asmussen and Klüppelberg (1995) in discrete times. Some applications of Theorem 20 are developed in Bertoin (1992-a,1995-a). Chaumont (1994, 1996) studied path decompositions in Lévy processes with positive jumps conditioned to stay positive.

VIII

Stable Processes and the Scaling Property

The scaling property is used to estimate the probability of certain basic events involving real-valued stable processes. These estimates are the key to the determination of the rates of growth, and to investigating further sample path properties. The scaling property also motivates the study of other processes indexed by the unit time interval, such as the stable bridge, normalized excursion and meander, which can be thought of as conditioned stable processes, and whose trajectories are connected with the trajectories of the stable process by simple path transformations.

1. Definition and probability estimates

The scaling property is a probabilistic analogue of the notion of self-similarity in the deterministic world. Here is a formal definition.

Definition (Scaling property) *Let $Y = (Y_t, t \geq 0)$ be a stochastic process taking values in $\mathbb{R}^d$. We say that Y has the scaling property of index $\alpha > 0$ if, for every $k > 0$, the rescaled process $(k^{-1/\alpha} Y_{kt}, t \geq 0)$ has the same finite-dimensional distributions as Y.*

A Lévy process X has the scaling property if and only if for each $t > 0$, the variables X_t and $t^{1/\alpha} X_1$ have the same law. In that case, one says that X is a *strictly stable process with index* α, but for the sake of brevity, we will omit the adverb 'strictly' in the sequel. Beware: the terminology 'stable process' is nowadays sometimes used to designate a wider class of processes analogous to Gaussian processes; see Ledoux and Talagrand (1991) and Samorodnitsky and Taqqu (1994). Recall that it follows from the Lévy-Khintchine formula that the range of the index α is $(0, 2]$.

Some properties of stable processes have already been presented in the preceding chapters as special cases of results for general Lévy processes. The purpose of the present chapter is to carry on their study, relying essentially on the scaling property. For simplicity, we will focus on the one-dimensional case, but several results have higher-dimensional analogues.

Henceforth, $(X, \mathbb{P})$ *will denote a real-valued stable process with index* $\alpha \in (0, 2]$. *The trivial case when* X *is a pure drift will be implicitly excluded.*

We start by collecting some standard results on stable laws (which are the one-dimensional distributions of stable processes) for which we refer to sections 9.8–11 in Breiman (1968) and section XVII.6 in Feller (1971). See also Zolotarev (1986) for an extensive account. The cases $\alpha = 1$ and $\alpha = 2$ are special and will be discussed last.

So assume first that $\alpha \in (0, 1) \cup (1, 2)$. The characteristic exponent Ψ has the form

$$\Psi(\lambda) = c|\lambda|^{\alpha} \left(1 - i\beta \mathrm{sgn}(\lambda) \tan(\pi\alpha/2)\right) , \qquad \lambda \in (-\infty, \infty), \tag{1}$$

where $c > 0$ and $\beta \in [-1, 1]$. The Lévy measure Π is absolutely continuous with respect to the Lebesgue measure, with density

$$\Pi(dx) = \begin{cases} c^+ x^{-\alpha-1} dx & \text{if } x > 0, \\ c^- |x|^{-\alpha-1} dx & \text{if } x < 0, \end{cases}$$

where c^+ and c^- are two nonnegative real numbers such that

$$\beta = (c^+ - c^-)/(c^+ + c^-).$$

The process has no positive jumps (respectively, no negative jumps) when $c^+ = 0$ (respectively, $c^- = 0$), or equivalently $\beta = -1$ (respectively, $\beta = 1$). The process is symmetric when $c^+ = c^-$, or equivalently when $\beta = 0$.

The case $\alpha = 2$ corresponds to a Gaussian law, $\Psi(\lambda) = c\lambda^2$ for some $c > 0$ and X is proportional to a Brownian motion. In particular, X is continuous and the Lévy measure is identically zero. Finally, the case $\alpha = 1$ corresponds to a symmetric Cauchy process with drift. The characteristic exponent has the form $\Psi(\lambda) = c|\lambda| + \mathrm{d}i\lambda$, where $\mathrm{d} \in (-\infty, \infty)$ is the drift coefficient and $c > 0$. The Lévy measure Π is then proportional to $|x|^{-2}dx$.

We see by Fourier inversion that for every $t > 0$, the stable law $\mathbb{P}(X_t \in \cdot)$ is absolutely continuous with respect to the Lebesgue measure and has a continuous density $p_t(\cdot)$. Except when X is a Brownian motion, a Cauchy process, or a stable subordinator with index $1/2$, there is no explicit formula for the density. When X is a subordinator, we have $p_t(x) = 0$ for $x \leq 0$ and $p_t(x) > 0$ for $x > 0$. When $|X|$ is not a subordinator, $p_t(\cdot)$ is positive everywhere. See section 2.7 in Zolotarev (1986).

By the scaling property, the quantity

$$\rho = \mathbb{P}(X_t \geq 0)$$

does not depend on $t > 0$. We call ρ the *positivity parameter* . Its value for an index $\alpha \neq 1, 2$ can be computed in terms of the parameter β,

$$\rho = \frac{1}{2} + (\pi\alpha)^{-1} \arctan\left(\beta \tan(\pi\alpha/2)\right).$$

See Zolotarev (1986), section 2.6. Note that for $0 < \alpha < 1$, ρ ranges over $[0, 1]$ (the boundary points $\rho = 1$ and $\rho = 0$ correspond to the cases when X is a subordinator and the negative of a subordinator, respectively), whereas for $1 < \alpha < 2$, ρ ranges over $[1 - 1/\alpha, 1/\alpha]$ (the boundary points $\rho = 1/\alpha$ and $\rho = 1 - 1/\alpha$ correspond to the cases when X has no positive jumps and no negative jumps, respectively). Finally, the positivity parameter necessarily equals $1/2$ when $\alpha = 2$, and ranges over the open unit interval for $\alpha = 1$.

Theoretically, one should be able to calculate all the analytic quantities that appear in fluctuation theory for stable processes in terms of the index and the positivity parameter. Unfortunately, it seems that this is seldom possible in practice, due to the absence of a closed expression for the stable densities. See, however, Doney (1987) for the case when $\rho + k = \ell/\alpha$, where k and ℓ are integers. In the next lemma, we simply determine the laws of the ladder processes separately, which will be sufficient for our purposes.

Lemma 1 *Suppose that $|X|$ is not a subordinator. The ladder time process L^{-1} is a stable subordinator of index ρ, and the ladder height process H is a stable process of index $\alpha\rho$.*

Proof According to Corollary VI.10, the Laplace exponent of L^{-1} is given for every $\lambda \geq 0$ by

$$\begin{aligned}\kappa(\lambda, 0) &= k \exp\left\{\int_0^\infty (e^{-t} - e^{-\lambda t}) t^{-1} \mathbb{P}(X_t \geq 0) dt\right\} \\ &= k\lambda^\rho \qquad \text{[by Frullani's integral]},\end{aligned}$$

which establishes our first assertion.

Corollary VI.10 also shows that the logarithm of the Laplace exponent of H is given by

$$\begin{aligned}\log \kappa(0,\lambda) &= \int_0^\infty t^{-1}\mathbb{E}(e^{-\lambda X_t} - e^{-t}, X_t \geq 0)dt \\ &= \int_0^\infty t^{-1}\mathbb{E}(e^{-X_t} - e^{-t\lambda^{-\alpha}}, X_t \geq 0)dt \qquad \text{[by scaling]} \\ &= \int_0^\infty t^{-1}\mathbb{E}(e^{-X_t} - e^{-t}, X_t \geq 0)dt \\ &\quad + \int_0^\infty t^{-1}(e^{-t} - e^{-t\lambda^{-\alpha}})\mathbb{P}(X_t \geq 0)dt \\ &= \log \kappa(0,1) + \alpha\rho \log \lambda \qquad \text{[by Frullani's integral]}.\end{aligned}$$

Thus $\kappa(0,\lambda) = \kappa(0,1)\lambda^{\alpha\rho}$ and our second assertion is proved. □

The second purpose of this section is to estimate the distributions of the unilateral and the bilateral suprema of a stable process,

$$S_t = \sup\{X_s : 0 \leq s \leq t\}\ ,\ X_t^* = \sup\{|X_s| : 0 \leq s \leq t\} \qquad (t > 0).$$

Observe that the increasing processes S and X^* satisfy the scaling property with index α; this will be used in the sequel without further mention. The case when X is a subordinator is somewhat degenerate, because then $X = S = X^*$ and estimates of their distribution functions either have been already obtained in chapter III, or can be deduced by application of Tauberian theorems. Therefore, we shall focus on the case when $|X|$ is not a subordinator.

In the next three propositions, k will denote a positive finite constant depending on the parameters of the stable process (its exact value is seldom explicitly known), which may change from result to result.

Proposition 2 *Suppose that $|X|$ is not a subordinator. Then there exists a constant $k > 0$ such that $\mathbb{P}(S_1 < x) \sim kx^{\alpha\rho}$ as $x \to 0+$.*

Proof We deduce from equation (8) of chapter VI (with $x = 1$) and Lemma 1 the estimate

$$1 - \mathbb{E}(\exp\{-qT_{(1,\infty)}\}) \sim kq^{\rho} \qquad (q \to 0+).$$

By the Tauberian theorem (cf. section O.7) this shows that

$$\mathbb{P}(S_t < 1) = \mathbb{P}(T_{(1,\infty)} \geq t) \sim kt^{-\rho} \qquad (t \to \infty).$$

On the other hand, the scaling property yields $\mathbb{P}(S_t < 1) = \mathbb{P}(S_1 < t^{-1/\alpha})$, which establishes our assertion. □

Proposition 3 *Suppose that $|X|$ is not a subordinator. Then there exists a constant $k > 0$ such that $\log \mathbb{P}(X_1^* < x) \sim -kx^{-\alpha}$ as $x \to 0+$.*

Proof By the scaling property, the estimate in the statement is equivalent to

$$\log \mathbb{P}(X_t^* < 1) \sim -kt \qquad (t \to \infty).$$

Consider the function $f(t) = \sup\{\mathbb{P}_x(X_t^* < 1) : |x| < 1\}$. The Markov property readily yields the inequality $f(t+s) \le f(t)f(s)$, thus $\log f$ is subadditive and there is a constant $k \in (0, \infty]$ such that

$$\lim_{t\to\infty} \frac{1}{t} \log f(t) = -k\,,$$

see e.g. Theorem 7.6.2 in Hille and Phillips (1957). In particular

$$\limsup_{t\to\infty} \frac{1}{t} \log \mathbb{P}(X_t^* < 1) \le -k.$$

Next, denote by $T_{a,b} = \inf\{t : X_t \notin [-a, b]\}$ the first exit time from $[-a, b]$, fix $\varepsilon > 0$ arbitrarily small and deduce by scaling that for every $t > 0$, there exists $y(t) \in [\varepsilon - 1, 1 - \varepsilon]$ such that

$$\begin{aligned} \mathbb{P}_{y(t)}(T_{1-\varepsilon,1-\varepsilon} > t) &\ge \frac{1}{2} \sup\{\mathbb{P}_x(T_{1-\varepsilon,1-\varepsilon} > t) : |x| < 1 - \varepsilon\} \\ &\ge \frac{1}{2} \exp\{-k(1+4\varepsilon)t\}. \end{aligned}$$

As a consequence,

$$\mathbb{P}_y(T_{1,1} > t) \ge \frac{1}{2} \exp\{-k(1+4\varepsilon)t\} \qquad \text{for every } y \in [y(t)-\varepsilon, y(t)+\varepsilon]. \quad (2)$$

We then observe that for every $x \in (-1, 1)$

$$\inf\left\{\mathbb{P}_x\left(X_1 \in [y-\varepsilon, y+\varepsilon], T_{1,1} > 1\right) : |y| \le 1 - \varepsilon\right\} > 0. \qquad (3)$$

Indeed, if (3) were false, then by a compactness argument, there would exist $y \in [\varepsilon - 1, 1 - \varepsilon]$ such that

$$\mathbb{P}_x\left(X_1 \in (y-\varepsilon, y+\varepsilon), T_{1,1} > 1\right) = 0,$$

which is absurd. (Typically, it is easy to see that the probability that the stable process enters $[y - \varepsilon/2, y + \varepsilon/2]$ before time $1 \wedge T_{1,1}$ and then remains inside $(y - \varepsilon, y + \varepsilon)$ during a unit amount of time is positive. Note that this argument fails for subordinators.) Combining now (2-3) and the Markov property, we deduce the converse inequality

$$\liminf_{t\to\infty} \frac{1}{t} \log \mathbb{P}_x(T_{1,1} > t) \ge -k(1+4\varepsilon)\,.$$

So all that is needed now is to check that $k < \infty$. To this end, recall that the one-dimensional stable law has a continuous positive density

$p_1(x)$ with respect to the Lebesgue measure (recall that $|X|$ is not a subordinator). As a consequence, we can pick $\eta, c, c' > 0$ such that

$$\mathbb{P}_x\left(X_1^* < c', |X_1| < c\right) > \eta \qquad \text{for all } |x| < c.$$

By the scaling property, we have for every integer $n > 0$

$$\inf_{|x|<cn^{-1/\alpha}} \mathbb{P}_x\left(X_{1/n}^* < c'n^{-1/\alpha}, |X_{1/n}| < cn^{-1/\alpha}\right) > \eta.$$

Then the Markov property yields the inequality

$$\begin{aligned}&\mathbb{P}\left(X_1^* < c'n^{-1/\alpha}\right)\\ &\geq \left(\inf_{|x|<cn^{-1/\alpha}} \mathbb{P}_x\left(X_{1/n}^* < c'n^{-1/\alpha}, |X_{1/n}| < cn^{-1/\alpha}\right)\right)^n \geq \eta^n,\end{aligned}$$

and this shows that k must be finite. □

The hypothesis in Propositions 2-3 that $|X|$ is not a subordinator is crucial. When X is a stable subordinator of index $\alpha \in (0,1)$, one obtains for instance from a Tauberian theorem of de Bruijn (Theorem 5.12.9 in Bingham, Goldie and Teugels (1987)) that $\log \mathbb{P}(X_1^* < x) \sim -kx^{-\alpha/(1-\alpha)}$ as $x \to 0+$.

Finally, we turn our attention to the behaviour of the tail distribution of the unilateral supremum. A related argument applies for the estimation of the tail of the bilateral supremum X_1^*, see Exercise 2.

Proposition 4 *Suppose that X possesses positive jumps, i.e. its Lévy measure does not vanish on $(0,\infty)$. Then there exists a constant $k > 0$ such that*

$$\mathbb{P}(X_1 > x) \sim \mathbb{P}(S_1 > x) \sim kx^{-\alpha} \quad (x \to \infty).$$

When X has no positive jumps, then either $\alpha < 1$, $-X$ is a subordinator and S_1 is identically zero, or $\alpha \in (1,2]$ and the exponential moments of S_1 are finite (see Corollary VII.2).

Proof The estimate $\mathbb{P}(X_1 > x) \sim kx^{-\alpha}$ is well known, see e.g. Zolotarev (1986) on page 95, or use Exercise I.1. It follows that

$$\liminf_{x\to\infty} \mathbb{P}(S_1 > x)x^{\alpha} \geq k.$$

Next, fix $\varepsilon > 0$ and note that

$$\begin{aligned}\mathbb{P}(X_1 > (1-\varepsilon)x) &\geq \mathbb{P}(S_1 > x, X_1 > (1-\varepsilon)x)\\ &\geq \int_0^1 \mathbb{P}(T_{(x,\infty)} \in dt)\mathbb{P}(X_{1-t} > -\varepsilon x) \qquad \text{[Markov]}\\ &\geq \mathbb{P}(S_1 > x)\mathbb{P}(X_1 > -\varepsilon x) \qquad \text{[scaling]}.\end{aligned}$$

Since $\mathbb{P}(X_1 > -\varepsilon x)$ tends to 1 as $x \to \infty$, we deduce that

$$\limsup_{x\to\infty} \mathbb{P}(S_1 > x)x^\alpha \le (1-\varepsilon)^{-\alpha}k,$$

and since ε can be chosen arbitrarily small, the proposition is proved. □

2. Some sample path properties

We are now going to apply the estimates of section 1 to deduce information on the sample path behaviour of stable processes. First, we consider the asymptotic rates of growth. For simplicity, we shall focus on the case of small times; results for large times are similar (simply replace 0+ by ∞ in the statements). The 'limsup' behaviour of stable processes is characterized by the following.

Theorem 5 *Let $f : (0,\infty) \to (0,\infty)$ be an increasing function.*

(i) *Suppose that X has positive jumps. Then*

$$\limsup_{t\to 0+} \frac{X_t}{f(t)} = 0 \text{ or } \infty \text{ a.s.}$$

according as the integral $\int_{0+} f(t)^{-\alpha} dt$ converges or diverges.

(ii) *Suppose that X has no positive jumps and that $\alpha > 1$. Then there exists a constant $k > 0$ such that*

$$\limsup_{t\to 0+} \frac{X_t}{t^{1/\alpha}\left(\log|\log t|\right)^{1-1/\alpha}} = k \qquad \text{a.s.}$$

(iii) *Suppose that $\alpha \ne 2$. Then*

$$\limsup_{t\to 0+} \frac{|X_t|}{f(t)} = 0 \text{ or } \infty \text{ a.s.}$$

according as the integral $\int_{0+} f(t)^{-\alpha} dt$ converges or diverges.

Proof (i) Suppose first that the integral converges and note that this forces $f(t)^{-\alpha} = o(t^{-1})$ as $t \to 0+$. For every integer $n > 0$, the scaling property and Proposition 4 imply

$$\mathbb{P}\left(S_{2^{-n}} > f(2^{-n-1})\right) = \mathbb{P}\left(S_1 > 2^{n/\alpha} f(2^{-n-1})\right) \sim k 2^{-n} f(2^{-n-1})^{-\alpha}.$$

The series $\sum 2^{-n} f(2^{-n-1})^{-\alpha}$ converges and the Borel-Cantelli lemma yields $S_{2^{-n}} < f(2^{-n-1})$ for all n large enough, a.s. A monotonicity argument then shows that $S_t \le f(t)$ for all $t > 0$ small enough a.s., and since the integral still converges when f is replaced by εf, we have $\lim_{t\to 0+}\left(S_t/f(t)\right) = 0$ a.s.

Suppose now that the integral diverges. Pick $a > 0$ and consider for every integer $n > 0$ the event

$$\Lambda_n = \left\{ |X_{2^{-n}}| \leq a2^{-n/\alpha}, X_{2^{-n+1}} - X_{2^{-n}} > f(2^{-n+1}) + a2^{-n/\alpha} \right\}.$$

Using the independence and homogeneity of the increments and the scaling property, we see that

$$\mathbb{P}(\Lambda_n) = \mathbb{P}(|X_1| \leq a)\mathbb{P}(X_1 > 2^{n/\alpha} f(2^{-n+1}) + a),$$

and we then deduce from Proposition 4 that the series $\sum \mathbb{P}(\Lambda_n)$ diverges. Of course, the events Λ_n are not independent, but using the inclusion

$$\Lambda_n \subseteq \Lambda'_n := \left\{ X_{2^{-n+1}} - X_{2^{-n}} > f(2^{-n+1}) + a2^{-n/\alpha} \right\}$$

and the independence of the Λ'_n, we see that for $n \neq m$

$$\mathbb{P}(\Lambda_m \cap \Lambda_n) \leq \mathbb{P}(\Lambda_m)\mathbb{P}(\Lambda_n)\mathbb{P}(|X_1| \leq a)^{-2}.$$

By a generalization of the Borel-Cantelli lemma for independent events (see for instance Spitzer (1964) on page 317), we have

$$\mathbb{P}(\limsup \Lambda_n) \geq \mathbb{P}(|X_1| \leq a)^2.$$

It follows that the probability that $\limsup_{t \to 0+} (X_t/f(t)) \geq 1$ is at least $\mathbb{P}(|X_1| \leq a)^2 > 0$, and thus equals 1 according to the Blumenthal zero-one law. Since the integral diverges when one replaces f by $\varepsilon^{-1} f$ as well, this shows that $\limsup_{t \to 0+} (X_t/f(t)) = \infty$ a.s.

(ii) This is a special case of Exercise VII.8(b).

(iii) With no loss of generality, we may suppose that X possesses negative jumps. If the integral diverges, then by (i), $\lim_{t \to 0+} (X_t/f(t)) = \infty$ a.s. and *a fortiori* $\lim_{t \to 0+} (|X_t|/f(t)) = \infty$ a.s.

Suppose now that the integral converges. When $|X|$ is a subordinator, we simply apply (i). When X possesses jumps of both signs, we deduce from (i) that $\lim_{t \to 0+} (X_t/f(t)) = \lim_{t \to 0+} (\widehat{X}_t/f(t)) = 0$, where $\widehat{X} = -X$, and our assertion follows. Finally, recall from section VII.1 that when X is a stable process with no positive jumps, the exponential moments of S_1 are finite. This forces $\mathbb{P}(S_1 > x) = o(e^{-x})$ as $x \to \infty$, and we then deduce from Proposition 4 that $\mathbb{P}(X_1^* > x) \sim kx^{-\alpha}$ as $x \to \infty$. The argument in the proof of (i) now applies and shows that $\lim_{t \to 0+} (X_t^*/f(t)) = 0$ a.s. □

We next turn our attention to the 'liminf' behaviour of the suprema processes.

Theorem 6 *Suppose that $|X|$ is not a subordinator.*

(i) *Let $f : (0,\infty) \to (0,\infty)$ be an increasing function. We have a.s.*

$$\liminf_{t\to 0+} \frac{S_t}{t^{1/\alpha} f(t)} = 0 \text{ or } \infty$$

according as the integral $\int_{0+} f(t)^{\alpha\rho} t^{-1} dt$ diverges or converges.

(ii) *Let $k > 0$ be the constant appearing in Proposition 3. We have a.s.*

$$\liminf_{t\to 0+} \frac{X_t^*}{\left[t/(\log|\log t|)\right]^{1/\alpha}} = k^{1/\alpha}.$$

Proof The proof of (i) is based on Proposition 2 and follows the same lines as that of Theorem 5(i); we skip the details. The proof of (ii) is close to that of Theorem III.11. It will be convenient to use the notation $X_t^* = X^*(t)$ and $f(t) = \left[t/(\log|\log t|)\right]^{1/\alpha}$ in the sequel.

First we prove the lower bound. Take $r < 1$ and $0 < c < c' < k^{1/\alpha}$ and note that for n large enough and r close enough to 1

$$\mathbb{P}(X^*(r^n) \leq cf(r^{n-1})) \leq \mathbb{P}(X^*(r^n) \leq c'f(r^n)) .$$

Applying Proposition 3 and the scaling property, we get

$$\log \mathbb{P}(X^*(r^n) \leq cf(r^{n-1})) \leq \log \mathbb{P}\left(X^*(1) \leq c'(\log(n\log(1/r)))^{-1/\alpha}\right)$$
$$\sim -kc'^{-\alpha} \log(n\log(1/r)) .$$

Since $kc'^{-\alpha} > 1$, we deduce that the series $\sum_n \mathbb{P}(X^*(r^n) \leq cf(r^{n-1}))$ converges. Therefore $X^*(r^n) \geq cf(r^{n-1})$ for all n large enough, a.s. An immediate monotonicity argument then shows that

$$\liminf_{t\to 0+} \left(X^*(t)/f(t)\right) \geq c \text{ a.s.}$$

Now we prove the converse upper bound. Take $c'' > k^{1/\alpha}$ and $r > 1$, and put $t_n = \exp\{-n^r\}$. Observe the inequality $t_n > 2t_{n+1}$ and deduce

$$\mathbb{P}\left(\sup\{|X(s) - X(t_{n+1})| : s \in [t_{n+1}, t_n]\} \leq x\right) \geq \mathbb{P}(X^*(t_n) \leq x) .$$

On the other hand, we see from Proposition 3 and the scaling property that

$$\log \mathbb{P}(X^*(t_n) \leq c'' f(t_n)) \sim -kc''^{-\alpha} r \log n .$$

Since $kc''^{-\alpha} < 1$, it follows that

$$\sum_n \mathbb{P}(\sup\{|X(s) - X(t_{n+1})| : s \in [t_{n+1}, t_n]\} \leq c'' f(t_n)) = \infty$$

provided that r has been chosen close enough to 1. According to the Borel-Cantelli lemma for independent events, we have

$$\liminf \frac{\sup\{|X(s) - X(t_{n+1})| : s \in [t_{n+1}, t_n]\}}{f(t_n)} \leq c'' \quad \text{a.s.}$$

We will prove in a while that

$$\lim \frac{X(t_{n+1})}{f(t_n)} = 0 \quad \text{a.s.} \tag{4}$$

We can then deduce from the foregoing that $\liminf_{t\to\infty} \left(X^*(t)/f(t)\right) \le c''$ a.s., so all that is needed now is to check (4).

Applying Proposition 4 in the second equality below, we have for every $\varepsilon > 0$

$$\begin{aligned}\mathbb{P}\left(|X(t_{n+1})| > \varepsilon f(t_n)\right) &= \mathbb{P}\left(|X_1| > \frac{\varepsilon \exp\{((n+1)^r - n^r)/\alpha\}}{(r\log n)^{1/\alpha}}\right) \\ &= O\left(\exp\{-(n+1)^r + n^r\}\right) \\ &= O\left(\exp\{-rn^{r-1}\}\right).\end{aligned}$$

This shows that the series $\sum \mathbb{P}\left(|X(t_{n+1})| > \varepsilon f(t_n)\right)$ converges. According to the Borel-Cantelli lemma, $\limsup\left(|X(t_{n+1})|/f(t_n)\right) \le \varepsilon$ a.s., which establishes (4) since $\varepsilon > 0$ can be chosen arbitrarily small. The proof is now complete. □

We now develop some applications of fluctuation theory based on Lemma 1. First, we obtain the following simple characterization of stable processes that possess increase times in terms of the positivity parameter (see section VI.5).

Proposition 7 *The probability that there exist increase times equals* 0 *if* $\rho \le 1/2$ *and* 1 *otherwise.*

Proof Of course, the positivity parameter of the dual process $\widehat{X}$ is $\widehat{\rho} = 1 - \rho$. Applying Corollaries VI.24-25 and Lemma 1, we see that the probability that there exist increase times equals 0 or 1 according as the integral

$$\int_{0+} x^{-\alpha(1-\rho)}\, d\left(x^{\alpha\rho}\right) = \alpha\rho \int_{0+} x^{2\alpha\rho-\alpha-1}\, dx$$

diverges or converges. □

It is remarkable that the criterion does not depend on the index α. Note also that the symmetric Cauchy process C has no increase times, whereas when one adds a drift $\mathrm{d} > 0$, $C_t + \mathrm{d}t$ possesses increase times.

Finally, we turn our attention to the way that the stable process hits a given point, say 1 for simplicity. For this purpose, only the case when the index α is greater than 1 is relevant, since otherwise single points are polar (see section II.5). Recall also that $T_{\{1\}}$ stands for the first passage time at 1 and that $T_{\{1\}} \in (0, \infty)$ a.s., since X is then recurrent.

Proposition 8 *Suppose that $\alpha > 1$.*

(i) *If X has no positive jumps, then X does not jump across 1 before time $T_{\{1\}}$, a.s.*
(ii) *If X has no negative jumps and $\alpha \neq 2$, then X makes exactly one jump across 1 before time $T_{\{1\}}$, a.s.*
(iii) *If X has both positive and negative jumps, then X makes infinitely many jumps across 1 before time $T_{\{1\}}$, a.s.*

Proof (i) is obvious.

(ii) In this case, the positivity parameter is $\rho = 1 - 1/\alpha$ and thus $\alpha\rho < 1$. By Lemma 1, the ladder height process has no drift, and therefore, according to Theorem VI.19, X jumps a.s. across 1 at its first passage time above 1, $T_{(1,\infty)}$. Immediately after time $T_{(1,\infty)}$, X is strictly above 1 and since it cannot jump downwards, $T_{(1,\infty)}$ is the instant of the only jump across 1 on the time interval $[0, T_{\{1\}}]$.

(iii) In that case, $\alpha\rho < 1$ and $\alpha(1-\rho) < 1$. By Lemma 1 and Theorem VI.19, we see that for every fixed level $a > 0$, X jumps a.s. across a at the first instant when it exceeds a. The same holds for the dual process $\widehat{X} = -X$. We conclude applying the Markov property at the successive instants when X passes above/below 1. □

3. Bridges

In section II.1, we considered the law of a Lévy process X conditioned on $X_t = b$, but this conditional law only makes sense for almost every b (with respect to the distribution of X_t), and not for a specific b. Our first aim in this section will be to give a rigorous definition of (standard) stable bridges, which are processes on the unit time interval which can be thought of as stable processes conditioned to be at level 0 at time 1. In the Brownian case, numerous constructions of the bridge are known (see in particular section IV.40 in Rogers and Williams (1987)), but most rely on the Gaussian property of Brownian motion and therefore do not extend to stable processes. Nonetheless, there exists a general construction of bridges for a wide class of Markov processes; we refer to Fitzsimmons, Pitman and Yor (1993) for a detailed account. We shall follow here a simple variation of this approach in our setting. It is obvious that the case when $|X|$ is a subordinator is degenerate for the problems we consider in this section, so it will be implicitly excluded in the sequel. In order to avoid heavy notation involving the parameters of the stable process, we shall also assume from now on that they have been fixed.

Since we are concerned with processes indexed by the unit interval, or more generally by a compact interval of time, it is convenient to introduce for every $t > 0$ the space of real-valued functions on $[0, t]$ which are continuous to the right and have limits to the left, $\Omega_t = D([0, t], \mathbb{R})$. This new canonical space is endowed with Skorohod's topology, and can be viewed as the restriction to the time interval $[0, t]$ of the paths $\omega \in \Omega$ with lifetime $\zeta(\omega) > t$.

Recall that for every $t > 0$, the law of X_t is absolutely continuous with respect to the Lebesgue measure and there is a continuous, everywhere positive version $p_t(\cdot)$ of the density. In particular, the law of X_t is equivalent to the Lebesgue measure, which allows us to use the expression *almost surely* without ambiguity. In this setting, the semigroup and scaling properties read respectively

$$p_t(x) = \int_{\mathbb{R}} p_{t-s}(y)p_s(x-y)dy \qquad (0 < s < t, x \in \mathbb{R})$$

and

$$p_t(x) = t^{-1/\alpha}p_1(t^{-1/\alpha}x) \qquad (0 < t, x \in \mathbb{R}).$$

The regular version of the conditional law of X_s given X_t is

$$\mathbb{P}(X_s \in da \mid X_t = b) = \frac{1}{p_t(b)}p_s(a)p_{t-s}(b-a)da \qquad (a, b \in \mathbb{R}). \qquad (5)$$

The way to get round the difficulty of giving a rigorous definition of the bridge stems from the inhomogeneous Markov property of conditional laws. Specifically, take an integer $n > 0$ and real numbers $0 < t_1 < \cdots < t_n$, and denote by $\mathbb{P}(\cdot \mid X_{t_1} = x_1, \cdots, X_{t_n} = x_n)$, $(x_1, \cdots, x_n) \in \mathbb{R}^n$, a version of the conditional law of $(X_u, 0 \le u \le t)$ given $(X_{t_1}, \cdots, X_{t_n})$ under $\mathbb{P}$. Write $t_0 = 0$ and $x_0 = 0$. We now claim the following.

Lemma 9 *For a.e. $(x_1, \cdots, x_n) \in \mathbb{R}^n$, under $\mathbb{P}(\cdot \mid X_{t_1} = x_1, \cdots, X_{t_n} = x_n)$, the processes*

$$\left(X_{t_i+u} - x_i, 0 \le u \le t_{i+1} - t_i\right), \qquad i = 0, \cdots, n-1,$$

are independent and have the laws $\mathbb{P}\left(\cdot \mid X_{t_{i+1}-t_i} = x_{i+1} - x_i\right)$, respectively.

Proof The case $n = 2$ follows immediately from the Markov property. We deduce the general case by iteration. □

Informally, Lemma 9 describes the pieces of paths obtained after splitting a bridge into n pieces, conditionally on the extremities of these pieces. The natural thing to do to define the bridge unambiguously is to splice back those pieces using (5). Here is the formal procedure.

Fix an arbitrary real number $t \in (0,1)$, and for a.e. $x \in \mathbb{R}$, consider two independent processes

$$Y^x = (Y_s^x, 0 \le s \le t) \quad \text{and} \quad Z^x = (Z_s^x, 0 \le s \le 1-t)$$

with laws $\mathbb{P}(\cdot \mid X_t = x)$ and $\mathbb{P}(\cdot \mid X_{1-t} = -x)$, respectively. Next put

$$X_s^x = \begin{cases} Y_s^x & \text{if } 0 \le s \le t, \\ x + Z_{s-t}^x & \text{if } t \le s \le 1, \end{cases}$$

and denote the law of X^x by $\mathbb{P}(\cdot \mid X_t = x, X_1 = 0)$. The probability measure on Ω_1 given by

$$\mathbb{Q}^t = \frac{1}{p_1(0)} \int_{-\infty}^{\infty} \mathbb{P}(\cdot \mid X_t = x, X_1 = 0) p_t(x) p_{1-t}(-x)\, dx \tag{6}$$

is now a natural candidate for being the law of the stable bridge. We check that it actually does not depend on t.

Lemma 10 *For all $0 < s < t < 1$, the laws $\mathbb{Q}^s$ and $\mathbb{Q}^t$ are the same.*

Proof We deduce first from Lemma 9 that the conditional laws $\mathbb{Q}^s(\cdot \mid X_s, X_t)$ and $\mathbb{Q}^t(\cdot \mid X_s, X_t)$ coincide a.s. So all that we have to verify is the identity

$$\mathbb{Q}^s(X_s \in da, X_t \in db) = \mathbb{Q}^t(X_s \in da, X_t \in db). \tag{7}$$

On the one hand, it follows from (5) that the left-hand side of (7) equals

$$\begin{aligned} &\mathbb{P}(X_s \in da \mid X_t = b) \frac{p_t(b) p_{1-t}(-b)}{p_1(0)} db \\ &= \frac{p_s(a) p_{t-s}(b-a)}{p_t(b)} \frac{p_t(b) p_{1-t}(-b)}{p_1(0)} dadb. \end{aligned}$$

Similarly, the right-hand side of (7) equals

$$\begin{aligned} &\mathbb{P}(X_{t-s} \in db - a \mid X_{1-s} = -a) \frac{p_s(a) p_{1-s}(-a)}{p_1(0)} db \\ &= \frac{p_{t-s}(b-a) p_{1-t}(-b)}{p_{1-s}(-a)} \frac{p_s(a) p_{1-s}(-a)}{p_1(0)} dadb. \end{aligned}$$

After simplification, we see that (7) holds. □

We are finally able to give a rigorous definition.

Definition (Bridge) *The probability measure* $\mathbb{Q}^t$ *on* Ω_1 *given by* (6) *is called the law of the stable bridge. It will be denoted henceforth by* $\mathbb{P}^{(br)}$.

It is now an easy matter to prove that the bridge appears in limits of elementary conditionings.

Proposition 11 *Fix an arbitrary* $t \in (0,1)$ *and take* $\Lambda \in \mathscr{F}_t$. *Then*

$$\mathbb{P}^{(br)}(\Lambda) = \lim_{\varepsilon \to 0+} \mathbb{P}(\Lambda \mid |X_1| < \varepsilon) .$$

Proof Applying the Markov property at time t, we get

$$\mathbb{P}(\Lambda \mid |X_1| < \varepsilon) = \frac{1}{\mathbb{P}(|X_1| < \varepsilon)} \int_{-\infty}^{\infty} \mathbb{P}(\Lambda \mid X_t = x)\mathbb{P}(|X_{1-t} + x| < \varepsilon)p_t(x)dx.$$

Then let ε tend to $0+$ and recall that the stable densities are bounded and continuous. On the one hand, $2\varepsilon/\mathbb{P}(|X_1| < \varepsilon)$ converges to $1/p_1(0)$, and on the other hand, $\mathbb{P}(|X_{1-t} + x| < \varepsilon)/2\varepsilon$ converges to $p_{1-t}(-x)$ and is bounded by $\sup\{p_{1-t}(y) : y \in \mathbb{R}\}$ uniformly in ε. It follows by dominated convergence that

$$\lim_{\varepsilon \to 0+} \mathbb{P}(\Lambda \mid |X_1| < \varepsilon) = \mathbb{Q}^t(\Lambda) = \mathbb{P}^{(br)}(\Lambda) . \qquad \square$$

It should be clear from the proof that, more generally, one can replace the conditioning on $\{|X_1| < \varepsilon\}$ in Proposition 11 by conditioning on any sensible approximation of the event $\{X_1 = 0\}$ (for instance, $\{0 \leq X_1 < \varepsilon\}$).

The very definition of the law of the stable bridge shows that it is absolutely continuous with respect to that of the stable process on every sigma-field $\mathscr{F}_t$, $0 < t < 1$. Specifically,

$$d\mathbb{P}^{(br)}_{|\mathscr{F}_t} = \frac{1}{p_1(0)} p_{1-t}(-X_t)\, d\mathbb{P}_{|\mathscr{F}_t}. \tag{8}$$

One sees in particular that the Radon-Nikodym derivative $p_{1-t}(-X_t)/p_1(0)$ is a $(\mathbb{P}, \mathscr{F}_t)$ martingale (recall also the semigroup property); specialists say that the law of the bridge is obtained by a space-time harmonic transform. Since finite-dimensional distributions determine laws on Ω_1, $\mathbb{P}^{(br)}$ is the unique probability measure on Ω_1 for which (8) holds and $X_1 = 0$ a.s.

One might think that it would have been simpler to define $\mathbb{P}^{(br)}$ directly using (8) and invoking Kolmogorov's projective theorem. However, this approach is far from being easy, because the projective limit is a probability measure on $\bigcup_{t<1} \Omega_t$. Proving that this law is actually supported by Ω_1 without using arguments similar to that appearing in Lemma 9 is quite a demanding task.

The scaling property had no rôle in the definition of the bridge, but we now conclude this section by presenting one of its applications in the case when single points are not polar, that is when the index α is greater than 1. It consists in a simple pathwise construction of the bridge from the stable process.

Theorem 12 *Suppose that $\alpha > 1$, denote by $g = \sup\{t < 1 : X_t = 0\}$ the last passage time at the origin on the unit time interval, and introduce the process*

$$X_t^{(br)} = g^{-1/\alpha} X_{gt}, \qquad t \in [0, 1].$$

Then under $\mathbb{P}$, $X^{(br)}$ and g are independent, $X^{(br)}$ is a bridge and g follows the generalized arcsine law with parameter $1 - 1/\alpha$.

One can give a rather short proof of Theorem 12 using arguments of excursion theory, see Chaumont (1996-a). We shall follow here an elementary approach which is slightly longer, but perhaps easier to follow for readers who are not familiar with excursion theory. It relies on the following formula for the distribution of the first passage time at a given real number.

Lemma 13 *For every $x \in \mathbb{R}$ and $t > 0$, one has*

$$\mathbb{P}\left(T_{\{x\}} < t\right) = \frac{1}{\Gamma(1-1/\alpha)\Gamma(1/\alpha)p_1(0)} \int_0^t s^{-1+1/\alpha} p_{t-s}(x) ds.$$

Proof Denote, for every $q > 0$, by $u^q(x) = \int_0^\infty e^{-qt} p_t(x) dt$, the continuous version of the q-resolvent density. Note by the scaling property of the semigroup that

$$u^q(0) = p_1(0) \int_0^\infty e^{-qt} t^{-1/\alpha} dt = q^{-1+1/\alpha} \Gamma(1-1/\alpha) p_1(0).$$

We then deduce from Corollary II.18 that

$$\int_0^\infty e^{-qt} \mathbb{P}\left(T_{\{x\}} < t\right) dt = q^{-1} u^q(x)/u^q(0) = \frac{q^{-1/\alpha} u^q(x)}{\Gamma(1-1/\alpha)p_1(0)}.$$

Using the identity

$$q^{-1/\alpha} = \frac{1}{\Gamma(1/\alpha)} \int_0^\infty e^{-qt} t^{1/\alpha - 1} dt,$$

we can invert the Laplace transform in the right-hand side and obtain the desired formula. □

The assertion in Theorem 12 that g has the generalized arcsine law with parameter $1 - 1/\alpha$ can be readily deduced from the feature that the inverse local time at 0 is a stable subordinator of index $1 - 1/\alpha$ (see section V.1) and Theorem III.6, using an argument similar to that in Exercise IV.5. Lemma 13 also yields the following direct proof. Applying the Markov property at time t, we can express $\mathbb{P}(g > t)$ as

$$\begin{aligned}
&\int_{-\infty}^{\infty} p_t(x)\mathbb{P}\left(T_{\{-x\}} < 1-t\right) dx \\
&= \frac{1}{\Gamma(1-1/\alpha)\Gamma(1/\alpha)p_1(0)} \int_{-\infty}^{\infty} p_t(x) \left(\int_0^{1-t} s^{-1+1/\alpha} p_{1-t-s}(-x) ds \right) dx \\
&= \frac{1}{\Gamma(1-1/\alpha)\Gamma(1/\alpha)p_1(0)} \int_0^{1-t} s^{-1+1/\alpha} p_{1-s}(0) ds
\end{aligned}$$

(by the semigroup property). Using the identity $p_{1-s}(0) = (1-s)^{-1/\alpha} p_1(0)$, we finally see this last quantity equals

$$\frac{1}{\Gamma(1-1/\alpha)\Gamma(1/\alpha)} \int_0^{1-t} s^{-1+1/\alpha}(1-s)^{-1/\alpha} ds .$$

So $1 - g$ has the generalized arcsine law with parameter $1/\alpha$, which is equivalent to our claim. We now prove Theorem 12.

Proof of Theorem 12 First, we rephrase the statement in terms of bridges of given length. Specifically, consider for every $t > 0$ the mapping $\mathcal{S}^{(t)} : \Omega_1 \to \Omega_t$ given by

$$\omega \to \mathcal{S}^{(t)}\omega \quad \text{where } \mathcal{S}^{(t)}\omega(s) = t^{1/\alpha}\omega(s/t) \qquad (s \in [0, t]).$$

The image of $\mathbb{P}^{(br)}$ under this mapping is denoted by $\mathbb{P}^{(br,t)}$ and is called the law of the stable bridge with length t. We thus see that the statement of Theorem 12 is equivalent to claiming that under $\mathbb{P}$, conditionally on g, the process $(X_s, s \leq g)$ is a stable bridge with length g.

Now fix an arbitrary $t \in (0,1)$ and $\Lambda \in \mathcal{F}_t$. Applying the Markov property at time t, we obtain

$$\begin{aligned}
&\mathbb{P}(\Lambda, g > t) \\
&= \mathbb{E}\left(\mathbb{P}_{X_t}\left(T_{\{0\}} < 1-t\right), \Lambda\right) \\
&= \frac{1}{\Gamma(1-1/\alpha)\Gamma(1/\alpha)p_1(0)} \mathbb{E}\left(\int_0^{1-t} s^{-1+1/\alpha} p_{1-t-s}(-X_t) ds \, , \, \Lambda \right) \\
&= \int_0^{1-t} \mathbb{P}(1-g \in ds) \frac{1}{p_{1-s}(0)} \mathbb{E}(p_{1-t-s}(-X_t), \Lambda) \, ,
\end{aligned}$$

where we used the identity $p_1(0) = (1-s)^{1/\alpha} p_{1-s}(0)$ and the fact that $1-g$ has the generalized arcsine law with parameter $1/\alpha$. To conclude, we simply observe that

$$\frac{1}{p_{1-s}(0)} \mathbb{E}(p_{1-t-s}(-X_t), \Lambda) = \mathbb{P}^{(br,1-s)}(\Lambda)$$

(by (8) and the scaling property). □

4. Normalized excursion and meander

We now turn our attention to the excursions of a stable process above its infimum, and their connections with the scaling property. The results presented in this section are essentially excerpts from Chaumont (1997) to whom we refer for a more complete account. Again we will assume that the parameters of the stable process have been fixed and that $|X|$ is not a subordinator.

Recall the notation $I_t = \inf\{X_s : 0 \le s \le t\}$ for the infimum up to time t, and that the (dual) reflected process $X - I = \widehat{S} - \widehat{X}$ has the Markov property. The origin is a regular instantaneous point in the classification of chapter IV, and we denote its local time by $\widehat{L}$. According to Lemma 1, the inverse local time $\widehat{L}^{-1}$ (that is the dual ladder time process) is a stable subordinator with index $\widehat{\rho} = 1 - \rho$. For simplicity, we decide to normalize the local time so that $\mathbb{E}(\exp\{-\lambda \widehat{L}^{-1}(t)\}) = \exp\{-t\lambda^{1-\rho}\}$, and then the tail of the Lévy measure $\widehat{\nu}$ of $\widehat{L}^{-1}$ is $\widehat{\nu}(t, \infty) = t^{\rho-1}/\Gamma(\rho)$.

We then consider the excursion measure of $X - I$ away from 0, $\widehat{n}$, which has been defined in section IV.4. In particular, the distribution of the lifetime ζ under $\widehat{n}$ is

$$\widehat{n}(\zeta > t) = \widehat{\nu}(t, \infty) = t^{\rho-1}/\Gamma(\rho)$$

and the conditional probability measure

$$\widehat{n}(\cdot \mid \zeta > t) = \widehat{n}(\cdot, \zeta > t)/\widehat{n}(\zeta > t)$$

is the law of the first excursion of $X - I$ with length $\ell > t$. The scaling property now immediately implies the following relation between these conditional laws. Recall that for every $\lambda > 0$, $\mathcal{S}^{(\lambda)} : \Omega \to \Omega$ denotes the mapping specified by

$$\mathcal{S}^{(\lambda)}\omega(s) = \begin{cases} \lambda^{1/\alpha}\omega(s/\lambda) & \text{if } s/\lambda < \zeta(\omega), \\ \partial & \text{otherwise.} \end{cases}$$

Lemma 14 *For every $\lambda > 0$ and $t > 0$, the image of $\widehat{n}(\cdot \mid \zeta > t)$ under $\mathcal{S}^{(\lambda)}$ is $\widehat{n}(\cdot \mid \zeta > t\lambda)$*

This elementary feature enables us to *normalize* the generic excursion by its length. Specifically one can associate to every path $\omega \in \Omega$ with lifetime $\zeta < \infty$ the path $\mathcal{S}^{(1/\zeta)}\omega \in \Omega$ which has lifetime $\zeta = 1$. Recall that the excursion measure $\widehat{n}$ assigns no mass to paths with infinite lifetime (because 0 is a recurrent point for $X - I$). Then taking into account Lemma 14, we see that the image of $\widehat{n}(\cdot \mid \zeta > t)$ under this transformation does not depend of $t > 0$, and this motivates the following definition.

Definition (Normalized excursion) *The image of $\widehat{n}(\cdot \mid \zeta > t)$ under $\mathcal{S}^{(1/\zeta)}$ is a probability measure on the subspace of Ω consisting of paths with lifetime* 1. *It is called the law of the stable normalized excursion and denoted by $\mathbb{P}^{(ex)}$.*

Observing that one can recover ω from $\mathcal{S}^{(1/\zeta)}\omega$ and ζ by an obvious inverse transformation, we deduce the following representation of the excursion measure $\widehat{n}$ in terms of the law of the normalized excursion:

$$\widehat{n}(\cdot) = \int_0^\infty \mathbb{P}^{(ex)}(\cdot \circ \mathcal{S}^{(t)})\,\widehat{n}(\zeta \in dt). \tag{9}$$

In particular, the law of the stable normalized excursion can also be viewed as the excursion measure conditioned on $\zeta = 1$.

We now present a simple pathwise construction of the normalized excursion which is reminiscent of that in Theorem 12. It will be convenient in the sequel to denote by $\underline{G}$ the set of left-end points of excursions intervals of $X - I$, and for every $s \in \underline{G}$, by $\ell(s)$ the length of the excursion interval starting at s.

Proposition 15 *Denote by $\underline{g} = \sup\{t < 1 : X_t - I_t = 0\}$ the (a.s. unique) instant when X attains its overall infimum on the unit time interval, and by $\ell = \ell(\underline{g})$ the length of the excursion interval straddling time* 1. *Under $\mathbb{P}$, the process*

$$X_t^{(ex)} = \ell^{-1/\alpha}(X_{\underline{g}+t\ell} - I_1), \qquad 0 \le t \le 1,$$

is a stable normalized excursion and is independent of ℓ.

Proof This is an easy application of excursion theory. For every left-end point $s \in \underline{G}$, denote by $\epsilon_s = ((X - I)_{s+t}, 0 \le t < \ell(s))$ the excursion of X above its minimum which starts at time s. Then take any bounded measurable functional F on Ω and apply the compensation formula for the excursion process (see Corollary IV.11) at the second equality below to get

$$\mathbb{E}(F(\epsilon_{\underline{g}})) = \mathbb{E}\left(\sum_{s\in \underline{G}} \mathbf{1}_{\{s<1\}}\mathbf{1}_{\{\ell(s)>1-s\}}F(\epsilon_s)\right)$$
$$= \mathbb{E}\left(\int_0^1 d\widehat{L}(s)\widehat{n}\left(F(\epsilon), \zeta > 1-s\right)\right),$$

where ϵ stands for the generic excursion and ζ for its lifetime. This shows that the law of $\epsilon_{\underline{g}}$ conditionally on ℓ is $\widehat{n}(\cdot \mid \zeta = \ell)$. By (9), this establishes our claim. □

The *meander* is a second process on the unit time interval which is naturally related to the excursions of the stable process above its infimum (see also Exercise 8 for a simple connection with the normalized excursion). Its law is that of the generic excursion restricted to the time interval $[0, 1]$ under $\widehat{n}(\cdot \mid \zeta > 1)$. Here is the formal definition.

Definition (Meander) *The image of $\widehat{n}(\cdot \mid \zeta > 1)$ under the mapping $\omega \to (\omega(s), 0 \leq s \leq 1)$ is a probability measure on Ω_1 denoted by $\mathbb{P}^{(me)}$ and called the law of the stable meander.*

Excursion theory and Lemma 14 yield the following pathwise construction of the meander, which is in the same vein as that described in Proposition 15.

Proposition 16 *Notation is as in Proposition 15. Under $\mathbb{P}$, the process*

$$X_t^{(me)} = (1-\underline{g})^{-1/\alpha}\left(X_{(1-\underline{g})t+\underline{g}} - I_1\right), \qquad 0 \leq t \leq 1,$$

is a stable meander independent of $\underline{g}$. Moreover the latter follows the generalized arcsine law with parameter $1-\rho$.

Proof The argument is similar to that in the proof of Proposition 15. The claim about the distribution of $\underline{g}$ follows from Theorem VI.13. □

We deduce now from the scaling property the following decomposition of the sample path of the stable process, which was noted first by Denisov (1984) in the Brownian case.

Corollary 17 *The notation is as in Proposition 16. Introduce the normalized reversed pre-infimum process*

$$\widehat{X}_t^{(me)} = \underline{g}^{-1/\alpha}\left(X_{\underline{g}(1-t)-} - I_1\right), \qquad 0 \le t \le 1.$$

Then under $\mathbb{P}$, $X^{(me)}$, $\widehat{X}^{(me)}$ *and* $\underline{g}$ *are independent, and* $\widehat{X}^{(me)}$ *has the law of the dual stable meander, viz.* $\widehat{\mathbb{P}}^{(me)}$.

Proof Again, this can be checked by standard arguments of excursion theory. The assertion that $\widehat{X}^{(me)}$ has the law of the dual stable meander is plain by time reversal. Alternatively, the scaling property enables us to replace the unit time interval by an interval of random length having an exponential distribution. According to Lemma VI.6(ii), the pair of processes obtained after splitting the path at the instant of its overall infimum are independent, and we can invoke Proposition 16. □

To conclude this section, we show that, informally, the meander can be thought of as a stable process on the unit time interval, conditioned to stay nonnegative. This was first discovered by Durrett, Iglehart and Miller (1977) in the Brownian case. For the sake of simplicity, we shall prove convergence in the sense of finite-dimensional distributions, but a closely related argument actually yields the convergence in the sense of Skorohod.

Theorem 18 *We have*

$$\mathbb{P}^{(me)} = \lim_{\varepsilon \to 0+} \mathbb{P}(\cdot \mid I_1 > -\varepsilon)$$

in the sense of finite-dimensional distributions.

The proof of Theorem 18 relies on the following result on the asymptotic distribution of $\underline{g}$, the instant of the minimum on $[0,1]$, conditionally on I_1.

Lemma 19 *For every* $t \in (0,1)$, *we have in the notation of Proposition 15*

$$\lim_{\varepsilon \to 0+} \mathbb{P}(\underline{g} > t \mid I_1 > -\varepsilon) = 0.$$

Proof. Applying the Markov property at time t, we see by a monotonicity argument that $\mathbb{P}(\underline{g} > t, I_1 > -\varepsilon)$ is bounded from above by

$$\begin{aligned}&\mathbb{P}(-\varepsilon < X_t \le 0, I_t > -\varepsilon)\mathbb{P}(I_{1-t} > -\varepsilon)\\ &\quad + \int_0^\infty \mathbb{P}(X_t \in dx, I_t > -\varepsilon)\mathbb{P}_x(I_{1-t} \in (-\varepsilon, 0))\\ &\le \mathbb{P}(I_t > -\varepsilon) \sup_{x\ge 0} \mathbb{P}(I_{1-t} \in [-x-\varepsilon, -x]).\end{aligned}$$

By Proposition 2 and the scaling property, there is a constant $c(t) > 0$ such that

$$\mathbb{P}(I_1 > -\varepsilon) \sim c(t)\mathbb{P}(I_{1-t} > -\varepsilon) \qquad (\varepsilon \to 0+),$$

and we thus only need to check that

$$\lim_{\varepsilon\to 0+} \sup_{x\ge 0} \mathbb{P}(I_{1-t} \in [-x-\varepsilon, -x]) = 0.$$

If this failed, then there would exist $\eta > 0$ and a sequence of real numbers $x_n \ge 0$ with $\mathbb{P}(|I_{1-t} - x_n| < 1/n) > \eta$ for all n. Such a sequence is necessarily bounded, so we may assume that it converges, say to $x \ge 0$. Then we would have $\mathbb{P}(I_{1-t} = x) \ge \eta$, which is absurd (apply the Markov property when X passes below $-x$ and recall that 0 is regular for the negative half-line). □

Finally we prove Theorem 18.

Proof of Theorem 18 Recall that $\underline{G}$ stands for the set of left-end points of the excursion intervals, and that for $s \in \underline{G}$, $\ell(s)$ denotes the length of the excursion interval starting at s. On the one hand, we deduce from the compensation formula of excursion theory,

$$\begin{aligned}\mathbb{P}(I_1 > -\varepsilon) &= \mathbb{E}\left(\sum_{s\in\underline{G}} \mathbf{1}_{\{s\le 1\}}\mathbf{1}_{\{I_s > -\varepsilon\}}\mathbf{1}_{\{\ell(s) > 1-s\}}\right)\\ &= \mathbb{E}\left(\int_0^1 d\widehat{L}(s)\mathbf{1}_{\{I_s > -\varepsilon\}}\widehat{n}(\zeta > 1-s)\right)\\ &\ge \widehat{n}(\zeta > 1)\mathbb{E}\left(\int_0^1 d\widehat{L}(s)\mathbf{1}_{\{I_s > -\varepsilon\}}\right).\end{aligned}$$

The same argument shows that for every $t \in (0,1)$

$$\begin{aligned}\mathbb{P}(I_1 > -\varepsilon, \underline{g} \le t) &= \mathbb{E}\left(\int_0^t d\widehat{L}(s)\mathbf{1}_{\{I_s > -\varepsilon\}}\widehat{n}(\zeta > 1-s)\right)\\ &\le \widehat{n}(\zeta > 1-t)\mathbb{E}\left(\int_0^1 d\widehat{L}(s)\mathbf{1}_{\{I_s > -\varepsilon\}}\right).\end{aligned}$$

According to Lemma 19, the left-hand side is equivalent to $\mathbb{P}(I_1 > -\varepsilon)$

as ε tends to $0+$, and since $t \in (0,1)$ can be chosen arbitrarily small, we deduce

$$\mathbb{P}(I_1 > -\varepsilon) \sim \widehat{n}(\zeta \geq 1)\mathbb{E}\left(\int_0^1 d\widehat{L}(s)\mathbf{1}_{\{I_s > -\varepsilon\}}\right) \qquad (\varepsilon \to 0+). \tag{10}$$

On the other hand, fix $t \in [0,1]$ and take a continuous function with compact support $f : \mathbb{R} \to [0,1]$. We deduce from Lemma 19, the right-continuity of the paths and the uniform continuity of f that

$$\lim_{\varepsilon \to 0+} \mathbb{E}\left(|f(X_t) - f(X_{t+\underline{g}} - I_1)| \mid I_1 > -\varepsilon\right) = 0.$$

Next, we apply the compensation formula and observe that

$$\begin{aligned}
&\mathbb{E}\left(f(X_{t+\underline{g}} - I_1), I_1 > -\varepsilon\right) \\
&\geq \mathbb{E}\left(f(X_{t+\underline{g}} - I_1), \ell(\underline{g}) > 1, I_1 > -\varepsilon\right) \\
&= \mathbb{E}\left(\sum_{s \in \underline{G}} \mathbf{1}_{\{s<1\}} f(X_{t+s} - I_s)\mathbf{1}_{\{I_s > -\varepsilon\}}\mathbf{1}_{\{\ell(s)>1\}}\right) \\
&= \mathbb{E}\left(\int_0^1 d\widehat{L}(s)\mathbf{1}_{\{I_s > -\varepsilon\}}\right)\widehat{n}(f(\epsilon(t)), \zeta > 1)
\end{aligned}$$

where ϵ denotes the generic excursion. It follows now from (10) that

$$\liminf_{\varepsilon \to 0+} \mathbb{E}(f(X_t) \mid I_1 > -\varepsilon) \geq \widehat{n}(f(\epsilon_t) \mid \zeta > 1) = \mathbb{E}^{(me)}(f(X_t)).$$

Replacing f by $1 - f$ yields the converse inequality for the limsup, and in conclusion we have proved that $\mathbb{P}(\cdot \mid I_1 > -\varepsilon)$ converges to $\mathbb{P}^{(me)}$ in the sense of the one-dimensional distributions. The proof for the convergence of finite-dimensional distributions is the same, but requires heavier notation. □

5. Exercises

1. *(Subordination of stable processes)* Prove that any stable process Y can be expressed in the form $Y = A \circ T$ where T is a stable subordinator and A is a completely asymmetric stable process (A has either no positive jumps or no negative jumps) independent of T.
2. *(Tail of the bilateral supremum)* (a) Suppose that X has no negative jumps, and let $k > 0$ be such that $\mathbb{P}(X_1 > x) \sim kx^{-\alpha}$ as x goes to ∞. Prove that

$$\mathbb{P}(X_1^* > x) \sim kx^{-\alpha} \qquad (x \to \infty),$$

where $X_1^* = \sup\{|X_s| : 0 \leq s \leq 1\}$.

(b) Suppose that X has both positive and negative jumps, and let $\hat{k} > 0$ be such that $\mathbb{P}(-X_1 > x) = \widehat{\mathbb{P}}(X_1 > x) \sim \hat{k}x^{-\alpha}$ as $x \to \infty$. Prove that

$$\mathbb{P}(X_1^* > x) \sim (k + \hat{k})x^{-\alpha} \qquad (x \to \infty).$$

3. *(Entrance law in a semi-finite interval)* Suppose that X has positive jumps. Prove that the distribution of X evaluated at its first entrance time in $[a, \infty)$, $T_{[a,\infty)} = \inf\{t \geq 0 : X_t \geq a\}$, is given for $0 < a < b$ by

$$\mathbb{P}(X_{T_{[a,\infty)}} > b) = \frac{1}{\Gamma(\alpha\rho)\Gamma(1-\alpha\rho)} \int_0^a x^{\alpha\rho-1}(b-x)^{-\alpha\rho}\, dx\,.$$

[Hint: use Lemma 1.]

4. *(Stable Ornstein-Uhlenbeck process)* Consider the process Y given by

$$Y(t) = \mathrm{e}^{-t/\alpha} X_{\exp t}\,, \qquad t \geq 0,$$

so that Y bears the same relationship to X as the Ornstein-Uhlenbeck process does to Brownian motion. Note that Y is stationary, more precisely $\mathbb{P}(Y(t) \in \cdot) = \mathbb{P}(X_1 \in \cdot)$ for every $t \geq 0$. Prove that Y is a Markov process with semigroup given by

$$\mathbb{P}(Y(t+u) \in dy \mid Y(t) = x)$$
$$= (\mathrm{e}^u - 1)^{-1/\alpha} \mathrm{e}^{u/\alpha} p_1\left((\mathrm{e}^u - 1)^{-1/\alpha}(\mathrm{e}^{u/\alpha}y - x)\right).$$

5. *(Reversed bridge)* (a) Prove that the law of the time-reversed process $(X_{(1-t)-}, 0 \leq t \leq 1)$ under $\mathbb{P}^{(br)}$ is that of the bridge of the dual process, viz. $\widehat{\mathbb{P}}^{(br)}$.

(b) Assume now that $\alpha \in (1, 2)$ and denote by $g = \sup\{t < 1 : X_t = 0\}$, the last zero of X on the unit interval. Let $f : (0, \infty) \to (0, \infty)$ be an increasing function. Deduce that

$$\limsup_{t \to 0+} \frac{|X_{g-t}|}{f(t)} = 0 \text{ or } \infty \qquad \mathbb{P}\text{-a.s.}$$

according as the integral $\int_{0+} f(t)^{-\alpha} dt$ converges or diverges. [Hint: use Theorems 5 and 12.]

6. *(Minimum of the bridge)* Check that the increments of the bridge are exchangeable, that is for every integer $n > 0$ and every permutation σ of $\{1, \cdots, n\}$, the variables

$$(X_{1/n} - X_0, \cdots, X_{n/n} - X_{(n-1)/n})$$

and

$$(X_{\sigma(1)/n)} - X_{(\sigma(1)-1)/n}, \cdots, X_{\sigma(n)/n} - X_{(\sigma(n)-1)/n})$$

have the same law under $\mathbb{P}^{(br)}$. Deduce that the (a.s. unique) instant when the bridge attains its overall infimum is uniformly distributed over the unit interval.

7. *(Pseudo-bridge)* Suppose that $\alpha \in (1,2]$ and put for $t \geq 0$

$$L(0,t) = \lim_{\varepsilon \to 0+} \frac{1}{2\varepsilon} \int_0^t \mathbf{1}_{\{|X_s|<\varepsilon\}} ds$$

whenever it makes sense. Recall that this happens $\mathbb{P}$-a.s. for all t according to Proposition V.2.
(a) Prove that $L(0,1)$ exists $\mathbb{P}^{(br)}$-a.s.
(b) We now work with respect to $\mathbb{P}$. We call a *pseudo-bridge* the process on the unit time interval,

$$\sigma^{-1/\alpha} X_{\sigma t} \qquad (t \in [0,1]),$$

where $\sigma = L^{-1}(0,1) = \inf\{t \geq 0 : L(0,t) = 1\}$. Prove that the law of the pseudo-bridge is absolutely continuous with respect to that of the (standard) bridge with density $kL(0,1)^{1/(\alpha-1)}$, where $k > 0$ is the constant of normalization.

8. *(Normalized excursion and meander conditioned on their end points)* For every path $\omega \in \Omega$ with lifetime 1, denote by $\widetilde{\omega}$ the path in Ω_1 given by $\omega = \widetilde{\omega}$ on $[0,1)$ and $\widetilde{\omega}(1) = \omega(1-)$. The image of $\mathbb{P}^{(ex)}$ under this transformation is a probability measure on Ω_1 denoted by $\mathbb{P}^{(\widetilde{ex})}$.
(a) Check that the law of $(X_t, t \in [0,1])$ under $\widehat{n}(\cdot \mid 1 < \zeta < 1+\varepsilon)$ converges to $\mathbb{P}^{(\widetilde{ex})}$ in the sense of the finite-dimensional distributions as $\varepsilon \to 0+$.
(b) Suppose now that there exist negative jumps. Deduce from Proposition 3 that $\mathbb{P}^{(\widetilde{ex})}$ is absolutely continuous with respect to $\mathbb{P}^{(me)}$ with density $kX_1^{-\alpha}$, where $k > 0$ is the constant of normalization.

9. *(Law of large numbers for crossings)* Suppose that X is a stable process of index $\alpha \in (0,1)$, but not completely asymmetric ($|X|$ is not a subordinator). Introduce the number of up-crossings across 0 made by X on the time interval $[1,t]$,

$$N_t = \text{Card}\{s \in [1,t] : X_{s-} < 0 \text{ and } X_s > 0\}, \qquad t \geq 1.$$

The purpose of this exercise is to determine the asymptotic behaviour for large t of N_t.
(a) Write $\overline{\Pi}(x) = \Pi((x,\infty))$, $x > 0$, for the tail of the Lévy measure. Using Exercise 4 and the ergodic theorem, check that a.s.

$$\int_1^t \mathbf{1}_{\{X_s<0\}} \overline{\Pi}(-X_s) ds \sim \mathbb{E}\left(\overline{\Pi}(-X_1), X_1 < 0\right) \log t \qquad (t \to \infty)$$

and that the expectation in the right-hand term is finite.

(b) By an argument based on the Borel-Cantelli lemma and the maximal inequality for compensated sums of section O.5, prove that a.s.

$$N_t \sim \int_1^t \mathbf{1}_{\{X_s<0\}} \overline{\Pi}(-X_s)ds \qquad (t \to \infty)$$

and conclude that

$$\lim_{t\to\infty} \frac{N_t}{\log t} = \mathbb{E}\left(\overline{\Pi}(-X_1), X_1 < 0\right) \quad \text{a.s.}$$

6. Comments

The importance of stable variables and processes stems from limit theorems for sums of independent random variables. See in particular Gnedenko and Kolmogorov (1949), Feller (1971), Ibragimov and Linnik (1971). A recent monograph by Janicki and Weron (1994) describes methods of approximation and simulation of stable processes and their applications.

We focused on strictly stable processes, and refer to Port and Stone (1969), Port (1989), Pruitt and Taylor (1983, 1985) and the references therein for information on the general case. More precise versions of Propositions 2 and 4 appear in Bingham (1973, 1973-a), and Proposition 3 is from Taylor (1967), see also Getoor (1961) for a sharper estimate in the symmetric case.

The second part of Theorem 5 was noted by Zolotarev (1964-a) and the third by Khintchine. Theorem 6(ii) is due to Taylor (1967) after Chung (1948) in the Brownian case; see also McKean (1955) and Pruitt and Taylor (1969). Extensions to fairly general Lévy processes have been obtained by Dupuis (1974) (in the symmetric case) and Wee (1988, 1990). Proposition 7 is from Bertoin (1994), but the present proof follows Doney (1995-a). Sharper properties of the set of increase times for stable processes have been established recently by Marsalle (1995). Proposition 8 is from Millar (1973). We refer to Takeuchi (1964, 1964-a), Takeuchi and Watanabe (1964), Jain and Pruitt (1968), Hendricks (1970, 1973), Mijnheer (1975), Fristedt (1979), Monrad and Silverstein (1979), Davis (1981) and Okoroafor and Ugbebor (1991, 1991-a) for further path properties of stable processes.

Bridges of general Lévy processes have been considered by Fitzsimmons and Getoor (1995) and Knight (1995), who proved analogues of the arcsine law for unconditioned processes. Theorem 12 and the whole of section 4 are essentially from Chaumont (1997) who also obtained interesting path transformations connecting the bridge, the normalized excursion and the meander, extending some of those known in the Brownian case.

Exercise 3 originates from Blumenthal, Getoor and Ray (1961), see also Port (1967), Pruitt and Taylor (1969-a) and Rogozin (1972). One can also calculate the entrance law in a semi-finite interval, see Ray (1958), and Pitman and Yor (1986) for the special case of the Cauchy process. Exercise 4 is from Breiman (1968), who used it to characterize the lower functions of stable subordinators. Exercise 6 is a special case of Knight (1995), and Exercises 7 and 8 are from Carmona, Petit and Yor (1994) and Chaumont (1996-a), respectively. We also refer to Donati-Martin, Song and Yor (1994) and Bertoin and Werner (1996) for further striking identities for symmetric stable processes.

References

O. Adelman (1985). Brownian motion never increases: a new proof to a result of Dvoretzky, Erdös and Kakutani. *Israel J. Math.* **50** 189–192.

E. Sparre Andersen (1953). On sums of symmetrically dependent random variables. *Scand. Aktuar. Tidskr.* **26** 123–138.

S. Asmussen and C. Klüppelberg (1995). Large deviation results in the presence of heavy tails, with applications to insurance risk. Preprint.

V. Bally and L. Stoica (1987). A class of Markov processes which admit local times. *Ann. Probab.* **15** 241–262.

M. T. Barlow (1981). Zero-one laws for the excursions and range of a Lévy process. *Z. Wahrscheinlichkeitstheorie verw. Gebiete* **55** 149–163.

M. T. Barlow (1985). Continuity of local times for Lévy processes. *Z. Wahrscheinlichkeitstheorie verw. Gebiete* **69** 23–35.

M. T. Barlow (1988). Necessary and sufficient conditions for the continuity of local time of Lévy processes. *Ann. Probab.* **16** 1389–1427.

M. T. Barlow and J. Hawkes (1985). Application de l'entropie métrique à la continuité des temps locaux des processus de Lévy. *C. R. Acad. Sci. Paris* **301** 237–239.

M. T. Barlow, E. A. Perkins and S. J. Taylor (1986). Two uniform intrinsic constructions for the local time of a class of Lévy processes. *Illinois J. Math.* **30** 19–65.

M. T. Barlow, E. A. Perkins and S. J. Taylor (1986–a). The behaviour and construction of local times for Lévy processes. *Seminar on Stochastic Processes 1984*, pp. 23–54. Birkhäuser, Boston, Mass.

R. Bass and D. Khoshnevisan (1992). Stochastic calculus and the continuity of local times of Lévy processes. *Séminaire de Probabilités XXVI*, pp. 1–10. Springer, Berlin.

G. Baxter and M. D. Donsker (1957). On the distribution of the supremum functional for processes with stationary independent increments. *Trans. Amer. Math. Soc.* **85** 73–87.

C. Berg and G. Forst (1975). *Potential theory on locally compact Abelian groups.* Springer, Berlin.

S. M. Berman (1969). Local times and sample function properties of stationary Gaussian processes. *Trans. Amer. Math. Soc.* **137** 277–299.

S. M. Berman (1986). The supremum of a process with stationary, independent and symmetric increments. *Stochastic Process. Appl.* **23** 281–290.

J. Bertoin (1990). Complements on the Hilbert transform and the fractional derivative of Brownian local times. *J. Math. Kyoto Univ.* **30** 651–670.

J. Bertoin (1991). Sur la décomposition de la trajectoire d'un processus de Lévy spectralement positif en son infimum. *Ann. Inst. Henri Poincaré* **27** 537–547.

J. Bertoin (1991–a). Increase of a Lévy process with no positive jumps. *Stochastics and Stochastics Reports* **37** 247–251.

J. Bertoin (1992). An extension of Pitman's theorem for spectrally positive Lévy processes. *Ann. Probab.* **20** 1464–1483.

J. Bertoin (1992–a). Factorizing Laplace exponents in a spectrally positive Lévy process. *Stochastic Process. Appl.* **42** 307–313.

J. Bertoin (1993). Splitting at the infimum and excursions in half-lines for random walks and Lévy processes. *Stochastic Process. Appl.* **47** 17–35.

J. Bertoin (1993–a). Lévy processes with no positive jumps at an increase time. *Probab. Theory Relat. Fields* **96** 123–135.

J. Bertoin (1994). Increase of stable processes. *J. Theoretic. Probab.* **7** 551–563.

J. Bertoin (1995). Some applications of subordinators to local times of Markov processes. *Forum Math.* **7** 629–644.

J. Bertoin (1995–a). On the local rate of growth of Lévy processes with no positive jumps. *Stochastic Process. Appl.* **55** 91–100.

J. Bertoin (1995–b). Lévy processes that can creep downwards never increase. *Ann. Inst. Henri Poincaré* **31** (2) 379–391.

J. Bertoin (1995–c). On the Hilbert transform of the local times of a Lévy process. *Bull. Sci. Math.* **119** (2) 147–156.

J. Bertoin (1995–d). Sample path behaviour in connection with generalized arcsine laws. *Probab. Theory Relat. Fields* **103** 317–327.

J. Bertoin (1997). Regularity of the half-line for Lévy processes. *Bull. Sci. Math.* **121** 345–354.

J. Bertoin (1998). *Subordinators: Examples and Applications.* To appear in *Ecole d'été de Probabilités de St-Flour XXVII*, Springer.

J. Bertoin and M. E. Caballero (1995). On the rate of growth of subordinators with slowly varying Laplace exponent. *Séminaire de Probabilités XXIX*, pp.125–132.

J. Bertoin and R. A. Doney (1994). On conditioning a random walk to stay positive. *Ann. Probab.* **22** 2152–2167.

J. Bertoin and R. A. Doney (1994–a). Cramér's estimate for Lévy processes. *Stat. Probab. Letters* **21** 363–365.

J. Bertoin and R. A. Doney (1997). Spitzer's condition for random walks and Lévy processes. *Ann. Inst. Henri Poincaré* **33** 167–178.

J. Bertoin and W. Werner (1996). Stable windings. *Ann. Probab.* **24** 1269–1279.

J. Bertoin and M. Yor (1996). Some independence results related to the arc-sine law. *J. Theoretic. Probab.* **9** 447–458.

Ph. Biane and M. Yor (1987). Valeurs principales associées aux temps locaux browniens. *Bull. Sci. Math.* **111** 23–101.

N. H. Bingham (1973). Maxima of sums of random variables and suprema of stable processes. *Z. Wahrscheinlichkeitstheorie verw. Gebiete* **26** 273–296.

N. H. Bingham (1973–a). Limit theorems in fluctuation theory. *Adv. Appl. Probab.* **5** 554–569.

N. H. Bingham (1975). Fluctuation theory in continuous time. *Adv. Appl. Probab.* **7** 705–766.

N. H. Bingham (1976). Continuous branching processes and spectral positivity, *Stochastic Process. Appl.* **4** 217–242.

N. H. Bingham, C. M. Goldie and J. L. Teugels (1987). *Regular variation.* Cambridge University Press, Cambridge.

R. M. Blumenthal (1992). *Excursions of Markov processes.* Birkhäuser, Boston, Mass.

R. M. Blumenthal and R. K. Getoor (1960). A dimension theorem for sample functions of stable processes. *Illinois J. Math.* **4** (3) 370–375.

R. M. Blumenthal and R. K. Getoor (1960–a). Some theorems on stable processes. *Trans. Amer. Math. Soc.* **95** (2) 263–273.

R. M. Blumenthal and R. K. Getoor (1961). Sample functions of stochastic processes with independent increments. *J. Math. Mech.* **10** 493–516.

R. M. Blumenthal and R. K. Getoor (1962). The dimension of the set of zeros and the graph of a symmetric stable process. *Illinois J. Math.* **6** 308–316.

R. M. Blumenthal and R. K. Getoor (1968). *Markov processes and potential theory.* Academic Press, New York.

R. M. Blumenthal and R. K. Getoor (1970). Dual processes and potential theory. *Proc. 12th. Biennial Sem. Canadian Math. Cong.* pp. 137–156.

R. M. Blumenthal, R. K. Getoor and D. B. Ray (1961). On the distribution of first hits for the symmetric stable processes. *Trans. Amer. Math. Soc.* **99** 540–554.

S. Bochner (1955). *Harmonic analysis and the theory of probability*. University of California Press, Berkeley.

A. A. Borovkov (1965). On the first passage time for one class of processes with independent increments. *Theory Probab. Appl.* **10** 331–334.

A. A. Borovkov (1970). Factorization identities and properties of the distribution of the supremum of sequential sums. *Theory Probab. Appl.* **15** 359–402.

A. A. Borovkov (1976). *Stochastic processes in queueing theory*. Springer, Berlin.

E. S. Boylan (1964). Local times for a class of Markov process. *Illinois J. Math.* **8** 19–39.

M. Braverman (1997). Suprema and sojourn times of Lévy processes with exponential tails. *Stochastic Process. Appl.* **68** 265–283.

M. Braverman and G. Samorodnitsky (1995). Functionals of infinitely divisible stochastic processes with exponential tails. *Stochastic Process. Appl.* **56** 207–231.

L. Breiman (1968). *Probability*. Addison-Wesley, Reading, Mass.

L. Breiman (1968–a). A delicate law of the iterated logarithm for non-decreasing stable processes. *Ann. Math. Stat.* **39** 1818–1824. [Correction id. (1970) Ibid. **41** 1126.]

J. Bretagnolle (1971). Résultats de Kesten sur les processus à accroissements indépendants. *Séminaire de Probabilités V*, pp. 21–36. Springer, Berlin.

J. Bretagnolle (1972). *p*-variation de fonctions aléatoires, 2ème partie: processus à accroissements indépendants. *Séminaire de Probabilités VI*, Lecture Notes in Math., pp. 64–71. Springer, Berlin.

J. Bretagnolle and D. Dacunha-Castelle (1968). Pointwise recurrence of processes with independent increments. *Probab. Theory Appl.* **13** 713–720.

K. Burdzy (1990). On nonincrease of Brownian motion. *Ann. Probab.* **18** 978–980.

Ph. Carmona, F. Petit and M. Yor (1994). Some extensions of the arc sine law as partial consequences of the scaling property of Brownian motion. *Probab. Theory Relat. Fields* **100** 1–29.

R. Carmona, W.C. Masters and B. Simon (1990). Relativistic Schrödinger operators: Asymptotic behavior of the eigenfunctions. *J. Funct. Anal.* **91** 117–142.

L. Chaumont (1994). Sur certains processus de Lévy conditionnés à rester positifs. *Stochastics and Stochastics Reports* **47** 1–20.

L. Chaumont (1996). Conditionings and path decompositions for Lévy processes. *Stochastic Process. Appl.* **64** 39–54.

L. Chaumont (1997). Excursion normalisée, méandre et pont pour des processus stables. *Bull. Sci. Math.* **121** 377–403.

Y. S. Chow (1986). On moments of ladder height variables. *Adv. Appl. Math.* **7** 46–54.

K. L. Chung (1948). On the maximum partial sums of sequences of independent random variables. *Trans. Amer. Math. Soc.* **64** 205–233.

K. L. Chung (1968). *A course in probability theory.* Academic Press, New York.

K. L. Chung (1988). Reminiscences of some of Paul Lévy's ideas in Brownian motion and in Markov chains. *Colloque Paul Lévy sur les processus stochastiques,* Astérisque **157–158** pp. 29–36. Société Mathématique de France, Paris.

K. L. Chung and W. H. J. Fuchs (1951). On the distribution of values of sums of random variables. *Memoirs Amer. Math. Soc.* **6** 1–12.

B. Davis (1981). On the paths of symmetric stable processes. *Trans. Amer. Math. Soc.* **281** 785–794.

C. Dellacherie, B. Maisonneuve and P. A. Meyer (1992). *Probabilités et potentiel,* vol. V. Processus de Markov, compléments de calcul stochastique. Hermann, Paris.

C. Dellacherie and P. A. Meyer (1975). *Probabilités et potentiel,* vol. I. Hermann, Paris.

C. Dellacherie and P. A. Meyer (1980). *Probabilités et potentiel,* vol. II. Théorie des martingales. Hermann, Paris.

C. Dellacherie and P. A. Meyer (1987). *Probabilités et potentiel,* vol. IV. Théorie du potentiel, processus de Markov. Hermann, Paris.

I. V. Denisov (1984). A random walk and a Wiener process near a maximum. *Theory Probab. Appl.* **28** 821–824.

C. Donati-Martin, S. Song and M. Yor (1994). Symmetric stable processes, Fubini's theorem, and some extensions of Ciesielski-Taylor identities in law. *Stochastics and Stochastics Reports* **50** 1–33.

R. A. Doney (1982). On the existence of the mean ladder height for random walk. *Z. Wahrscheinlichkeitstheorie verw. Gebiete* **59** 373–382.

R. A. Doney (1982–a). On the exact asymptotic behaviour of the distribution of ladder epochs. *Stochastics Process. Appl.* **12** 203–214.

R. A. Doney (1987). On the Wiener-Hopf factorisation and the distribution of extrema for certain stable processes. *Ann. Probab.* **15** 1352–1362.

R. A. Doney (1989). On the asymptotic behaviour of first passage times for transient random walks. *Z. Wahrscheinlichkeitstheorie verw. Gebiete* **81** 239–246.

R. A. Doney (1991). Hitting probabilities for spectrally positive Lévy processes. *J. London Math. Soc.* **44** 566–576.

R. A. Doney (1993). A path decomposition for Lévy processes. *Stochastic Process. Appl.* **47** 167–181.

R. A. Doney (1995). Spitzer's condition and ladder variables in random walks. *Probab. Theory Relat. Fields* **101** 577–580.

R. A. Doney (1995–a). Increase of Lévy processes. *Ann. Probab.*

J. L. Doob (1957). Conditional Brownian motion and the boundary limits of harmonic functions. *Bull. Soc. Math. France* **85** 431–458.

J. L. Doob (1984). *Classical potential theory and its probabilistic counterpart.* Springer, Berlin.

R. M. Dudley (1973). Sample functions of the Gaussian process. *Ann. Probab.* **1** 66–103.

C. Dupuis (1974). Mesure de Hausdorff de la trajectoire de certains processus à accroissements indépendants et stationnaires. *Séminaire de Probabilités VIII*, pp. 37–77. Springer, Berlin.

R. T. Durrett, D. L. Iglehart and D. R. Miller (1977). Weak convergence to Brownian meander and Brownian excursion. *Ann. Probab.* **5** 117–129.

A. Dvoretzky, P. Erdös and S. Kakutani (1961). On nonincrease everywhere of the Brownian motion process. *Proc. 4th Berkeley Symp. Math. Stat. Probab.* II, pp. 103–116.

E. B. Dynkin (1965). *Markov processes*, I, II. Springer, Berlin.

D. J. Emery (1973). Exit problem for a spectrally positive process. *Adv. Appl. Probab.* **5** 498–520.

K. B. Erickson (1973). The strong law of large numbers when the mean is undefined. *Trans. Amer. Math. Soc.* **185** 371–381.

S. N. Evans (1987). Multiple points in the sample path of a Lévy process. *Probab. Theory Relat. Fields* **76** 359–367.

S. N. Evans (1989). The range of a perturbed Lévy process. *Probab. Theory Relat. Fields* **81** 555–557.

S. N. Evans (1994). Multiplicities of a random sausage. *Ann. Inst. Henri Poincaré* **30** 501–518.

W. E. Feller (1971). *An introduction to probability theory and its applications*, 2nd edn, vol. 2. Wiley, New York.

B. de Finetti (1929). Sulle funzioni a incremento aleatorio. *Rend. Accad. Lincei* **6** (10) 163–168.

P. J. Fitzsimmons, B. E. Fristedt and B. Maisonneuve (1985). Intersections and limits of regenerative sets. *Z. Wahrscheinlichkeitstheorie verw. Gebiete* **70** 157–173.

P. J. Fitzsimmons and R. K. Getoor (1992). On the distribution of the Hilbert transform of the local time of a symmetric Lévy process. *Ann. Probab.* **20** 1484–1497.

P. J. Fitzsimmons and R. K. Getoor (1992–a). Limit theorems and variation properties for fractional derivatives of the local time of a stable process. *Ann. Inst. Henri Poincaré* **28** 311–333.

P. J. Fitzsimmons and R. K. Getoor (1995). Occupation time distributions for Lévy bridges and excursions. *Stochastic Process. Appl.* **58** 73–89.

P. J. Fitzsimmons and M. Kanda (1992). On Choquet's dichotomy of capacity for Markov processes. *Ann. Probab.* **20** 342–349.

P. J. Fitzsimmons, J. W. Pitman and M. Yor (1993). Markovian bridges: construction, Palm interpretation, and splicing. *Seminar on Stochastic Processes 1992*, pp. 102–133. Birkhäuser, Boston, Mass.

P. J. Fitzsimmons and S. C. Port (1991). Local times, occupation times, and the Lebesgue measure of the range of a Lévy process. *Seminar on Stochastic Processes 1989*, pp. 59–73. Birkhäuser, Boston, Mass.

P. J. Fitzsimmons and T.S. Salisbury (1989). Capacity and energy for multiparameter Markov processes. *Ann. Inst. Henri Poincaré* **25** 325–350.

S. Fourati (1998). Points de croissance des processus de Lévy et théorie générale des processus. *Probab. Theory Relat. Fields* **110** 13–49.

B. E. Fristedt (1964). The behavior of increasing stable processes for both small and large times. *J. Math. Mech.* **13** 849–856.

B. E. Fristedt (1967). Sample function behaviour of increasing processes with stationary independent increments. *Pac. J. Math.* **21** 21–33.

B. E. Fristedt (1967–a). An extension of a theorem of S. J. Taylor concerning the multiple points of the symmetric stable process. *Z. Wahrscheinlichkeitstheorie verw. Gebiete* **9** 62–64.

B. E. Fristedt (1971). Upper functions for symmetric processes with stationary independent increments. *Indiana Univ. Math. J.* **21** 177–185. [Correction id. (1973) Ibid. **23** 445].

B. E. Fristedt (1974). Sample functions of stochastic processes with stationary, independent increments. *Advances in Probability* **3**, pp. 241–396. Dekker, New York.

B. E. Fristedt (1979). Uniform local behavior of stable subordinators. *Ann. Probab.* **7** 1003–1013.

B. E. Fristedt and P. E. Greenwood (1972). Variation of processes with stationary independent increments. *Z. Wahrscheinlichkeitstheorie verw. Gebiete* **23** 171–186.

B. E. Fristedt and W. E. Pruitt (1971). Lower functions for increasing random walks and subordinators. *Z. Wahrscheinlichkeitstheorie verw. Gebiete* **18** 167–182.

B. E. Fristedt and W. E. Pruitt (1972). Uniform lower functions for subordinators. *Z. Wahrscheinlichkeitstheorie verw. Gebiete* **24** 63–70.

B. E. Fristedt and S. J. Taylor (1973). Strong variation for the sample functions of a stable process. *Duke Math. J.* **40** 259–278.

B. E. Fristedt and S. J. Taylor (1983). Construction of local time for a Markov process. *Z. Wahrscheinlichkeitstheorie verw. Gebiete* **62** 73–112.

B. E. Fristedt and S. J. Taylor (1992). The packing measure of a general subordinator. *Probab. Theory Relat. Fields* **92** 493–510.

M. Fukushima, Y. Oshima and M. Takeda (1994). *Dirichlet forms and Markov processes.* De Gruyter, Berlin.

I. M. Gel'fand and G. E. Shilov (1964). *Generalized functions 1, properties and operations.* Academic Press, New York.

D. Geman and J. Horowitz (1980). Occupation densities. *Ann. Probab.* **8** 1–67.

R. K. Getoor (1961). First passage times for symmetric stable processes in space. *Trans. Amer. Math. Soc.* **101** (1) 75–90.

R. K. Getoor (1965). Some asymptotic formulas involving capacity. *Z. Wahrscheinlichkeitstheorie verw. Gebiete* **4** 248–252.

R. K. Getoor and H. Kesten (1972). Continuity of local times for Markov processes. *Compos. Math.* **24** 277–303.

R. K. Getoor and M. J. Sharpe (1994). On the arc sine laws for Lévy processes. *J. Appl. Probab.* **31** 76–89.

R. K. Getoor and M. J. Sharpe (1994–a). Local times on rays for a class of planar Lévy processes. *J. Theoretic. Probab.* **7** 799–811.

I.I. Gihman and A. V. Skorohod (1975). *The theory of stochastic processes* II. Springer, Berlin.

B. V. Gnedenko and A. N. Kolmogorov (1949). *Limit distributions for sums of independent random variables.* Addison-Wesley, Reading, Mass.

P. E. Greenwood (1969). The variation of a stable path is stable. *Z. Wahrscheinlichkeitstheorie verw. Gebiete* **14** 140–148.

P. E. Greenwood (1975). Wiener-Hopf methods, decompositions, and factorisation identities for maxima and minima of homogeneous random processes. *Advances in Appl. Probability* **7** 767–785.

P. E. Greenwood (1975–a). Extreme time of stochastic processes with stationary independent increments. *Ann. Probab* **3** 664–676.

P. E. Greenwood (1976). Wiener–Hopf decomposition of random walks and Lévy processes. *Z. Wahrscheinlichkeitstheorie verw. Gebiete* **34** 193–198.

P. E. Greenwood and A. A. Novikov (1986). One-sided boundary crossing for processes with independent increments. *Theory Probab. Appl.* **31** 221–232.

P. E. Greenwood and J. W. Pitman (1980). Fluctuation identities for Lévy processes and splitting at the maximum. *Adv. Appl. Probab.* **12** 893–902.

P. E. Greenwood and J. W. Pitman (1980–a). Construction of local time and Poisson point processes from nested arrays. *J. London Math. Soc.* **22** 182–192.

R. Griego (1967). Local times and random time changes. *Z. Wahrscheinlichkeitstheorie verw. Gebiete* **8** 325–331.

D. V. Gusak (1969). On the joint distribution of the first exit time and exit value for homogeneous processes with independent increments. *Theory Probab. Appl.* **14** 14–23.

D. V. Gusak and V. S. Korolyuk (1969). On the joint distribution of a process with stationary independent increments and its maximum. *Theory Probab. Appl.* **14** 400–409.

J. Hawkes (1970). Polar sets, regular points and recurrent sets for the symmetric and increasing stable processes. *Bull. London Math. Soc.* **2** 53–59.

J. Hawkes (1971). A lower Lipschitz condition for the stable subordinator. *Z. Wahrscheinlichkeitstheorie verw. Gebiete* **17** 23–32.

J. Hawkes (1971–a). On the Hausdorff dimension of the intersection of the range of a stable process with a Borel set. *Z. Wahrscheinlichkeitstheorie verw. Gebiete* **19** 90–102.

J. Hawkes (1974). Local times and zero sets for processes with infinitely divisible distributions. *J. London Math. Soc.* **8** 517–525.

J. Hawkes (1975). On the potential theory of subordinators. *Z. Wahrscheinlichkeitstheorie verw. Gebiete* **33** 113–132.

J. Hawkes (1977). Intersection of Markov random sets. *Z. Wahrscheinlichkeitstheorie verw. Gebiete* **37** 243–251.

J. Hawkes (1978). Multiple points for symmetric Lévy processes. *Math. Proc. Cambridge Phil. Soc.* **83** 83–90.

J. Hawkes (1979). Potential theory of Lévy processes. *Proc. London Math. Soc.* **38** 335–352.

J. Hawkes (1984). Some geometric aspects of potential theory. *Stochastic analysis and applications*, Lecture Notes in Math. 1095, pp. 130–154. Springer, Berlin.

J. Hawkes (1985). Local times as stationary processes. *From local times to global geometry*, Pitman Research Notes in Math. 150, pp. 111–120. Longman, Chicago.

J. Hawkes and W. E. Pruitt (1974). Uniform dimension results for processes with independent increments. *Z. Wahrscheinlichkeitstheorie verw. Gebiete* **28** 277–288.

W. J. Hendricks (1970). Lower envelopes near zero and infinity for processes with stable components. *Z. Wahrscheinlichkeitstheorie verw. Gebiete* **16** 261–278.

W. J. Hendricks (1973). A dimension theorem for the sample functions of processes with stable components. *Ann. Probab.* **1** 849–853.

E. Hille and R. S. Phillips (1957). *Functional analysis and semigroups.* Amer. Math. Soc., Providence, RI.

J. Hoffmann-Jørgensen (1969). Markov sets. *Math. Scand.* **24** 145–166.

J. Horowitz (1968). The Hausdorff dimension of the sample path of a subordinator. *Israel J. Math.* **6** 176–182.

J. Horowitz (1972). Semilinear Markov processes, subordinators and renewal theory. *Z. Wahrscheinlichkeitstheorie verw. Gebiete* **24** 167–193.

X. Hu and S. J. Taylor (1997). The multifractal structure of stable occupation measure. *Stochastic Process. Appl.* **66** 283–299.

B. Huff (1969). The strict subordination of a differential process. *Sankhya Sera. A* **31** 403–412.

G. A. Hunt (1958). Markov processes and potentials III. *Illinois J. Math.* **2** 151–213.

I. A. Ibragimov and Yu. V. Linnik (1971). *Independent and stationary sequences of random variables.* Wolters-Noordhoff, Groningen, Netherlands.

K. Itô (1942). On stochastic processes. I. (Infinitely divisible laws of probability). *Japan J. Math.* **18** 261–301.

K. Itô (1961). *Lectures on stochastic processes.* Tata Institute of Fundamental Research. Springer, Berlin.

K. Itô (1970). Poisson point processes attached to Markov processes. *Proc. 6th Berkeley Symp. Math. Stat. Probab.* III, pp. 225–239.

K. Itô and H. P. McKean (1965). *Diffusion processes and their sample paths.* Springer, Berlin.

J. Jacod (1979). *Calcul stochastique et problèmes de martingales.* Lecture Notes in Math. 714. Springer, Berlin.

J. Jacod and A. N. Shiryaev (1987). *Limit theorems for stochastic processes.* Springer, Berlin.

S. Jaffard (1996). Sur la nature multifractale des processus de Lévy. *C. R. Acad. Sci. Paris Sér. I Math.* **323** 1059–1064.

N. C. Jain and W. E. Pruitt (1968). The correct measure function for the graph of a transient stable process. *Z. Wahrscheinlichkeitstheorie verw. Gebiete* **9** 131–138.

N. C. Jain and W. E. Pruitt (1987). Lower tail probabilities estimates for subordinators and nondecreasing random walks. *Ann. Probab.* **15** 75–102.

A. Janicki and A. Weron (1994). *Simulation and chaotic behavior of α-stable stochastic processes.* Dekker, New York.

O. Kallenberg (1974). Path properties of processes with independent and interchangeable increments. *Z. Wahrscheinlichkeitstheorie verw. Gebiete* **28** 257–271.

O. Kallenberg (1992). Some time change representations of stable integrals, via predictable transformations of local martingales. *Stochastic Process. Appl.* **40** 199–223.

M. Kanda (1976). Two theorems on capacity for Markov processes with stationary independent increments. *Z. Wahrscheinlichkeitstheorie verw. Gebiete* **35** 159–165.

M. Kanda (1978). Characterization of semipolar sets for processes with stationary independent increments. *Z. Wahrscheinlichkeitstheorie verw. Gebiete* **42** 141–154.

M. Kanda (1983). On the class of polar sets for a class of Lévy processes on the line. *J. Math. Soc. Japan* **35** 221–242.

J. Keilson (1963). The first passage time density for homogeneous skip-free walks on the continuum. *Ann. Math. Stat.* **34** 1003–1011.

D. G. Kendall (1957). Some problems in the theory of dams. *J. Roy. Stat. Soc. B* **19** 207–212.

H. Kesten (1969). Hitting probabilities of single points for processes with stationary independent increments. *Memoirs Amer. Math. Soc.* **93.**

H. Kesten (1976). Lévy processes with a nowhere dense range. *Indiana Univ. Math. J.* **25** 45–64.

A. Ya. Khintchine (1938). Zwei Sätze über stochastische Prozesse mit stabilen Verteilungen. *Math. Sbornik.* **3** 577–584.

A. Ya. Khintchine (1939). Sur la croissance locale des processus stochastiques homogènes à accroissements indépendants. *Izv. Akad. Nauk SSSR* **3** 487–508.

D. Khoshnevisan (1995). The rate of convergence in the ratio ergodic theorem for Markov processes. Preprint.

D. Khoshnevisan (1997). Escape rates for Lévy processes. *Studia Sci. Math. Hungar.* **33** 177–183.

F. B. Knight (1996). The uniform law for exchangeable and Lévy bridges. Hommage à P. A. Meyer et J. Neveu. *Astérisque* **236** 171–188.

V. S. Korolyuk, V. N. Suprun and V. M. Shurenkov (1976). Method of potential in boundary problems for processes with independent increases and jumps of the same sign. *Theory. Probab. Appl.* **21** , 243–249.

M. Ledoux and M. Talagrand (1991). *Probability in Banach spaces.* Springer, Berlin.

M. L. T. Lee and G. A. Whitmore (1990). Subordinated processes: a survey. *Proceedings 1990 Taipei Symp. Stat.* pp. 395–405. Institute of Statistical Science, Taipei.

J. F. Le Gall (1987). Temps locaux d'intersection et points multiples des processus de Lévy. *Séminaire de Probabilités XXI*, Lecture Notes in Math. 1247, pp. 341–374. Springer, Berlin.

J. F. Le Gall and Y. Le Jan (1995). Arbres aléatoires et processus de Lévy. *C. R. Acad. Sci. Paris* **321** 1241–1244.

J. F. Le Gall, J. Rosen and N. R. Shieh (1989). Multiple points of Lévy processes. *Ann. Probab.* **17** 503–515.

P. Lévy (1934). Sur les intégrales dont les éléments sont des variables aléatoires indépendantes. *Ann. Scuola Norm. Pisa* **3** 337–366 and **4** 217–218.

P. Lévy (1939). Sur certains processus stochastiques homogènes. *Compos. Math.* **7** 283–339.

P. Lévy (1954). *Théorie de l'addition des variables aléatoires.* 2nd edn. Gauthier-Villars, Paris.

P. Lévy (1965). *Processus stochastiques et mouvement brownien.* 2nd edn. Gauthier-Villars, Paris.

M. Loève (1963). *Probability theory,* 3rd edn. Van Nostrand, Princeton, NJ.

B. Maisonneuve (1971). Ensembles régénératifs, temps locaux et subordinateurs. *Séminaire de Probabilités V,* Lecture Notes in Math. 191, pp. 147–169. Springer, Berlin

B. Maisonneuve (1975). Exit systems. *Ann. Probab.* **3** 395–411

M. B. Marcus and J. Rosen (1992). Sample path properties of the local times of strongly symmetric Markov processes via Gaussian processes. *Ann. Probab.* **20** 1603–1684.

M. B. Marcus and J. Rosen (1992–a). p-variation of the local times of symmetric stable processes and of Gaussian processes with stationary increments. *Ann. Probab.* **20** 1685–1713.

M. B. Marcus and J. Rosen (1992–b). Moduli of continuity of local times of strongly symmetric Markov processes via Gaussian processes. *J. Theoretic. Probab.* **5** 791–825.

M. B. Marcus and J. Rosen (1993). ϕ-variation of the local times of symmetric Lévy processes and stationary Gaussian processes. *Seminar on Stochastic Processes 1992*, pp. 209–220. Birkhäuser, Boston, Mass.

M. B. Marcus and J. Rosen (1994). Laws of the iterated logarithm for the local times of symmetric Lévy processes and recurrent random walks. *Ann. Probab.* **22** 626–658.

M. B. Marcus and J. Rosen (1994–a). Laws of the iterated logarithm for the local times of recurrent random walks on Z^2 and of Lévy processes and recurrent random walks in the domain of attraction of Cauchy random variables. *Ann. Inst. Henri Poincaré* **30** 467–499.

M. B. Marcus and J. Rosen (1994–b). Exact rates of convergence to the local times of symmetric Lévy processes. *Séminaire de Probabilités XXVIII*, Lecture Notes in Math. 1583, pp. 102–109. Springer, Berlin.

L. Marsalle (1995). Hausdorff measures and capacities for increase times of stable processes. To appear in *Potential Analysis.*

P. McGill (1989). Computing the overshoot of a Lévy process. *Stochastic analysis, path integration and dynamics.* Pitman Research Notes in Math. 200, pp. 165–196. Longman, Chicago.

P. McGill (1989–a). Wiener-Hopf factorisation of Brownian motion. *Probab. Theory Relat. Fields* **83** 355–389.

H. P. McKean (1955). Sample functions of stable processes. *Ann. Math.* **61** 564–579.

H. P. McKean (1963). Excursions of a non-singular diffusion. *Z. Wahrscheinlichkeitstheorie verw. Gebiete* **1** 230–239.

P. A. Meyer (1969). Processus à accroissements indépendants et positifs. *Séminaire de Probabilités III*, Lecture Notes in Math. 88, pp. 175–189. Springer, Berlin.

P. A. Meyer (1970). Ensembles régénératifs, d'après Hoffmann-Jørgensen. *Séminaire de Probabilités IV*, Lecture Notes in Math. 124, pp. 133–140. Springer, Berlin.

J. L. Mijnheer (1975). *Sample path properties of stable processes.* Math. Centre Tracts 59, Amsterdam.

P. W. Millar (1971). Path behaviour of processes with stationary independent increments. *Z. Wahrscheinlichkeitstheorie verw. Gebiete* **17** 53–73.

P. W. Millar (1973). Exit properties of stochastic processes with stationary independent increments. *Trans. Amer. Math. Soc.* **178** 459–479.

P. W. Millar (1977). Zero-one laws and the minimum of a Markov process. *Trans. Amer. Math. Soc.* **226** 365–391.

P. W. Millar (1978). A path decomposition for Markov processes. *Ann. Probab.* **6** 345–348.

P.W. Millar (1981). Comparison theorems for sample function growth. *Ann. Probab.* **9** 330–334.

P. W. Millar and L. T. Tran (1974). Unbounded local times. *Z. Wahrscheinlichkeitstheorie verw. Gebiete* **30** 87–92.

D. Monrad and M. L. Silverstein (1979). Stable processes: Sample function growth at a local minimum. *Z. Wahrscheinlichkeitstheorie verw. Gebiete* **49** 177–210.

I. Monroe (1972). On the γ-variation of processes with stationary independent increments. *Ann. Math. Stat.* **43** 1213–1220.

T. Mountford and S. Port (1991). The range of a Lévy process. *Ann. Probab.* **19** 221–225.

M. Nagasawa (1964). Time reversion of Markov processes. *Nagoya J. Math.* **24** 177–204.

M. Nagasawa (1993). *Schrödinger equations and diffusion theory.* Monographs in Mathematics, Birkhäuser, Boston, Mass.

A. C. Okoroafor and O. O. Ugbebor (1991). Lower asymptotic behavior of the sojourn time for a stable process. *Contemporary stochastic analysis*, pp. 109–126. World Scientific, Singapore.

A. C. Okoroafor and O. O. Ugbebor (1991–a). Upper rates of escape for stable processes. *Contemporary stochastic analysis*, pp. 127–147. World Scientific, Singapore.

S. Orey (1967). Polar sets for processes with stationary independent increments. *Markov processes and potential theory*, pp. 117–126. Wiley, New York.

E. A. Pecherskii (1974). Some identities related to the exit of a random walk out of a segment or a semi-finite interval. *Theory Probab. Appl.* **19** 104–119.

E. A. Pecherskii and B. A. Rogozin (1969). On the joint distribution of random variables associated with fluctuations of a process with independent increments. *Theory Probab. Appl.* **14** 410–423.

M. Perman (1993). Order statistics for jumps of normalized subordinators. *Stochastic Process. Appl.* **46** 267–281.

M. Perman, J. W. Pitman and M. Yor (1992). Size-biased sampling of Poisson point processes, and excursions. *Probab. Theory Relat. Fields* **92** 21–40.

J. W. Pitman (1975). One-dimensional Brownian motion and the three-dimensional Bessel process. *Adv. Appl. Probab.* **7** 511–526.

J. W. Pitman and L. C. G. Rogers (1981). Markov functions. *Ann. Probab.* **9** 573–581.

J. W. Pitman and M. Yor (1986). Level crossings of a Cauchy process. *Ann. Probab.* **14** 780–792.

J. W. Pitman and M. Yor (1992). Arc sine laws and interval partitions derived from a stable subordinator. *Proc. London Math. Soc.* **65** 326–356.

S. C. Port (1963). An elementary approach to fluctuation theory. *J. Math. Anal. Appl.* **6** 109–151.

S. C. Port (1967). Hitting times and potentials for recurrent stable processes. *J. Anal. Math.* **20** 371–395.

S. C. Port (1989). Stable processes with drift on the line. *Trans. Amer. Math. Soc.* **313** 805–841.

S. C. Port (1990). Asymptotic expansions for the expected volume of the stable sausage. *Ann. Probab.* **18** 319–341.

S. C. Port and C. J. Stone (1969). The asymmetric Cauchy process on the line. *Ann. Math. Stat.* **40** 137–143.

S. C. Port and C. J. Stone (1971). Infinitely divisible processes and their potential theory I, II. *Ann. Inst. Fourier* **21** (2) 157–275, and **21** (4) 179–265.

S. C. Port and C. J. Stone (1978). *Brownian motion and classical potential theory.* Academic Press, New York.

N. U. Prabhu (1970). Ladder variables for a continuous time stochastic process. *Z. Wahrscheinlichkeitstheorie verw. Gebiete* **16** 157–164.

N. U. Prabhu (1972). Wiener-Hopf factorization for convolution semi-groups. *Z. Wahrscheinlichkeitstheorie verw. Gebiete* **23** 103–113.

N. U. Prabhu (1981). *Stochastic storage processes, queues, insurance risk and dams.* Springer, Berlin.

N. U. Prabhu and M. Rubinovitch (1973). Further results for ladder processes in continuous time. *Stochastic Process. Appl.* **1** 151–168.

P. Protter (1990). *Stochastic integration and stochastic differential equations: A new approach.* Springer, Berlin.

W. E. Pruitt (1969). The Hausdorff dimension of the range of a process with stationary independent increments. *J. Math. Mech.* **19** 371–378.

W. E. Pruitt (1981). The growth of random walks and Lévy processes. *Ann. Probab.* **9** 948–956.

W. E. Pruitt (1991). An integral test for subordinators. *Random walks, Brownian motion and iteracting particle systems: A Festschrift in honor of Frank Spitzer*, pp. 389–398. Birkhäuser, Boston, Mass.

W. E. Pruitt and S. J. Taylor (1969). Sample path properties of processes with stable components. *Z. Wahrscheinlichkeitstheorie verw. Gebiete* **12** 267–289.

W. E. Pruitt and S. J. Taylor (1969–a). The potential kernel and hitting probabilities for the general stable process in R^n. *Trans. Amer. Math. Soc.* **146** 299–321.

W. E. Pruitt and S. J. Taylor (1983). The behavior of asymmetric Cauchy processes for large time. *Ann. Probab.* **11** 302–327.

W. E. Pruitt and S. J. Taylor (1985). The local structure of the sample paths of asymmetric Cauchy processes. *Z. Wahrscheinlichkeitstheorie verw. Gebiete* **70** 535–561.

W. E. Pruitt and S. J. Taylor (1996). Packing and covering indices for a general Lévy process. *Ann. Probab.* **24** 971–986.

M. Rao (1977). On a result of M. Kanda. *Z. Wahrscheinlichkeitstheorie verw. Gebiete* **41** 35–37.

M. Rao (1987). On polar sets for Lévy processes. *J. London Math. Soc. (2)* **35** 569–576.

M. Rao (1988). Hunt's hypothesis for Lévy processes. *Proc. Amer. Math. Soc.* **104** 621–624.

D. B. Ray (1958). Stable processes with an absording barrier. *Trans. Amer. Math. Soc.* **89** 16–24.

D. Revuz (1984). *Markov chains.* 2nd edn. North-Holland, Amsterdam.

D. Revuz and M. Yor (1994). *Continuous martingales and Brownian motion*, 2nd edn. Springer, Berlin.

F. Rezakhanlou and S. J. Taylor (1988). The packing measure of the graph of a stable process. *In: Colloque Paul Lévy sur les Processus Stochastiques. Astérisque* **157–158** 341–362.

C. A. Rogers (1970). *Hausdorff measure.* Cambridge University Press, Cambridge.

L. C. G. Rogers (1984). A new identity for real Lévy processes. *Ann. Inst. Henri Poincaré* **20** 21–34

L. C. G. Rogers (1989). Multiple points of Markov processes in a complete metric space. *Séminaire de Probabilités XXIII*, pp. 186–197. Springer, Berlin.

L. C. G. Rogers (1990). The two-sided exit problem for spectrally positive Lévy processes. *Adv. Appl. Probab.* **22** 486–487.

L. C. G. Rogers and D. Williams (1987). *Diffusions, Markov processes, and martingales vol. 2: Itô calculus.* Wiley, New York.

L. C. G. Rogers and D. Williams (1994). *Diffusions, Markov processes, and martingales vol. 1: Foundations.* [1st ed. by D. Williams, 1979.] Wiley, New York.

B. A. Rogozin (1965). On some class of processes with independent increments. *Theory Probab. Appl.* **10** 479–483.

B. A. Rogozin (1966). On the distribution of functionals related to boundary problems for processes with independent increments. *Theory Probab. Appl.* **11** 580–591.

B. A. Rogozin (1968). Local behavior of processes with independent increments. *Theory Probab. Appl.* **13** 482–486.

B. A. Rogozin (1971). The distribution of the first ladder moment and height and fluctuation of a random walk. *Theory Probab. Appl.* **16** 575–595.

B. A. Rogozin (1972). The distribution of the first hit for stable and asymptotically stable walks in an interval. *Theory Probab. Appl.* **17** 332–338.

M. Rubinovitch (1971). Ladder phenomena in stochastic processes with stationary independent increments. *Z. Wahrscheinlichkeitstheorie verw. Gebiete* **20** 58–74.

G. Samorodnitsky and M. S. Taqqu (1994). *Stable non-Gaussian random processes, stochastic models with infinite variance.* Chapman & Hall New York.

K. Sato (1972). Potential for Markov processes. *Proc. 6th. Berkeley Symp. Math. Stat. Probab.* III, pp. 193–211.

K. Sato (1990). *Stochastic processes with stationary independent increments* (in Japanese). Kinokuniya, Tokyo.

K. Sato (1995). *Lévy processes on the Euclidean spaces.* University of Zurich.

M. J. Sharpe (1969). Zeroes of infinitely divisible densities. *Ann. Math. Stat.* **40** 1503–1505.

M. J. Sharpe (1989). *General theory of Markov processes.* Academic Press, New York.

E. S. Shtatland (1965). On local properties of processes with independent increments. *Theory Probab. Appl.* **10** 317–322.

E. S. Shtatland (1966). The distribution of the time the maximum is achieved for processes with independent increments. *Theory Probab. Appl.* **11** 637–642.

M. L. Silverstein (1980). Classification of coharmonic and coinvariant functions for a Lévy process. *Ann. Probab.* **8** 539–575.

A. V. Skorohod (1965). *Studies in the theory of random processes.* Addison-Wesley, Reading, Mass.

A. V. Skorohod (1991). *Random processes with independent increments.* Kluwer, Dordrecht, Netherlands.

F. Spitzer (1956). A combinatorial lemma and its applications to probability theory. *Trans. Amer. Math. Soc.* **82** 323–339.

F. Spitzer (1964). *Principles of random walk.* Van Nostrand, Princeton, NJ.

E. M. Stein (1970). *Singular integrals and differentiability properties of functions.* Princeton University Press.

F. W. Steutel (1970). *Preservation of infinite divisibility under mixing, and related topics.* Math. Centre Tracts 33, Amsterdam.

V. N. Suprun (1976). Problem of destruction and resolvent of terminating process with independent increments. *Ukrainian Math. J.* **28** 39–45.

V. N. Suprun and V. M. Shurenkov (1980). Limit distribution of the position at the time of exit from an interval of a semi-continuous process with independent increments with zero mean and infinite variance. *Ukrainian Math. J.* **32** 170–172.

L. Takács (1966). *Combinatorial methods in the theory of stochastic processes.* Wiley, New York.

J. Takeuchi (1964). On the sample paths of the symmetric stable processes in space. *J. Math. Soc. Japan.* **16** 109–127.

J. Takeuchi (1964–a). A local asymptotic law for the transient stable process. *Proc. Japan Acad.* **40** 141–144.

J. Takeuchi and S. Watanabe (1964). Spitzer's test for the Cauchy process on the line. *Z. Wahrscheinlichkeitstheorie verw. Gebiete* **3** 204–210.

S. J. Taylor (1966). Multiple points for the sample paths of the symmetric stable process. *Z. Wahrscheinlichkeitstheorie verw. Gebiete* **5** 247–264.

S. J. Taylor (1967). Sample path properties of a transient stable process. *J. Math. Mech.* **16** 1229–1246.

S. J. Taylor (1973). Sample path properties of processes with stationary independent increments. *Stochastic analysis*, pp. 387–414. Wiley, London.

S. J. Taylor and J. G. Wendel (1966). The exact Hausdorff measure of the zero set of a stable process. *Z. Wahrscheinlichkeitstheorie verw. Gebiete* **6** 170–180.

H. F. Trotter (1958). A property of Brownian motion paths. *Illinois J. Math.* **2** 425–433.

H. G. Tucker (1965). On a necessary and sufficient condition that an infinitely divisible distribution be absolutely continuous. *Trans. Amer. Math. Soc.* **118** 316–330.

H. G. Tucker (1975). The support of infinitely divisible distribution functions. *Proc. Amer. Math. Soc.* **49** 436–440.

I. S. Wee (1988). Lower functions for processes with stationary independent increments. *Probab. Theory Relat. Fields* **77** 551–566.

I. S. Wee (1990). Lower functions for asymmetric Lévy processes. *Probab. Theory Relat. Fields* **85** 469–488.

I. S. Wee (1991). The law of the iterated logarithm for local time of a Lévy process. *Gaussian random fields*, pp. 384–395. World Scientific, Singapore.

J. G. Wendel (1960). Order statistics for partial sums. *Ann. Math. Stat.* **31** 1034–1044.

E. Willekens (1987). On the supremum of an infinitely divisible process. *Stochastic Process. Appl.* **26** 173–175.

D. Williams (1974). Path decomposition and continuity of local time for one-dimensional diffusions. *Proc. London Math. Soc.* **28** 738–768.

D. Williams (1979). *Diffusions, Markov processes, and martingales vol. 1: Foundations* [2nd edn jointly with L.C.G. Rogers (1994).] Wiley, New York.

T. Yamada (1985). On the fractional derivative of Brownian local times. *J. Math. Kyoto Univ.* **25** 49–58.

T. Yamada (1986). On some limit theorems for occupation times of one dimensional Brownian motion and its continuous additive functionals locally of zero energy. *J. Math. Kyoto Univ.* **26** 309–322.

M. Yamazato (1978). Unimodality of infinitely divisible distributions of class L. *Ann. Probab.* **6** 523–531.

M. Yamazato (1982). On strongly unimodal infinitely divisible distributions. *Ann. Probab.* **10** 589–601.

M. Yor (1982). Sur la transformée de Hilbert des temps locaux browniens et une extension de la formule d'Itô. *Séminaire de Probabilités XVI*, Lecture Notes in Math. 920, pp. 238–242, Springer, Berlin.

M. Yor (1995). Some aspects of Brownian motion, Part II: some new martingale problems.

J. Zabczyk (1970). Sur la théorie semi-classique du potentiel pour les processus à accroissements indépendants. *Studia Math.* **35** 227–247.

J. Zabczyk (1975). A note on semipolar sets for processes with independent increments. *Probability winter school, Karpacz, Poland, 1975.* Lecture Notes in Math. 472, pp. 277–283. Springer, Berlin.

V. M. Zolotarev (1964). The first-passage time of a level and the behavior at infinity for a class of processes with independent increments. *Theory Probab. Appl.* **9** 653–664.

V. M. Zolotarev (1964–a). Analog of the iterated logarithm law for semi-continuous stable processes. *Theory Probab. Appl.* **9** 512–513.

V. M. Zolotarev (1986). *One-dimensional stable distributions.* Amer. Math. Soc., Providence, RI.

List of symbols

$\mathscr{C}_0$	space of continuous functions that tend to 0 at infinity
$C^q(B)$	q-capacity of a set B
∂	cemetery point
d	drift coefficient of a Lévy process with bounded variation
Δ, ΔX	jump process
δ_x	Dirac point mass at x
d, $d_n(a)$	right-end point of an excursion interval
$e = (e_t, t \geq 0)$	excursion process
ϵ	generic excursion
$\mathscr{F}_t$	$\mathbb{P}$-complete sigma-field generated by $(X_s, s \leq t)$
$(\mathscr{G}_t)$	some filtration
g, $g_n(a)$	left-end point of an excursion interval
G_t	last zero of the reflected process before time t
H, $H(t)$	ladder height process
$H(x,t)$	Hilbert transform of the local times
I	infimum process
$\Im(z)$	imaginary part of a complex number z
k_t	killing operator
$\kappa(\lambda)$	Laplace exponent of the inverse local time
$\kappa(\alpha, \beta)$	bivariate Laplace exponent of the ladder point process
ℓ, $\ell_n(a)$	length of an excursion interval

L	local time of a Markov process, of the reflected process
L^{-1}	inverse local time, ladder time process
$L(x,t)$	local time of a Lévy process at level x and time t
μ_B^q	q-capacitary measure of a set B
N^B	process of the number of visits of a set B for a Poisson point process
ω	generic right-continuous path having limits to the left
Ω	space right-continuous path having limits to the left
$p_t(x)$	continuous version of the density of a stable law
$\mathbb{P}$, $\mathbb{P}_0$	law of the Lévy process started at 0
$\mathbb{P}_x$	law of the Lévy process started at $x \in \mathbb{R}^d$
$\mathbb{P}^{\uparrow}$	law of the Lévy process conditioned to stay positive
$\mathbb{P}^{\natural}$	law of the Lévy process conditioned to drift to ∞
Φ	Laplace exponent of a subordinator
Ψ	characteristic exponent of a Lévy process
ψ	Laplace exponent of a Lévy process with no positive jumps
Π	Lévy measure
$\overline{\Pi}$	tail of the Lévy measure
Q	Gaussian coefficient
$\Re(z)$	real part of a complex number z
ρ	positivity parameter
S	supremum process
σ	inverse of the local time at 0
$\varsigma(x)$	last passage time below x
T_B	first passage time in a set B
T'_B	first hitting time of a set B
$T(x)$, T_x	first passage time above x
τ, $\tau(q)$	exponential time of parameter q, independent of X
θ_t	shift operator
U^q, U	q-resolvent operator, potential operator
$U(dx)$	potential measure of a subordinator
$\mathcal{U}$	renewal function of a subordinator
u^q, u	q-resolvent density, potential density
$\mathcal{V}$	renewal function of the ladder height process
W	scale function

X	canonical process
X_{t-}	left limit of X at time t
$\widehat{X}$	dual process ($= -X$)
X^{*}	absolute supremum of X
$\mathscr{Z}$	zero set of a Markov process
ζ	lifetime of a path

Functional notation

$f = o(g)$	$\lim (f/g) = 0$
$f = O(g)$	$\limsup (f/g) < \infty$
$f \sim g$	$\lim (f/g) = 1$
$f \asymp g$	f/g is bounded away from 0 and ∞
$f * g$	convolution of f and g
$f \circ g$	composition of f and g
$\mathscr{F}\mu$, $\mathscr{F}f$	Fourier transform of a measure μ, of a function f
$\mathscr{L}\mu$, $\mathscr{L}f$	Laplace transform of a measure μ, of a function f

Index

Printed in the United States
By Bookmasters